T0344689

OPERATIONS RESEARCH FOR UNMANNED SYSTEMS

OPERATIONS RESEARCH FOR UNMANNED SYSTEMS

Edited by

Jeffrey R. Cares
Captain, US Navy (Ret.)
Alidade Inc.
USA

and

John Q. Dickmann, Jr.
Sonalysts Inc.
USA

This edition first published 2016
© 2016 John Wiley & Sons, Ltd.

Registered Office
John Wiley & Sons, Ltd, The Atrium, Southern Gate, Chichester, West Sussex, PO19 8SQ, United Kingdom

For details of our global editorial offices, for customer services and for information about how to apply for permission to reuse the copyright material in this book please see our website at www.wiley.com.

The right of the author to be identified as the author of this work has been asserted in accordance with the Copyright, Designs and Patents Act 1988.

All rights reserved. No part of this publication may be reproduced, stored in a retrieval system, or transmitted, in any form or by any means, electronic, mechanical, photocopying, recording or otherwise, except as permitted by the UK Copyright, Designs and Patents Act 1988, without the prior permission of the publisher.

Wiley also publishes its books in a variety of electronic formats. Some content that appears in print may not be available in electronic books.

Designations used by companies to distinguish their products are often claimed as trademarks. All brand names and product names used in this book are trade names, service marks, trademarks or registered trademarks of their respective owners. The publisher is not associated with any product or vendor mentioned in this book.

Limit of Liability/Disclaimer of Warranty: While the publisher and author have used their best efforts in preparing this book, they make no representations or warranties with respect to the accuracy or completeness of the contents of this book and specifically disclaim any implied warranties of merchantability or fitness for a particular purpose. It is sold on the understanding that the publisher is not engaged in rendering professional services and neither the publisher nor the author shall be liable for damages arising herefrom. If professional advice or other expert assistance is required, the services of a competent professional should be sought.

Library of Congress Cataloging-in-Publication Data

Names: Cares, Jeffrey R., editor. | Dickmann, Jr., John Q., editor.
Title: Operations research for unmanned systems / edited by Jeffrey R. Cares and John Q. Dickmann, Jr.
Description: Chichester, UK ; Hoboken, NJ : John Wiley & Sons, 2016. | Includes bibliographical references and index. |
    Description based on print version record and CIP data provided by publisher; resource not viewed.
Identifiers: LCCN 2015036918 (print) | LCCN 2015033015 (ebook) | ISBN 9781118918913 (Adobe PDF) |
    ISBN 9781118918920 (ePub) | ISBN 9781118918944 (cloth)
Subjects: LCSH: Autonomous vehicles–Industrial applications. | Drone aircraft–Industrial applications. |
    Vehicles, Remotely piloted–Industrial applications.
Classification: LCC TL152.8 (print) | LCC TL152.8 .O64 2016 (ebook) | DDC 629.04/6–dc23
LC record available at http://lccn.loc.gov/2015036918

A catalogue record for this book is available from the British Library.

Set in 10/12pt Times by SPi Global, Pondicherry, India
Printed and bound in Singapore by Markono Print Media Pte Ltd

1   2016

Contents

About the Contributors   xiii

Acknowledgements   xix

1   Introduction   **1**
   1.1   Introduction   1
   1.2   Background and Scope   3
   1.3   About the Chapters   4
   References   6

**2   The In-Transit Vigilant Covering Tour Problem for Routing
    Unmanned Ground Vehicles**   **7**
   2.1   Introduction   7
   2.2   Background   8
   2.3   CTP for UGV Coverage   9
   2.4   The In-Transit Vigilant Covering Tour Problem   9
   2.5   Mathematical Formulation   11
   2.6   Extensions to Multiple Vehicles   14
   2.7   Empirical Study   15
   2.8   Analysis of Results   21
   2.9   Other Extensions   24
   2.10 Conclusions   25
   Author Statement   25
   References   25

**3   Near-Optimal Assignment of UAVs to Targets Using
      a Market-Based Approach   27**
      3.1   Introduction   27
      3.2   Problem Formulation   29
             3.2.1   Inputs   29
             3.2.2   Various Objective Functions   29
             3.2.3   Outputs   31
      3.3   Literature   31
             3.3.1   Solutions to the MDVRP Variants   31
             3.3.2   Market-Based Techniques   33
      3.4   The Market-Based Solution   34
             3.4.1   The Basic Market Solution   36
             3.4.2   The Hierarchical Market   37
                      3.4.2.1   Motivation and Rationale   37
                      3.4.2.2   Algorithm Details   40
             3.4.3   Adaptations for the Max-Pro Case   41
             3.4.4   Summary   41
      3.5   Results   42
             3.5.1   Optimizing for Fuel-Consumption (Min-Sum)   43
             3.5.2   Optimizing for Time (Min-Max)   44
             3.5.3   Optimizing for Prioritized Targets (Max-Pro)   47
      3.6   Recommendations for Implementation   51
      3.7   Conclusions   52
      Appendix 3.A   A Mixed Integer Linear Programming (MILP) Formulation   53
             3.A.1   Sub-tour Elimination Constraints   54
      References   55

**4   Considering Mine Countermeasures Exploratory Operations Conducted
      by Autonomous Underwater Vehicles   59**
      4.1   Background   59
      4.2   Assumptions   61
      4.3   Measures of Performance   62
      4.4   Preliminary Results   64
      4.5   Concepts of Operations   64
             4.5.1   Gaps in Coverage   64
             4.5.2   Aspect Angle Degradation   64
      4.6   Optimality with Two Different Angular Observations   65
      4.7   Optimality with N Different Angular Observations   66
      4.8   Modeling and Algorithms   67
             4.8.1   Monte Carlo Simulation   67
             4.8.2   Deterministic Model   67
      4.9   Random Search Formula Adapted to AUVs   68
      4.10   Mine Countermeasures Exploratory Operations   70
      4.11   Numerical Results   71
      4.12   Non-uniform Mine Density Distributions   72
      4.13   Conclusion   74
      Appendix 4.A   Optimal Observation Angle between Two AUV Legs   75
      Appendix 4.B   Probabilities of Detection   78
      References   79

5   Optical Search by Unmanned Aerial Vehicles: Fauna Detection Case Study   81
  5.1   Introduction   81
  5.2   Search Planning for Unmanned Sensing Operations   82
        5.2.1   Preliminary Flight Analysis   84
        5.2.2   Flight Geometry Control   85
        5.2.3   Images and Mosaics   86
        5.2.4   Digital Analysis and Identification of Elements   88
  5.3   Results   91
  5.4   Conclusions   92
  Acknowledgments   94
  References   94

6   A Flight Time Approximation Model for Unmanned Aerial Vehicles:
    Estimating the Effects of Path Variations and Wind   95
  Nomenclature   95
  6.1   Introduction   96
  6.2   Problem Statement   97
  6.3   Literature Review   97
        6.3.1   Flight Time Approximation Models   97
        6.3.2   Additional Task Types to Consider   98
        6.3.3   Wind Effects   99
  6.4   Flight Time Approximation Model Development   99
        6.4.1   Required Mathematical Calculations   100
        6.4.2   Model Comparisons   101
        6.4.3   Encountered Problems and Solutions   102
  6.5   Additional Task Types   103
        6.5.1   Radius of Sight Task   103
        6.5.2   Loitering Task   105
  6.6   Adding Wind Effects   108
        6.6.1   Implementing the Fuel Burn Rate Model   110
  6.7   Computational Expense of the Final Model   111
        6.7.1   Model Runtime Analysis   111
        6.7.2   Actual versus Expected Flight Times   113
  6.8   Conclusions and Future Work   115
  Acknowledgments   117
  References   117

7   Impacts of Unmanned Ground Vehicles on Combined Arms Team
    Performance   119
  7.1   Introduction   119
  7.2   Study Problem   120
        7.2.1   Terrain   120
        7.2.2   Vehicle Options   122
        7.2.3   Forces   122
              7.2.3.1   Experimental Force   123
              7.2.3.2   Opposition Force   123
              7.2.3.3   Civilian Elements   123
        7.2.4   Mission   124

7.3   Study Methods   125
      7.3.1   *Closed-Loop Simulation*   125
      7.3.2   *Study Measures*   126
      7.3.3   *System Comparison Approach*   128
7.4   Study Results   128
      7.4.1   *Basic Casualty Results*   128
              7.4.1.1   *Low Density Urban Terrain Casualty Only Results*   128
              7.4.1.2   *Dense Urban Terrain Casualty-Only Results*   130
      7.4.2   *Complete Measures Results*   131
              7.4.2.1   *Low Density Urban Terrain Results*   131
              7.4.2.2   *Dense Urban Terrain Results*   132
              7.4.2.3   *Comparison of Low and High Density Urban Results* 133
      7.4.3   *Casualty versus Full Measures Comparison*   135
7.5   Discussion   136
References   137

**8   Processing, Exploitation and Dissemination: When is Aided/Automated
      Target Recognition "Good Enough" for Operational Use?**   **139**
8.1   Introduction   139
8.2   Background   140
      8.2.1   *Operational Context and Technical Issues*   140
      8.2.2   *Previous Investigations*   141
8.3   Analysis   143
      8.3.1   *Modeling the Mission*   144
      8.3.2   *Modeling the Specific Concept of Operations*   145
      8.3.3   *Probability of Acquiring the Target under the Concept of Operations* 146
      8.3.4   *Rational Selection between Aided/Automated Target Recognition
              and Extended Human Sensing*   147
      8.3.5   *Finding the Threshold at which Automation is Rational*   148
      8.3.6   *Example*   148
8.4   Conclusion   149
Acknowledgments   151
Appendix 8.A   151
      Ensuring $\mathbb{E}[Q_*]$ Decreases as ζ_* Increases   152
References   152

9   Analyzing a Design Continuum for Automated Military Convoy Operations **155**
9.1   Introduction   155
9.2   Definition Development   156
      9.2.1   *Human Input Proportion (H)*   156
      9.2.2   *Interaction Frequency*   157
      9.2.3   *Complexity of Instructions/Tasks*   157
      9.2.4   *Robotic Decision-Making Ability (R)*   157
9.3   Automation Continuum   157
      9.3.1   *Status Quo (SQ)*   158
      9.3.2   *Remote Control (RC)*   158
      9.3.3   *Tele-Operation (TO)*   158

		9.3.4	Driver Warning (DW)	158
		9.3.5	Driver Assist (DA)	158
		9.3.6	Leader-Follower (LF)	159
			9.3.6.1 Tethered Leader-Follower (LF1)	159
			9.3.6.2 Un-tethered Leader-Follower (LF2)	159
			9.3.6.3 Un-tethered/Unmanned/Pre-driven Leader-Follower (LF3)	159
			9.3.6.4 Un-tethered/Unmanned/Uploaded Leader-Follower (LF4)	159
		9.3.7	Waypoint (WA)	159
			9.3.7.1 Pre-recorded "Breadcrumb" Waypoint (WA1)	160
			9.3.7.2 Uploaded "Breadcrumb" Waypoint (WA2)	160
		9.3.8	Full Automation (FA)	160
			9.3.8.1 Uploaded "Breadcrumbs" with Route Suggestion Full Automation (FA1)	160
			9.3.8.2 Self-Determining Full Automation (FA2)	160
	9.4	Mathematically Modeling Human Input Proportion (H) versus System Configuration		161
		9.4.1	Modeling H versus System Configuration Methodology	161
		9.4.2	Analyzing the Results of Modeling H versus System Configuration	165
		9.4.3	Partitioning the Automation Continuum for H versus System Configuration into Regimes and Analyzing the Results	168
	9.5	Mathematically Modeling Robotic Decision-Making Ability (R) versus System Configuration		169
		9.5.1	Modeling R versus System Configuration Methodology	169
		9.5.2	Mathematically Modeling R versus System Configuration When Weighted by H	171
		9.5.3	Partitioning the Automation Continuum for R (Weighted by H) versus System Configuration into Regimes	175
		9.5.4	Summarizing the Results of Modeling H versus System Configuration and R versus System Configuration When Weighted by H	177
	9.6	Mathematically Modeling H and R		178
		9.6.1	Analyzing the Results of Modeling H versus R	178
	9.7	Conclusion		180
	9.A	System Configurations		180
10	**Experimental Design for Unmanned Aerial Systems Analysis: Bringing Statistical Rigor to UAS Testing**			**187**
	10.1	Introduction		187
	10.2	Some UAS History		188
	10.3	Statistical Background for Experimental Planning		189
	10.4	Planning the UAS Experiment		192
		10.4.1	General Planning Guidelines	192
		10.4.2	Planning Guidelines for UAS Testing	193
			10.4.2.1 Determine Specific Questions to Answer	194
			10.4.2.2 Determine Role of the Human Operator	194

		10.4.2.3	Define and Delineate Factors of Concern for the Study	195
		10.4.2.4	Determine and Correlate Response Data	196
		10.4.2.5	Select an Appropriate Design	196
		10.4.2.6	Define the Test Execution Strategy	198
	10.5	Applications of the UAS Planning Guidelines		199
		10.5.1	Determine the Specific Research Questions	199
		10.5.2	Determining the Role of Human Operators	199
		10.5.3	Determine the Response Data	200
		10.5.4	Define the Experimental Factors	200
		10.5.5	Establishing the Experimental Protocol	201
		10.5.6	Select the Appropriate Design	202
		10.5.6.1	Verifying Feasibility and Practicality of Factor Levels	202
		10.5.6.2	Factorial Experimentation	202
		10.5.6.3	The First Validation Experiment	203
		10.5.6.4	Analysis: Developing a Regression Model	204
		10.5.6.5	Software Comparison	204
	10.6	Conclusion		205
	Acknowledgments			205
	Disclaimer			205
	References			205

11   Total Cost of Ownership (TOC): *An Approach for Estimating UMAS Costs*   207
	11.1	Introduction		207
	11.2	Life Cycle Models		208
		11.2.1	DoD 5000 Acquisition Life Cycle	208
		11.2.2	ISO 15288 Life Cycle	208
	11.3	Cost Estimation Methods		210
		11.3.1	Case Study and Analogy	210
		11.3.2	Bottom-Up and Activity Based	211
		11.3.3	Parametric Modeling	212
	11.4	UMAS Product Breakdown Structure		212
		11.4.1	Special Considerations	212
		11.4.1.1	Mission Requirements	214
		11.4.2	System Capabilities	214
		11.4.3	Payloads	214
	11.5	Cost Drivers and Parametric Cost Models		215
		11.5.1	Cost Drivers for Estimating Development Costs	215
		11.5.1.1	Hardware	215
		11.5.1.2	Software	218
		11.5.1.3	Systems Engineering and Project Management	218
		11.5.1.4	Performance-Based Cost Estimating Relationship	220
		11.5.1.5	Weight-Based Cost Estimating Relationship	223
		11.5.2	Proposed Cost Drivers for DoD 5000.02 Phase Operations and Support	224
		11.5.2.1	Logistics–Transition from Contractor Life Support (CLS) to Organic Capabilities	224

	11.5.2.2	Training	224
	11.5.2.3	Operations – Manned Unmanned Systems Teaming (MUM-T)	225
11.6	Considerations for Estimating Unmanned Ground Vehicle Costs		225
11.7	Additional Considerations for UMAS Cost Estimation		230
	11.7.1	Test and Evaluation	230
	11.7.2	Demonstration	230
11.8	Conclusion		230
Acknowledgments			231
References			231

12 Logistics Support for Unmanned Systems — **233**

12.1	Introduction		233
12.2	Appreciating Logistics Support for Unmanned Systems		233
	12.2.1	Logistics	234
	12.2.2	Operations Research and Logistics	236
	12.2.3	Unmanned Systems	240
12.3	Challenges to Logistics Support for Unmanned Systems		242
	12.3.1	Immediate Challenges	242
	12.3.2	Future Challenges	242
12.4	Grouping the Logistics Challenges for Analysis and Development		243
	12.4.1	Group A – No Change to Logistics Support	243
	12.4.2	Group B – Unmanned Systems Replacing Manned Systems and Their Logistics Support Frameworks	244
	12.4.3	Group C – Major Changes to Unmanned Systems Logistics	247
12.5	Further Considerations		248
12.6	Conclusions		251
References			251

13 Organizing for Improved Effectiveness in Networked Operations — **255**

13.1	Introduction	255
13.2	Understanding the IACM	256
13.3	An Agent-Based Simulation Representation of the IACM	259
13.4	Structure of the Experiment	260
13.5	Initial Experiment	264
13.6	Expanding the Experiment	265
13.7	Conclusion	269
Disclaimer		270
References		270

14 An Exploration of Performance Distributions in Collectives — **271**

14.1	Introduction	271
14.2	Who Shoots How Many?	272
14.3	Baseball Plays as Individual and Networked Performance	273
14.4	Analytical Questions	275

14.5   Imparity Statistics in Major League Baseball Data   277
         14.5.1   *Individual Performance in Major League Baseball*   278
         14.5.2   *Interconnected Performance in Major League Baseball*   281
14.6   Conclusions   285
Acknowledgments   286
References   286

**15   Distributed Combat Power: The Application of Salvo Theory
        to Unmanned Systems**   **287**
15.1   Introduction   287
15.2   Salvo Theory   288
         15.2.1   *The Salvo Equations*   288
         15.2.2   *Interpreting Damage*   289
15.3   Salvo Warfare with Unmanned Systems   290
15.4   The Salvo Exchange Set and Combat Entropy   291
15.5   Tactical Considerations   292
15.6   Conclusion   293
References   294

Index   **295**

About the Contributors

Rajan Batta
Associate Dean for Research and Graduate Education, School of Engineering
and Applied Sciences
SUNY Distinguished Professor, Department of Industrial & Systems Engineering
University at Buffalo (State University of New York), NY
USA

Fred D. J. Bowden
Land Capability Analysis Branch
Joint and Operations Analysis Division
DSTO-E, Edinburgh, SA
Australia

Shannon R. Bowling
Dean of the College of Engineering Technology and Computer Science
Bluefield State College
Bluefield, WV
USA

Juan José Vales Bravo,
Environmental Information Network of Andalusia
Environment and Water Agency
Johan Gutenberg, Seville
Spain

Elena María Méndez Caballero
Environmental Information Network of Andalusia
Environment and Water Agency
Johan Gutenberg, Seville
Spain

Jeffrey R. Cares
Captain, US Navy (Ret.)
Alidade Inc.
USA

Francisco Cáceres Clavero
Regional Ministry of Environment and Spatial Planning
Avenida Manuel Siurot, Seville
Spain

Kelly Cohen
Department of Aerospace Engineering and Engineering Mechanics
College of Engineering
University of Cincinnati
Cincinnati, OH
USA

Andrew W. Coutts
Land Capability Analysis Branch
Joint and Operations Analysis Division
DSTO-E, Edinburgh, SA
Australia

Agamemnon Crassidis
The Kate Gleason College of Engineering
Rochester Institute of Technology
Rochester, NY
USA

Sean Deller
Joint and Coalition Warfighting (J7)
Washington, DC
USA

Richard M. Dexter
Land Capability Analysis Branch
Joint and Operations Analysis Division
DSTO-E, Edinburgh, SA
Australia

John Q. Dickmann, Jr.
Sonalysts Inc.
USA

Fernando Giménez de Azcárate
Regional Ministry of Environment and Spatial Planning
Avenida Manuel Siurot, Seville
Spain

Luke Finlay
Land Capability Analysis Branch
Joint and Operations Analysis Division
DSTO-E, Edinburgh, SA
Australia

Matthew J. Henchey
Herren Associates
Washington, DC
USA

Patrick Chisan Hew
Defence Science and Technology Organisation
HMAS Stirling
Garden Island, WA
Australia

Raymond R. Hill
Department of Operational Sciences
US Air Force Institute of Technology/ENS
Wright-Patterson AFB
Dayton, OH
USA

David Hopkin
Section Head
Maritime Asset Protection
Defence Research and Development Canada – Atlantic
Dartmouth, NS
Canada

Keirin Joyce
University of New South Wales at the Australian Defence Force Academy
  (UNSW Canberra)
Australia

Mark Karwan
Department of Industrial and Systems Engineering
University at Buffalo (State University of New York), NY
USA

Elad Kivelevitch
Department of Aerospace Engineering and Engineering Mechanics
College of Engineering
University of Cincinnati
Cincinnati, OH
USA

Manish Kumar
Department of Mechanical, Industrial & Manufacturing Engineering
University of Toledo
Toledo, OH
USA

José Manuel Moreira Madueño
Regional Ministry of Environment and Spatial Planning
Avenida Manuel Siurot, Seville
Spain

David M. Mahalak
Applied Logistics Integration Consulting
Dallas, PA
USA

Gregoria Montoya Manzano
Environmental Information Network of Andalusia
Environment and Water Agency
Johan Gutenberg, Seville
Spain

Raquel Prieto Molina
Environmental Information Network of Andalusia
Environment and Water Agency
Johan Gutenberg, Seville
Spain

Bao Nguyen
Senior Scientist
Defence Research and Development Canada
Centre for Operational Research and Analysis
Ottawa, ON
Canada

Ben Pietsch
Land Capability Analysis Branch
Joint and Operations Analysis Division
DSTO-E, Edinburgh, SA
Australia

Ghaith Rabadi
Department of Engineering Management & Systems Engineering (EMSE)
Old Dominion University
Norfolk, VA
USA

Laura Granado Ruiz
Environmental Information Network of Andalusia
Environment and Water Agency
Johan Gutenberg, Seville
Spain

Thomas R. Ryan, Jr. (Tommy Ryan)
Department of Systems Engineering
United States Military Academy
West Point, NY
USA

Irene Rosa Carpintero Salvo
Environmental Information Network of Andalusia
Environment and Water Agency
Johan Gutenberg, Seville
Spain

Isabel Pino Serrato
Environmental Information Network of Andalusia
Environment and Water Agency
Johan Gutenberg, Seville
Spain

Denis R. Shine
Land Capability Analysis Branch
Joint and Operations Analysis Division
DSTO-E, Edinburgh, SA
Australia

Brian B. Stone
Department of Operational Sciences, US Air Force Institute of Technology/ENS
Wright-Patterson AFB, OH
USA

Huang Teng Tan
Department of Operational Sciences
US Air Force Institute of Technology/ENS
Wright-Patterson AFB
Dayton, OH
USA

Andreas Tolk
The MITRE Corporation
USA

Ricardo Valerdi
Department of Systems and Industrial Engineering
University of Arizona
Tucson, AZ
USA

Handson Yip
Supreme Allied Command Transformation
Staff Element Europe
SHAPE, Mons
Belgium

Acknowledgements

At the end of any major literary undertaking such as writing this book, authors take time to thank the many people who support their work. In this case, however, the editors are supporting, and the main effort, the real work that went into this book, was accomplished by the contributing authors. We justifiably thank them for the time and effort spent to create each chapter, and the patience they displayed with a lengthy submission and editing process. There is no doubt, moreover, that each of them has their own thanks to convey, but it would have made the process much longer and more complicated to include their own thanks here. To those who also supported the various authors in their independent efforts, we are likewise truly grateful.

We are also grateful for the support provided by the Military Operations Research Society (MORS). Four of the chapters, those by Han and Hill; Nguyen, Hopkin and Yip; Henchey, Batta, Karwan and Crassidis; and Deller, Rabadi, Tolk and Bowling first appeared in their research publication, *MORS Journal*. It is with their kind permission that they are reprinted here. It speaks loudly for the *MORS Journal* that some of the cutting edge research on this important topic was first presented in their pages.

1

Introduction

Jeffrey R. Cares[1] and John Q. Dickmann, Jr.[2]

[1] Captain, US Navy (Ret.), Alidade Inc., USA
[2] Sonalysts Inc., USA

1.1 Introduction

Given all the attention and investment recently bestowed on unmanned systems, it might seem surprising that this book does not already exist. Even the most cursory internet search on this topic will show professional journal articles, industry symposia proceedings, and technical engineering texts conveying broad interest, substantial investment, and aggressive development in unmanned systems. Yet an internet bookstore or library search for "operations research" combined with "unmanned systems" will come up blank. This book will indeed be the first of its kind.

Historians of military innovation would not be surprised. In fact, they point to a recurring tendency of the study of *usage* to lag *invention*. Such a hyper-focus on engineering and production might be perfectly understandable (for program secrecy, to work the "bugs" out of early production models, or simply because of the sheer novelty of radically new devices) but the effect is often the same: a delayed understanding of how operators could use new hardware in new ways. As the preeminent World War II scientist P. M. S. Blackett observed of the innovations of his time, "relatively too much scientific effort has been expended hitherto on the *production* of new devices and too little in the *proper use* of what we have got." [1] It is ironic that a study of usage is one of the best ways to understand how to develop and improve a new technology; but engineering, not usage, gets the most attention early in an innovation cycle.

One does not need to be a student of military innovation to know that the study of usage is not the engineer's purview. Blackett's counterparts across the Atlantic, Morse and Kimball, noted that the "the branches ... of engineering ... are involved in the construction and production of equipment, whereas operations research is involved in its use. The engineer is the consultant to the builder, the producer of equipment, whereas the operations research worker

Operations Research for Unmanned Systems, First Edition. Edited by
Jeffrey R. Cares and John Q. Dickmann, Jr.
© 2016 John Wiley & Sons, Ltd. Published 2016 by John Wiley & Sons, Ltd.

is the consultant to the *user* of the equipment." [2] Engineering tells you how to build things and operations research tells you how things should be used. In development of new military hardware, however, engineering nearly always has a head start over operations research.

Three of the many ways that engineering overshadows usage early in unmanned systems development have delayed a book such as this from reaching professional bookshelves. The first is that most engineers have not yet recognized that unmanned systems can be so much more than merely systems without a human onboard. This *anthropomorphism* – creating in our own image – was the first fertile ground for engineers, and early success with this approach made it seem unnecessary to conceive of unmanned operations as any different than those studied by operations researchers for decades.

The second reason is that since engineers build *things*, not *operations*, the engineer's approach to improving operations is to refine the vehicles. Such engineering-centered solutions have already been observed in existing unmanned programs, driving up vehicle complexity and cost – without regard to how modifying operational schemes might be a better way to increase operational performance.

The third reason is that since humans are the most expensive "total cost of ownership" (TOC) components of modern military systems, the military and defense industries have been content to lean on "manpower cost avoidance" as the overriding value proposition for unmanned systems. For now, unmanned systems are convincingly sold on cost alone – there is no reason for program managers to answer questions about operational value that no one is yet asking. The engineer's present task is to keep development and production costs lower than equivalent manned systems for a given level of performance – not to explore the performance–cost trade space.

The historian of military innovation would be quick to clarify that usage lags invention mostly in the *initial* phases of maturation. Engineers and program managers pre-occupied with production can indeed be quite successful. In the case of unmanned vehicle development, second- and third-generation variants have already replaced prototypes and initial production models in the fleet, field, and flightline. Major acquisition programs (such as the Global Hawk and Predator systems) are already out of adolescence. Now that well-engineered platforms are employed on a much larger scale, a growing cadre of operations research analysts are at last being asked to answer operational questions – questions of usage.

While the three reasons cited above are among those that have heretofore preempted this book, they also constitute an initial set of topics for the operations researcher. What we might now call "operations research for unmanned systems" is emerging with three main themes:

- *The Benefits of "Unmanning":* While the challenges of removing humans from platforms are still manifold and rightfully deserve our attention, operations researchers are now looking past the low hanging fruit of "unmanning" these systems – such as less risk to humans, longer sortie duration, higher *g*-force tolerance – to develop entirely new operations for unmanned systems and to discern new ways of measuring effectiveness.
- *Improving Operations:* The introduction of large numbers of unmanned vehicles into a legacy order of battle may transform warfare in profound ways. Some authors in the defense community have coined the term "Age of Robotics" to refer to this transformation, but from an analytical perspective, this term (like "Network Centric Warfare" and others of their ilk) is still more rubric than operational concept. While a full appreciation of such a new age may remain elusive, operations researchers are approaching the study

of unmanned collectives in a more modest way. Through careful study and operational experimentation with smaller groups of vehicles, these analysts are starting to build evidence for claims of increasing returns and show why and how they may be possible (or, just as importantly, not).

- *The True Costs of Unmanned Systems:* The only "unmanned" part of today's unmanned systems are the vehicles – the humans have been moved somewhere else in the system. The life-cycle cost savings accrue to the platforms, but is the overall system cheaper? In some systems, centralized human control and cognition may be a much more costly approach, requiring substantially more technological investment, greater manning, and networks with much higher capacity than legacy manned systems. Analyzing this trade space is an area of new growth for operations research.

1.2   Background and Scope

As recent as the late 1990s, unmanned vehicles were still seen as a threat to the legacy defense investments of the world's leading defense establishments. Even the mildest endorsements of their value to the warfighter for anything but the most mundane military tasks were met with derision, suspicion, and resistance. At the same time, more modest militaries and their indigenous industries – unconstrained by the need to perpetuate big-ticket, long-term acquisition strategies – began to develop first-generation unmanned platforms and capabilities that could no longer be denied by their bigger counterparts.

Concurrently and independently, innovations in secure, distributed networking and high-speed computing – the two most basic building blocks of advanced unmanned systems – began to achieve the commercial successes that made unmanned military vehicles seem more viable as a complement to legacy platforms in the fleet, field, and flight-line. But while the war on terror has seen focused employment of surveillance drones and explosive ordinance disposal robots, defense budget reductions are spurring a more widespread use of unmanned military systems more for the cost savings they provide than for the capabilities they deliver.

The five-year future of unmanned systems is uncertain, except in one respect: every new operational concept or service vision produced by the world's leading militaries *expect* that unmanned vehicles will be a major component of future force structures. The details of this expectation – which platforms will garner the most investment, what technological breakthrough will have the most impact or where unmanned systems will have their first, game-changing successes – are the subject of intense speculation. This book will be successful if it helps bring some operational focus to the current debate.

While it is common to assert that increasing returns must surely accrue as more unmanned hardware is connected to a larger "network-enabled" systems of systems, engineers still concentrate on the robotic vehicles, unable to conceive of how unmanned collectives might indeed perform better than merely the sum total of all the vehicles' individual performance. Without better analyses of group operations, the engineer's solution to improving the performance of a collective is simply to engineer better performance into each vehicle of the group. Network engineers have been the loudest advocates for "networked effects," but like the hardware engineers they have largely ignored operations research, devoting their efforts to engineering architectural standards and interconnection protocols. To make matters worse, in many cases the *process* (engineering activity) has become the *product*.

This book will benefit readers by providing them with a new perspective on how to use and value unmanned systems. Since there is no other place where these types of analyses are yet assembled, this book will serve as a seminal reference, establishing the context in which operations research should be applied to unmanned systems, catalyzing additional research into the value of unmanned platforms, and providing critical initial feedback to the unmanned systems engineering community.

Good operations research analysis is at once digestible by operators and informative to specialists, so we have attempted to strike a balance between the two. Fortunately, nearly all defense community operators have a solid technical education and training (albeit somewhat dated), and can follow the main arguments from college-level physics, statistics, and engineering. Chapters in this book should briefly refresh their education and bring it into operational context. Defense engineers, by contrast, are expert at their applicable "hard science," but must be informed of operational context. This book should confirm for a technical audience that the writers understand the most important technical issues, and then show how the technical issues play out in an operational context. Both will buy this book expecting to learn something more than they already know about unmanned systems; this book will have to approach this learning experience from both of these perspectives.

1.3   About the Chapters

Fourteen chapters follow this introduction. Considering that unmanned vehicle systems development is by nature multidisciplinary, there are certainly many ways that these chapters might be appropriately arranged. The editors opted to arrange the chapters on a continuum from individual problems to analyses of vehicle groups, then to organizational issues, and finally to broad theoretical questions of command and control. Some of the topics may be new intellectual ground for many readers. For this reason, the editors have tried to ensure each chapter has enough basic context for a general audience with some mathematical background to digest each chapter, no matter what the subject. They also hope that this will satisfy readers who come to this book for, say, unmanned vehicle routing techniques, to stay for a discussion of Test and Evaluation or TOC.

Huang Teng Tan and Dr. Raymond R. Hill of the US Air Force Institute of Technology provided the first chapter, *The In-Transit Vigilant Covering Tour Problem for Routing Unmanned Ground Vehicles*. One might rightly wonder why Air Force researchers care about unmanned ground vehicles, but the answer is simple: the US Air Force has a significant Force Protection mission at its many bases worldwide, and the total ownership costs of human sentries are high. Unmanned sentries can augment and replace humans at lower costs. This chapter provides a formal discussion of how to efficiently address the covering tour problem, or in other words, what is the best way for a robotic sentry to "make its rounds." There are obvious border patrol and civilian security applications of this research.

The next chapter, *Near-Optimal Assignment of UAVs to Targets Using a Market-Based Approach* by Dr. Elad Kivelevitch, Dr. Kelly Cohen, and Dr. Manish Kumar, is an application of "market-based" optimization to sensor–target pairings. This family of optimization techniques is inspired by economic markets, such as in this example where unmanned vehicles act as rational economic agents and bid for targets using a valuation and trading scheme. The authors show the benefits and limits of this approach to obtaining a fast, reliable optimization under conditions of high uncertainty.

A chapter discussing naval applications of unmanned underwater vehicles comes next. In *Considering Mine Countermeasures Exploratory Operations Conducted by Autonomous Underwater Vehicles*, Dr. Bao Nguyen, David Hopkin, and Dr. Handson Yip look at ways to evaluate the performance of Commercial Off-The-Shelf (COTS) unmanned underwater vehicles in searches for underwater mines. They present and discuss measures of effectiveness and compare and contrast different search patterns.

Optical Search by Unmanned Aerial Vehicles: Fauna Detection Case Study, by Raquel Prieto Molina *et al.*, is a very interesting chapter that harkens back to some of the very early operations research work from World War II. Readers familiar with Koopman's *Search and Screening*[3], for example, will note the strong parallel between this chapter and World War II research on lateral range curves and the inverse cube law. Both were trying to describe the basic physics of visual detection (Prieto *et al.*, are, of course dealing with artificial visual detection), and how it impacts search patterns and detection probabilities.

There are many cases in modern military operations where a clever scheme or algorithm devised *in silico* unravels when it is placed in operation in a real environment. Recognizing this, Dr. Matthew J. Henchey, Dr. Rajan Batta, Dr. Mark Karwan, and Dr. Agamemnon Crassidis show how algorithms might compensate for environmental effects in *Flight Time Approximation Model for Unmanned Aerial Vehicles: Estimating the Effects of Path Variations and Wind*. While crafted for air vehicles, this research could be adapted for any unmanned vehicles operations where delay or resistance are encountered (such as set and drift at sea, reduced trafficability on land, or interruption by an adversary in any medium).

For many militaries and corporations, unmanned vehicles are now major acquisition programs, requiring high-level analyses of alternatives (AOAs) not just between vehicles, but between human–vehicle hybrid systems. Fred D. J. Bowden, Andrew W. Coutts, Richard M. Dexter, Luke Finlay, Ben Pietsch, and Denis R. Shine present a template for these types of studies in *Impacts of Unmanned Ground Vehicles on Combined Arms Team Performance*. While specific to Australian Army trade-off analyses, this chapter is certainly useful for other AOA analyses in both cabinet departments and corporate executive suites.

With respect to human–vehicle hybrids, how much human work should robots actually do? All senior military officials will insist that a human must always be in the loop, but in many cases this is just to confirm an automated solution before weapons are employed. But how much was this human involved in the automated solution, and how reliable is the robot's "thinking"? Patrick Chisan Hew's chapter, *Processing, Exploitation, and Dissemination: When is Aided/Automated Target Recognition "Good Enough" for Operational Use?*, offers a formal mathematical treatment to this and similar questions of the operational and ethical impact of automated cognition.

Also exploring the man–machine trade-space is *Analyzing a Design Continuum for Automated Military Convoy Operations,* by David M. Mahalak. This chapter used logistics convoy operations to show how automated control can supplant human control in a continuum of increasingly automated convoys. This is yet another chapter that applies to a broader range of vehicles and operations.

Continuing along the continuum to higher levels of organizational problems, another chapter from Dr. Raymond R. Hill (this time with Brian B. Stone, also of the US Air Force Institute of Technology), *Experimental Design for Unmanned Aerial Systems Analysis: Bringing Statistical Rigor to UAS Testing,* addresses new operational test and evaluation issues wrought by the introduction of unmanned systems.

It has long been assumed that automated systems are cheaper than manned systems. As this introduction has stated, however, this has not been as well investigated as investments in unmanned vehicle systems should warrant. Dr. Ricardo Valerdi and Captain Thomas R. Ryan, Jr., US Army, address this issue and provide costing techniques in *Total Cost of Ownership (TOC): An Approach for Estimating UMAS Costs.*

Part of the TOC of any system are the costs associated with logistics and maintenance. Major Keirin Joyce, Australian Army, discusses modeling techniques for logistics operations with a focus on how well current logistics models can support unmanned vehicle operations. In a very important section of this chapter, *Logistics Support for Unmanned Systems,* Major Joyce extrapolates from current models and operations to address logistics support challenges for future unmanned systems.

As more systems are automated and dispersed throughout the battlespace or commercial work environment, there is an increasing need to understand how networks of collectives are effectively operated and controlled. *Organizing for Improved Effectiveness in Networked Operations* by Dr. Sean Deller, Dr. Ghaith Rabadi, Dr. Andreas Tolk, and Dr. Shannon R. Bowling combines concepts of interaction patterns in biochemistry with modern agent-based modeling techniques to explore a general model of command and control in a distributed, networked system.

Two chapters addressing theoretical topics complete the volume. *An Exploration of Performance Distributions in Collectives* by Jeffrey R. Cares compares individual and collective performance in competition, using baseball as a proxy. In *Distributed Combat Power: The Application of Salvo Theory to Unmanned Systems*, the same author shows how Hughes' Salvo Equations might be modified to evaluate outcomes from missile combat between large platforms when advanced unmanned vehicles are employed.

References

1. P. M. S. Blackett, originally in *Operational Research Section Monograph*, "The Work of the Operational Research Section," Ch. 1, p. 4., quoted from Samuel E. Morison, *The Two Ocean War*, Little Brown, Boston, 1963, p. 125.
2. P. Morse & G. Kimball, *Methods of Operations Research*, John Wiley & Sons, Inc., New York, 1951.
3. B. O. Koopman, Pergamon Press, New York, 1980.

2

The In-Transit Vigilant Covering Tour Problem for Routing Unmanned Ground Vehicles*

Huang Teng Tan and Raymond R. Hill
Department of Operational Sciences, US Air Force Institute of Technology/ENS, Wright-Patterson AFB, Dayton, OH, USA

2.1 Introduction

The Maximization of Observability in Navigation for Autonomous Robotic Control (MONARC) project within the Air Force Institute of Technology (AFIT) has an overarching goal to develop an autonomous robotic, network-enabled, Search, Track, ID, Geo-locate, and Destroy (Kill Chain) capability, effective in any environment, at any time. One area of interest in the MONARC project is mission planning for base security protection of Key Installations (KINs) from adversarial intrusions using autonomous Unmanned Ground Vehicles (UGVs). This UGV mission planning task is multifaceted and requires the consolidation of intelligence, management of system readiness, centralized operational planning and dissemination of Command and Control (C2) information. Sensory data from various locations around the KINs are fused into a Recognized Ground Situation Picture (RGSP) and augmented with intelligence from various agencies. A centralized C2 center consolidates and manages the real-time system serviceability and readiness state of the UGVs. The mission planners input the security requirements, such as key surveillance points, potential intrusion spots, and Rules of Engagement (ROE) into a Ground Mission Planning System which provides a surveillance approach for use by the team of UGVs.

The protection of a large KIN, such as a military airbase, requires a team of UGVs patrolling along certain routes to effectively cover numerous intrusion spots. The surveillance approach of the security defense task can be formulated as a combinatorial optimization model, in which

---

*Contribution originally published in MORS Journal, Vol. 18, No. 4, 2013. Reproduced with permission from Military Operations Research Society.

Operations Research for Unmanned Systems, First Edition. Edited by
Jeffrey R. Cares and John Q. Dickmann, Jr.
© 2016 John Wiley & Sons, Ltd. Published 2016 by John Wiley & Sons, Ltd.

multiple security entities must visit multiple locations, while covering certain adversarial locations, in the shortest total distance traveled for all entities. When adversarial locations are sensed by UGVs at their route location, the problem is a Covering Tour Problem (CTP) [1–3]. The CTP is a Traveling Salesman Problem (TSP) with a Set Covering Problem (SCP) structure. The multiple vehicle variant is a natural extension. Not addressed in prior CTP research, but quite applicable in the current context, is the vehicle covering capability while transiting along edges between route locations. This in-transit vigilance component is important to the current mission planning environment. A new variant of the multiple vehicle Covering Tour Problem (mCTP) model called the in-transit Vigilant Covering Tour Problem (VCTP) is used as a mission planning tool for base security applications. The single vehicle version is evaluated to assess the viability of the VCTP.

2.2   Background

The CTP model has been applied extensively in the health care industry, especially in planning deployments of mobile health care units traveling in developing countries [4]. Mobile health care units have access to a limited number of villages due to factors such as infrastructure restrictions, unit capacity, and cost. Therefore, it is infeasible to travel to all villages. Instead, a tour route is planned so that the unvisited villages are within reasonable walking distance of the visited villages thereby providing greater access for those in need of health care. The vehicle routes of the health care units are efficiently planned to reduce the amount of travel required, but to visit enough villages to provide sufficient overall medical coverage. A real-life problem associated with the planning of mobile health care units was in the Suhum district, Ghana [5] and solved by Hachicha *et al.* [6] as a CTP.

Another important application of the CTP is the placement of mailbox locations to reduce the traveling distance of the postal delivery service while ensuring maximum coverage [7]. Good locations of mailboxes to cover a region of users and an optimal route for mail distribution are constructed. Alternatively, this approach applies to the management of centralized post offices, that is, post offices are centralized in towns with larger populations while the smaller nearby towns are covered by the centralized post offices.

The CTP model has also been applied to the transportation industry such as in the design of bi-level, hierarchical transportation networks [8]. For an overnight mail delivery service provider, such as DHL, FedEx, and so on, the optimal tour route represents the route taken by the primary vehicle (aircraft) to the distribution centers and the coverage radius is represented by the maximum distance traveled by the delivery trucks from the distribution centers to the customers. The route ensures that the overall distribution cost is minimized and provides the required delivery service to its customers.

The mCTP [6] is defined as a complete undirected graph $G = (V, E)$ where V of size $n + 1$, and indexed as $v_0, ..., v_n$, is the vertex set, and $E = \{(v_i, v_j) \mid v_i, v_j \in V, i \neq j\}$ is the edge set. Vertex v_0 is a depot (base station), $U \subseteq V$ is the subset of vertices that must be visited ($v_0 \in T$), and W is the set of targets to cover of size p and indexed as $w_1, ..., w_p$. Each element of V and W have a location given by their x and y coordinates. A distance matrix $C = (c_{ij})$ provides the edge length for each element in E. A final parameter is c, the pre-defined maximum size of the cover. A solution to the mCTP consists in defining a set of m vehicle routes of minimum total length, all starting and ending at the depot such that every target in W is covered. Target coverage is satisfied if it lies within distance c from a vertex in V in a tour route. Each vehicle uniquely visits a selected vertex within its route but vertices may overlap among the individual vehicle routes. Targets need only be covered one time although multiple coverage is permitted.

2.3   CTP for UGV Coverage

One defined MONARC scenario relates to the protection of KINs from potential adversarial intrusions. A team of UGVs are tasked to protect a critical installation. Their surveillance capabilities are augmented by static sensors located throughout the installation. An RGSP is available to a mission planner to assist in finding UGV tour routes. The UGVs should only patrol routes that cover all the required checkpoints and the overall route length should be minimized. All UGVs originate from a base station, the depot. There are certain checkpoints that the UGVs must visit (these checkpoints are usually critical ones requiring compulsory surveillance) and there are also checkpoints that may be visited. There are also potential spots where the adversary may appear and these spots must be covered by visiting a checkpoint that is within a fixed proximity distance. Each checkpoint is visited by a UGV and all UGVs return to the base station.

Coverage of targets by a UGV at a visited checkpoint (vertex) is defined as the circular area of a fixed radius, where any target within the area is covered by that checkpoint. The circular area of coverage is analogous to the effective range of a weapon or sensor system onboard the UGVs. Each UGV is modeled as an individual vehicle traveling on different routes of minimum-length tours. During the route, the UGV covers targets when at the vertices and all targets must be covered for there to be a feasible solution to the overall problem.

There are some key assumptions and limitations made in modeling the base defense security scenario as an mCTP:

1. All UGVs are homogeneous and have equal capabilities in movement and coverage.
2. UGVs can uniquely visit as many vertices as needed and transit as long as required.
3. UGVs travel in a straight line between vertices.
4. Potential adversarial spots are known at a specific point of time or are pre-defined and are thus part of the problem structure.

In reality, UGVs, or for that matter any sensor craft, can sense while traveling. Thus, coverage only at vertices is artificially limiting and coverage while in transit between vertices is reasonable. The CTP model is thus extended to include target coverage via traveled edges. This new variant of the CTP for a generic base security defense scenario, which considers *coverage by both visited vertices and traveled edges*, is described in the next section. The extension of the VCTP to an mVCTP (multiple vehicle Vigilant Covering Tour Problem) is discussed in the subsequent section.

2.4   The In-Transit Vigilant Covering Tour Problem

The CTP can be used to model a UGV assigned to protect a critical installation. However, a scenario may exist in which a potential adversarial spot is not covered by any vertex. In this case, the CTP is infeasible. Figure 2.1 illustrates a single vehicle example which is infeasible.

The tour in Figure 2.1 is a minimum-length tour constructed with all required vertices visited. However, the solution is infeasible since there is an uncovered target. We note the UGV could sense the target while in transit. The CTP model is modified to allow coverage of such targets by the en route UGV.

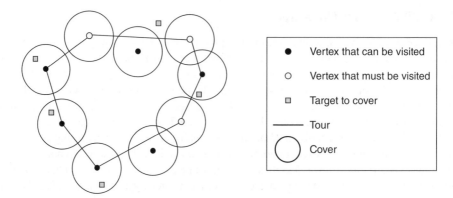

Figure 2.1 Possible solution for a CTP. *Source*: Tan, Huang Teng; Hill, Raymond R., "The In-Transit Vigilant Covering Tour Problem for Routing Unmanned Ground Vehicles," *MORS Journal*, Vol. 18, No. 4, 2013, John Wiley and Sons, LTD.

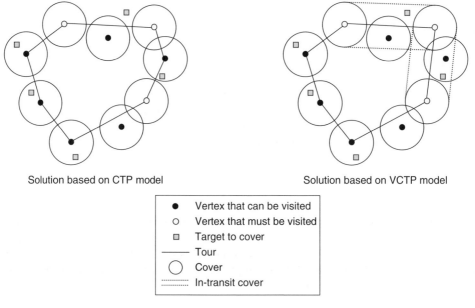

Figure 2.2 Optimal solution for a CTP and a VCTP. *Source*: Tan, Huang Teng; Hill, Raymond R., "The In-Transit Vigilant Covering Tour Problem for Routing Unmanned Ground Vehicles," *MORS Journal*, Vol. 18, No. 4, 2013, John Wiley and Sons, LTD.

Considering coverage *during transit* is a logical assumption for a base security defense problem, that is, a UGV can cover a potential adversarial spot during its movement between checkpoints. Thus, while a UGV is traveling along the route and transiting between check-points, it could pass within some fixed proximity distance and detect (or cover) the adversarial spot. Figure 2.2 compares the infeasible result in Figure 2.1 with a solution based on the VCTP.

There are three distinct differences between the CTP and VCTP models. First, note that the vertex not covered in the CTP is now covered in the VCTP during a route transition with no change of route required. Thus, the revised model effectively increases the amount of coverage as both the traveled vertices and edges provide coverage. Second, we can shorten the tour length, as one of the visited vertices is not required in the VCTP tour since a tour edge provides the requisite coverage. Lastly, the solution based on the VCTP model is feasible.

As illustrated in Figure 2.2, the coverage of a target by a vertex visit changes to coverage by an edge transit, thus if the same vertices are considered, the solution of the VCTP model will involve an equal or lesser number of vertices compared to the CTP model. Thus, the optimal tour length of the VCTP model provides a lower bound for the CTP. Since target coverage is by both edge and/or vertex, the VCTP provides an upper bound for the CTP on targets covered.

2.5   Mathematical Formulation

The mathematical development of the VCTP is presented in this section. The basic Vehicle Routing Problem(VRP) model [9] was used as a basis for the VCTP model. We utilize the two-index vehicle flow formulation in the single vehicle variant of the VCTP model [10]; it is extended into the three-index vehicle flow formulation for the mVCTP. The two-index vehicle flow formulation uses $O(n^2)$ binary variables a_{ij} and $O(n)$ binary variables b_i, where a_{ij} and b_i are defined as:

$$a_{ij} = \begin{cases} 1, & \text{edge}\left(v_i, v_j\right)\text{is part of the tour} \\ 0, & \text{otherwise} \end{cases}$$

$$b_i = \begin{cases} 1, & \text{vertex } v_i \text{ is part of the tour} \\ 0, & \text{otherwise} \end{cases}$$

An important component of the VCTP lies in the introduction of two pre-processed matrices, α_{ij}^k and β_i^k, which are defined as the in-transit edge coverage and vertex coverage matrices, respectively.

For the in-transit edge coverage matrix α_{ij}^k, for target, k, an i by j matrix is formulated to determine if edge (v_i, v_j) can provide coverage of a target. Figure 2.3 illustrates in-transit vigilant coverage of target w_k by edge (v_i, v_j) as it lies within the pre-determined perpendicular distance c from the edge.

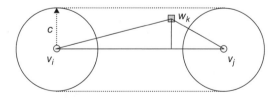

Figure 2.3   In-transit vigilant coverage by edge (v_i, v_j) on target w_k. *Source*: Tan, Huang Teng; Hill, Raymond R., "The In-Transit Vigilant Covering Tour Problem for Routing Unmanned Ground Vehicles," *MORS Journal*, Vol. 18, No. 4, 2013, John Wiley and Sons, LTD.

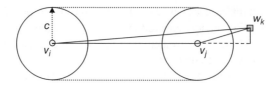

Figure 2.4 Lack of coverage by edge (v_i, v_j) on target w_k. *Source*: Tan, Huang Teng; Hill, Raymond R., "The In-Transit Vigilant Covering Tour Problem for Routing Unmanned Ground Vehicles," *MORS Journal*, Vol. 18, No. 4, 2013, John Wiley and Sons, LTD.

The construction of the matrix should be carefully considered as a target could lie within the perpendicular distance c from the edge, but fall outside the perpendicular boundaries of edge (v_i, v_j) as shown in Figure 2.4. The α_{ij}^k will have a value of 0 in such a case.

Thus, if we consider the triangle bounded by vertices v_i, v_j, and w_k, the following two conditions must hold for the target to be covered by the edge:

1. Target w_k must lie within the pre-determined perpendicular distance c from edge (v_i, v_j).
2. Angles at vertices v_i and v_j must be equal or less than 90 degrees.

Therefore the value of α_{ij}^k is defined as:

$$\alpha_{ij}^k = \begin{cases} 1, & \text{edge}\left(v_i, v_j\right) \text{covers } w_k \\ 0, & \text{otherwise} \end{cases}$$

Algorithm 2.1 defines the construction of the α_{ij}^k matrix:

Algorithm 2.1   Construction of α_{ij}^k Matrix (Edge Covering Matrix).

```
Given: Set of V vertices and set of W targets with their
   coordinates, x and y, and distance matrix C
Initialize matrix α_ij^k = [0]
for all i from 1 to n, j from 1 to n (i ≠ j) and k from 1 to p do
         Construct triangle with corners having corresponding
coordinates (x_i, y_i), (x_j, y_j) & (x_k, y_k)
             Let the opposite side lengths be x, y & z where
             x = c_jk,
             y = c_ik,
             z = c_ij
             Let s = (x + y + z)/2
```
$$h = \frac{2}{c_{ij}} \sqrt{s(s-x)(s-y)(s-z)}$$
```
         Let
             if h ≤ c, do
```
$$Angle\, i = arcos\left(\frac{y^2 + z^2 - x^2}{2yz}\right)$$
```
             Let
```
$$Angle\, j = arcos\left(\frac{x^2 + z^2 - y^2}{2xz}\right)$$

```
                               Let
                    if angle i ≤ 90 degs AND
    angle j ≤ 90 degs, then
                                         α_ij^k = 1
                         else, α_ij^k = 0
                         end
               else, α_ij^k = 0
               end
           Update α_ij^k
end
```

The vertex covering matrix, β_i^k, is also formulated as a binary matrix. Element (i, k) takes a value of 1 if target w_k is within distance c of vertex v_i[3].

The single vehicle VCTP therefore involves the undirected graph G, with vertex set V and edge set E with edge length matrix C, target set W, and the new covering indicator matrices α_{ij}^k and β_i^k. To prevent the formation of subtours, a Subtour Elimination (STE) constraint is added [11].

The formulation of the integer linear program (LP) of the VCTP model is as follows.

Sets	
V	Set of vertices indexed using i and j
W	Set of targets to cover, indexed using k
E	Set of edges $\left\{ \left(v_i, v_j \right) \mid v_i, v_j \in V, i \neq j \right\}$
Data	
c_{ij}	Distance of edge (v_i, v_j)
α_{ij}^k	1 if edge (v_i, v_j) covers target k, 0 otherwise
β_i^k	1 if vertex v_i covers target k, 0 otherwise
Binary Decision Variables	
a_{ij}	1 if edge (v_i, v_j) is part of the tour, 0 otherwise
b_i	1 if node v_i is part of the tour, 0 otherwise

$$\text{Minimize} \sum_{(v_i, v_j) \in E} c_{ij} a_{ij} \tag{2.1}$$

Subject to

$$\sum_{i \in V} a_{ij} = b_j \ \forall \, j \in V \tag{2.2}$$

$$\sum_{i \in V} \beta_i^k b_i + \sum_{(v_i, v_j) \in E} \alpha_{ij}^k a_{ij} \geq 1 \, \forall \, k \in W \tag{2.3}$$

$$\sum_{i \in V} a_{ij} = \sum_{l \in V} a_{jl} \ \forall \, j \in V \tag{2.4}$$

$$\sum_{j=1}^{n} a_{1j} = 1, \sum_{i=1}^{n} a_{i1} = 1 \tag{2.5}$$

$$\sum_{i \in S} \sum_{j \in S} a_{ij} \leq |S| - 1, S \subset V, 2 \leq |S| \leq |V| - 1 \tag{2.6}$$

The objective function (2.1) minimizes the tour cost. Constraint (2.2) sets the vertices that are on tour. Constraint (2.3) ensures all targets are covered, either by a vertex or during transit of an edge. Constraint (2.4) balances flow through each vertex. Constraint (2.5) ensures that the tour starts and ends at the depot. Constraint (2.6) presents the STE constraint.

2.6   Extensions to Multiple Vehicles

We next extend the VCTP model into an mVCTP. Similar to the VRP, there are m available identical vehicles based at the depot. The mVCTP involves designing a set of minimum total length vehicle routes satisfying the following constraints:

1. There are at most m vehicle routes and each start and end at the depot, v_0.
2. Each vertex of V belongs to at most one route.
3. Each target in W must be covered by an edge or vertex in the routes.

This formulation explicitly indicates the vehicle that traverses an edge, in order to impose more constraints on the routes and overcome some of the drawbacks associated with the two-index model. We use the *three-index vehicle flow formulation* which uses $O\ (n^2m)$ binary variables a_{hij} and $O\ (nm)$ binary variables b_{hi}, where a_{hij} and b_{hi} are as defined:

$$a_{hij} = \begin{cases} 1, & \text{edge}\left(v_i, v_j\right) \text{in tour by vehicle } h \\ 0, & \text{otherwise} \end{cases}$$

$$b_{hi} = \begin{cases} 1, & \text{vertex } v_i \text{ in tour by vehicle } h \\ 0, & \text{otherwise} \end{cases}$$

The formulation of the integer linear program of the mVCTP model is as follows.

Sets and Data
The notation for the sets and data are the same as the VCTP model

Binary Decision Variables
a_{hij} 1 if edge (v_i, v_j) in tour by vehicle h, 0 otherwise
b_{hi} 1 if node v_i in tour by vehicle h, 0 otherwise

$$\text{Minimize} \sum_{h=1}^{m} \sum_{\left(v_i, v_j\right) \in E} c_{ij} a_{hij} \tag{2.7}$$

Subject to

$$\sum_{h=1}^{m} \sum_{i \in V} a_{hij} = \sum_{h=1}^{m} b_{hj} \left(\forall j \in V\right) \tag{2.8}$$

$$\sum_{h=1}^{m} \sum_{i \in V} \beta_i^k b_{hi} + \sum_{h=1}^{m} \sum_{\left(v_i, v_j\right) \in E} \alpha_{ij}^k a_{hij} \geq 1 \left(\forall k \in W\right) \tag{2.9}$$

$$\sum_{i \in V} a_{hij} = \sum_{l \in V} a_{hjl} \left(\forall j \in V; h = 1, \ldots, m\right) \tag{2.10}$$

$$\sum_{j=1}^{n} a_{h1j} = 1, \sum_{i=1}^{n} a_{hi1} = 1 \left(h = 1, \ldots, m \right) \tag{2.11}$$

$$\sum_{i \in S} \sum_{j \in S} a_{hij} \leq |S| - 1, \left(S \subset V; 2 \leq |S| \leq |V| - 1; h = 1, \ldots, m \right) \tag{2.12}$$

The objective function and the constraints for the mVCTP are similar to the VCTP model, with the inclusion of indices for the multiple vehicles.

2.7   Empirical Study

The VCTP and the mVCTP provide tour cost and target coverage benefits over the CTP and mCTP, respectively. Unfortunately, the benefits come at a computational cost. The empirical study is designed and conducted to focus on examining the benefits and costs of the VCTP approach. The focus is on exact solutions leaving heuristics search methods for follow-on work. The study varies the number of vertices and number of targets while focusing on the single vehicle version and fixing the sensor range. This provides a focus on the particulars of the operational problem, not the UGV platform.

The integer linear program described was coded in [12,13] and tested on randomly generated test problems. Unlike many combinatorial optimization problems where test data and their accompanying optimal solutions are available, there is no existing database for CTP models. Thus, our test data were derived from the various Solomon [14,15] data sets, which are VRP with Time Windows data sets using Euclidean distances between vertices. These routing data sets are classified into randomly generated data points set R1 and clustered data points set C1. Any clustering of the vertices is the clustering present in the original Solomon problem; we do not do any additional clustering. As each data set contains 101 points, we randomly select vertices to visit and vertices designated as targets in the test problem.

For the MONARC area of operations, we can classify the potential adversarial locations into random and clustered data points to agree with the test problem structure. For a large battlefield with undefined boundaries, we assume that the adversaries appear in a homogeneous fashion and thus the randomly generated data points provide a good approximation. In a battlefield with some high value assets scattered throughout the area of operations, the threats can be identified into certain clusters and the clustered data set is a reasonable fit. Thus, we can make a reasonable case for using each type of problem.

An unattractive feature of the STE constraints are the exponential increase in the number of constraints with the number of N vertices to approximately 2^N constraints. Miller, Tucker, and Zemlin [16] introduced another formulation of the STE constraint which adds n variables to the model, but dramatically decreases the number of constraints to approximately n^2. However, the Dantzig formulation is much tighter than the Miller-Tucker-Zemlin (MTZ) formulation as shown by Nemhauser et al. [17]. Desrochers et al. [18] strengthened the MTZ formulation by lifting the MTZ constraints into facets of the TSP polytope. Thus, the Desrochers' STE constraint is LINGO code as it provides a good compromise between the number of constraints and their tightness. Additionally, as the motivation here is to demonstrate the viability of the VCTP integer linear model, we will only examine problems of small data size, hence circumventing the computational concerns with the exponential increase of STE constraints and the increase in computational time.

The sets V and W were defined by randomly choosing $n + 1$ and p points from the first $n + p + 1$ points, respectively from the Solomon data sets. The first point in V is chosen as the depot, vertex v_0. The c_{ij} coefficients were computed as the Euclidean distance between these points. The value of the covering distance c was set to a value of 10. For comparison, the data points occupied an approximate 80×80 grid. The α_{ij}^k and β_i^k matrices were pre-processed with Microsoft Excel and read into LINGO.

Tests were run for various combinations of n and p on the random and clustered data sets. Specifically, the following values were tested: $n=20, 30, p=5, 10$. The CTP and VCTP models were used to examine their robustness on the varied data sets. With a 2^2 design, using random and clustered data sets equally, for 10 random data sets each, a total of 80 test problems were used.

For comparison, the optimal tour taken, tour length, and computational effort were recorded for both models. The number of targets covered by vertices and by edges were also collected for the VCTP model.

The results are summarized in Tables 2.1–2.4 with the headings defined as:

Tour length	:	Optimal tour length
Iterations	:	Number of iterations required by LINGO 11.0
Tour vertices	:	Number of vertices visited (including depot)
Vertex coverage	:	Number of targets covered by vertices
Edge coverage	:	Number of targets covered by edges

Table 2.1 Results for $n=20$ and $p=5$.

Problem	CTP model			VCTP model				
	Tour length	Tour vertices	Iterations	Tour length	Tour vertices	Iterations	Vertex coverage	Edge coverage
R101	124.566	4	65 062	124.566	4	87 032	5	0
R102	Infeasible	—	—	118.6011	5	51 037	2	3
R103	Infeasible	—	—	85.16839	3	14 520	1	4
R104	Infeasible	—	—	133.1632	5	162 410	3	2
R105	177.1577	5	120 020	176.0813	5	178 926	4	1
R106	140.7824	5	1 783	136.6277	5	7 233	4	1
R107	Infeasible	—	—	134.9327	5	118 764	3	2
R108	Infeasible	—	—	121.5904	5	71 216	3	2
R109	Infeasible	—	—	Infeasible	—	—	—	—
R110	Infeasible	—	—	139.4547	6	28 749	4	1
R1 avg	**147.50**	**4.67**	**62 288**	**145.76**	**4.78**	**79 987**	**3.22**	**1.78**
C101	153.886	5	143 137	153.6151	4	100 712	4	1
C102	Infeasible	—	—	120.069	3	2 708	1	4
C103	192.1403	5	16 689	192.0509	4	31 158	4	1
C104	177.3902	6	159 729	173.2213	5	45 544	3	2
C105	170.9239	5	238 430	162.565	4	489 711	4	1
C106	88.36234	4	10 351	85.09786	4	44 258	4	1
C107	Infeasible	—	—	179.8654	5	29 998	3	2
C108	Infeasible	—	—	Infeasible	—	—	—	—
C109	Infeasible	—	—	Infeasible	—	—	—	—
C110	143.7321	4	114 153	143.7321	4	114 153	5	0
C1 avg	**154.41**	**4.83**	**113 748**	**151.71**	**4.13**	**107 280**	**3.50**	**1.50**

Table 2.2 Results for $n = 20$ and $p = 1$.

Problem	CTP model			VCTP model				
	Tour length	Tour vertices	Iterations	Tour length	Tour vertices	Iterations	Vertex coverage	Edge coverage
R101	133.0123	7	162318	127.0204	5	46268	8	2
R102	Infeasible	—	—	157.3024	5	6687	5	5
R103	Infeasible	—	—	184.0486	6	93927	6	4
R104	Infeasible	—	—	162.241	5	363884	4	6
R105	Infeasible	—	—	Infeasible	—	—	—	—
R106	Infeasible	—	—	182.347	6	9322	8	2
R107	Infeasible	—	—	213.2143	8	55003	5	5
R108	Infeasible	—	—	186.6751	5	100927	4	6
R109	Infeasible	—	—	Infeasible	—	—	—	—
R110	Infeasible	—	—	148.0407	6	11094	8	2
R1 avg	**133.01**	**7.00**	**162318**	**127.02**	**5.75**	**85889**	**6.00**	**4.00**
C101	220.2951	7	6131	216.273	5	7416	7	3
C102	Infeasible	—	—	227.401	6	141213	3	5
C103	229.1829	7	120279	223.3114	5	49836	7	3
C104	221.4393	7	225792	214.0772	5	206575	5	5
C105	Infeasible	—	—	170.9239	5	67945	7	3
C106	Infeasible	—	—	192.1624	7	35178	7	3
C107	Infeasible	—	—	179.8654	5	80666	5	5
C108	Infeasible	—	—	Infeasible	—	—	—	—
C109	Infeasible	—	—	Infeasible	—	—	—	—
C110	Infeasible	—	—	Infeasible	—	—	—	—
C1 avg	**223.64**	**7.00**	**117401**	**217.89**	**5.43**	**84118**	**5.86**	**3.86**

Table 2.3 Results for $n = 30$ and $p = 5$.

Problem	CTP model			VCTP model				
	Tour length	Tour vertices	Iterations	Tour length	Tour vertices	Iterations	Vertex coverage	Edge coverage
R101	124.566	4	2110132	124.566	4	4582202	5	0
R102	81.31466	4	71233	81.31466	4	85620	5	0
R103	Infeasible	—	—	85.16839	3	90267	1	4
R104	116.5963	6	273980	115.5935	5	457159	4	1
R105	165.3386	5	8075691	165.3386	5	24509374	5	0
R106	140.679	5	65905	136.6277	5	185666	4	1
R107	Infeasible	—	—	117.1561	5	141753	2	3
R108	117.1231	5	4815096	117.1231	5	11537324	5	0
R109	Infeasible	—	—	102.6575	5	341768	2	3
R110	147.754	5	1632075	136.8977	6	5601861	3	2
R1 avg	**127.62**	**4.86**	**2434873**	**125.35**	**4.70**	**4753299**	**3.60**	**1.40**
C101	150.652	5	12639563	150.652	5	12496739	5	0
C102	116.0504	5	71723	109.0399	4	188004	3	2
C103	191.3442	6	10465191	191.3354	4	16372338	4	1
C104	174.7859	6	21694919	173.1091	5	20295879	3	2
C105	170.9239	5	22827889	162.565	4	16574573	4	1
C106	72.82747	3	238103	72.82747	3	400078	5	0
C107	Infeasible	—	—	175.747	5	4769732	2	3
C108	163.7361	6	6327886	163.7307	5	5056208	4	1
C109	171.6362	6	29558883	166.9823	3	35535858	2	3
C110	120.7869	4	75859	120.7869	4	120336	5	0
C1 avg	**148.08**	**5.11**	**11544446**	**145.67**	**4.20**	**11180975**	**3.70**	**1.30**

Table 2.4 Results for $n=30$ and $p=10$.

Problem	CTP model			VCTP model				
	Tour length	Tour vertices	Iterations	Tour length	Tour vertices	Iterations	Vertex coverage	Edge coverage
R101	132.5805	7	120 200	127.0204	5	262 229	8	2
R102	160.3054	8	138 887	157.3024	5	352 454	5	5
R103	Infeasible			178.5465	6	461 902	6	4
R104	177.7494	10	35 179 360	162.241	5	37 613 680	4	6
R105	Infeasible			183.5834	5	1 435 766	5	5
R106	162.0383	7	185 151	159.0077	6	81 644	7	3
R107	Infeasible			189.0305	8	411 934	5	5
R108	138.9221	8	12 843 343	134.9712	7	32 451 923	8	2
R109	Infeasible			132.0172	6	357 727	7	3
R110	161.7819	9	294 380	148.0407	7	390 280	8	2
R1 avg	**155.56**	**8.17**	**8 126 887**	**148.10**	**6.00**	**7 381 954**	**6.30**	**3.70**
C101	215.9099	7	8 112 801	212.0746	5	17 819 282	7	3
C102	216.4693	8	4 427 765	204.0802	6	1 556 739	5	5
C103	226.3285	7	12 494 718	220.4994	5	8 249 504	7	3
C104	218.2831	7	69 249 875	214.0772	5	78 678 647	5	5
C105	Infeasible			170.9239	5	2 965 931	7	3
C106	Infeasible			170.5045	5	2 292 541	7	3
C107	Infeasible			175.747	5	9 761 083	4	6
C108	163.9464	7	2 493 496	163.7307	5	5 356 536	5	5
C109	207.3529	8	55 206 267	194.5455	5	25 891 461	3	7
C110	180.1757	7	2 300 252	177.4336	4	7 855 860	5	5
C1 avg	**204.07**	**7.29**	**22 040 739**	**198.06**	**5.00**	**16 042 758**	**5.50**	**4.50**

We notice the following general observations from Tables 2.1 to 2.4:

1. For data sets permitting feasible solutions to both CTP and VCTP models, the VCTP tours are never longer.
2. The vertex coverage dominates but edge coverage is utilized for VCTP models.
3. More problems are feasible when edge coverage is exploited.

Graphical examples comparing tour solutions from the CTP and VCTP models, for data sets C101 and R101, are shown in Figures 2.5 and 2.6, respectively.

Figure 2.5 provides a scatter plot of cluster data set C101 with $n = 30$, $p = 10$ with solutions from the CTP and VCTP models that clearly illustrate the difference in coverage. We observe that the CTP model requires seven traveled vertices and an overall longer tour length to cover all targets. The VCTP model requires only five vertices for full coverage with a shorter tour length.

For the random data set R101 shown in Figure 2.6, with $n = 30$, $p = 10$, we see a significant difference in the optimal tour route between the CTP and VCTP model. The original model requires seven traveled vertices to cover all 10 targets. The VCTP model utilizes just five vertices for target coverage.

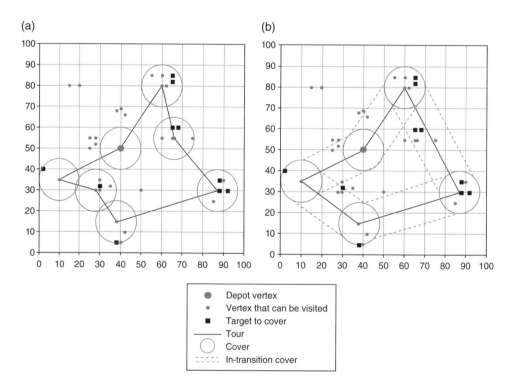

Figure 2.5 Comparison of CTP (a) and VCTP (b) solution for data set C101. *Source*: Tan, Huang Teng; Hill, Raymond R., "The In-Transit Vigilant Covering Tour Problem for Routing Unmanned Ground Vehicles," *MORS Journal*, Vol. 18, No. 4, 2013, John Wiley and Sons, LTD.

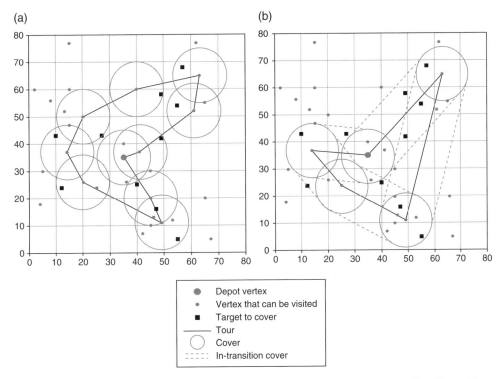

Figure 2.6 Comparison of CTP (a) and VCTP (b) solution for data set R101. *Source*: Tan, Huang Teng; Hill, Raymond R., "The In-Transit Vigilant Covering Tour Problem for Routing Unmanned Ground Vehicles," *MORS Journal*, Vol. 18, No. 4, 2013, John Wiley and Sons, LTD.

2.8 Analysis of Results

From the raw results, we calculated the following Measures of Performance (MOPs) to further compare the performance of both models:

1. Number of times the VCTP model has a shorter tour length than the original CTP model.
2. Average percentage tour length savings.
3. Average percentage of targets covered by edges.
4. Computational efficiency (in number of iterations).

Generally, the VCTP model performed better than the CTP model in an operational sense not considering the computational burden. The number of infeasible solutions for each combination of test runs is shown in Table 2.5.

Of the 80 problems, there were only 8 infeasible problems using the VCTP model. However, there were 38 infeasible problems using the CTP model. The main bulk of infeasibilities occurred for the $n = 20$, $p = 10$ scenario, which is reasonable as the number of available vertices for travel may not be sufficient to cover the proportionally large number of targets, given the fixed coverage distance.

We also compared results, based on the optimal tour length generated. A shorter tour length equates to better performance. The raw results showed that the VCTP model performed better

Table 2.5 Number of infeasible solutions for each combination of data sets.

Problem	n	p	CTP model	VCTP model
R1	20	5	7	1
C1			4	2
R1		10	9	2
C1			7	3
R1	30	5	3	0
C1			1	0
R1		10	4	0
C1			3	0
Total			38	8

Table 2.6 Comparison of performance between the CTP and VCTP models.

Problem	n	p	Better performance	Identical performance	Infeasibilities on both models
R1	20	5	8	1	1
C1			7	1	2
R1		10	8	0	2
C1			7	0	3
R1	30	5	6	4	0
C1			7	3	0
R1		10	10	0	0
C1			10	0	0
Total			63	9	8

in 63 instances than the CTP model. The number of times of VCTP superior performance, identical performance, and infeasibilities of the two models are tabulated in Table 2.6.

While the values in Table 2.6 indicate VCTP improvement over CTP, we can use a non-parametric test to make a stronger, statistically based statement of the results. A non-parametric binomial test is conducted with the null hypothesis of same performance by both models and alternate hypothesis of difference performance based on the Table 2.6 results.

$$H_0 : Performance_{CTP} = Performance_{VCTP}$$

$$H_1 : Performance_{CTP} \neq Performance_{VCTP}$$

For $\alpha = 0.01$, the critical region lies outside the confidence interval of $72\left(\frac{1}{2}\right) \pm 2.326\sqrt{72\left(\frac{1}{2}\right)\left(\frac{1}{2}\right)} = [26.1, 45.9]$; we see 63 instances of better performance meaning the null hypothesis is rejected at the 99% significant level and we conclude that the VCTP model performs better.

We also observed that the VCTP model performed better as the number of targets increases. This is attributed to the complimentary capability of vertex and edge coverage; as the CTP is unable to cover as many targets based on vertex coverage alone.

The average tour length for each solved problem was computed to compare the CTP and VCTP performance. The average tour length of the original CTP model was used as the basis and Table 2.7 shows the percentage of average tour length savings.

The overall average tour length savings was 2.64%. For each problem set, we observed that the optimal tour length of the VCTP model is always less than or equal to the original CTP model. This agrees with our claim that the VCTP sets the lower-bound tour length for the original CTP solution. Again, we observe that the VCTP model yielded a higher percentage of average tour length savings when the number of targets was higher.

The next MOP examines the improvement in coverage due to the edge covering capability of the VCTP model. We calculated the percentage of targets covered by edges. The results are tabulated in Table 2.8.

The overall percentage of targets covered by edges is significant at 35.2%. In the presence of higher number of targets, the VCTP model provided more coverage via the edges. As expected, the edge covering capability is exploited by the route construction.

Computational efficiency of the two models was compared using the number of iterations (branch and bound nodes) required. Only instances where both the original CTP and VCTP model gave feasible solutions were compared. The percentage of computational efficiency was calculated as the difference in the number of iterations between the two models divided by the number of iterations used by the CTP model. The overall percentage in computational

Table 2.7 Percentage of savings in average tour lengths.

Problem	n	p	Average tour length savings (%)
R1	20	5	1.18
C1			1.74
R1		10	4.50
C1			2.57
R1	30	5	1.78
C1			1.63
R1		10	4.80
C1			2.94
Total			2.64

Table 2.8 Percentage of targets covered by edges in the VCTP model.

Problem	n	p	Average number of targets covered by edges (%)
R1	20	5	35.6
C1			30.0
R1		10	40.0
C1			39.7
R1	30	5	28.0
C1			26.0
R1		10	37.0
C1			45.0
Total			35.2

Table 2.9 Number of iterations by both models and percentage comparison.

Problem	n	p	Average number of iterations		Improvement of computational efficiency (%)
			CTP model	VCTP model	
R1	20	5	62288	91064	−46.20
C1			113748	137589	−20.96
R1		10	162318	46268	71.50
C1			117401	87942	25.09
R1	30	5	2434873	6708458	−175.52
C1			11544446	11893335	−3.02
R1		10	8126887	11858702	−45.92
C1			22040739	20772576	5.75
Overall					−38.30

efficiency is a weighted average based on the number of instances across each problem set. The comparison of the computational efficiency is shown in Table 2.9.

The VCTP model requires significantly more computational effort to generate optimal tour lengths. This is expected as the VCTP model uses n^2p more parameters due to the additional α_{ij}^k related variables for edge coverage computation. This likely produces more options for the solver to consider in the VCTP. We do not have computational times associated with solving each LP, but fully expect the times to be comparable between the CTP and the VCTP as the number of constraints and variables remain the same in each model. Thus, we can focus computational differences using just the number of iterations. For comparison, in an $n=30$, $p=10$ scenario, there are 970 and 9970 variables and parameters in the CTP and VCTP models, respectively.

2.9 Other Extensions

The VCTP is a baseline model for a generic base security defense scenario. It is natural to discuss other extensions to improve the computational efficiency and the problem formulation. Some areas for further research are as listed:

1. *Heuristic Method:* The VCTP is non-deterministic polynomial-time (NP)-hard. Furthermore, our results show the problems are hard in practice. A heuristic is a polynomial time algorithm that produces optimal or near optimal solutions on some input instances [19]. For a relatively small instance of $n=30$, $p=10$, we observed that 78 million iterations by LINGO 11.0 are required. For a real-world scenario where n could go quite large, exact solvers are impractical and heuristic methods should be developed. As the VCTP is a TSP with SCP structure, a potential heuristic approach is the combination of the GENIUS heuristic [20] for TSP with a modified version of the PRIMAL1 set covering heuristic [21] to account for the additional edge coverage capability.
2. *Multiple Vehicle Variant:* The TSP has received a lot of research attention; the multiple TSP is more adequate to model real-world applications [22]. Similarly, further research should be conducted on the multiple vehicle variant of the VCTP model.

3. *Dynamic Routing:* Based on the assumptions made for the VCTP, it is a *static* and *deterministic* problem, where all inputs are known beforehand and routes do not change during execution [23]. Real-world applications often include two important dimensions: evolution and quality of information. Evolution implies that information may change during execution of routes and quality reflects possible uncertainty on the available data. Thus, for a *dynamic* and *stochastic* VCTP, the tour route can be redefined in an ongoing fashion based on changing travel vertices and appearance of pop-up targets.

2.10 Conclusions

An application of the newly defined VCTP is used to model a UGV assigned to base security protection. The VCTP has the novel additional edge coverage, to model the sensing capability of the UGV while traveling.

The empirical study showed that the VCTP model performed better across all combinations of scenarios. Specifically, it performed significantly better when more targets need to be covered. For the same problem data sets, the VCTP is more robust yielding more feasible solutions (72 out of 80) as compared to the CTP (42 out of 80). All VCTP optimal tour lengths were also equal or shorter than the CTP model, with an average tour length savings of 2.64%. The edge coverage capability of the VCTP accounted for 35.2% of target coverage. However, the VCTP required 38.3% more computational effort for tour length generation.

Further research areas, such as development of a high quality and quick running heuristic solver, are essential for large sized problems. The multiple vehicle and dynamic routing variants can be explored to better mimic selected real-world situations. The main contribution of this work is to provide a nascent model showcasing the importance and usefulness of both vertex and edge coverage, as well as utilizing it as a first cut mission planning optimizer tool for the MONARC base security problem.

Author Statement

The views expressed in this chapter are those of the authors and do not reflect the official policy or position of the United States Air Force, Department of Defense, or the United States Government.

References

1. Current, J. R. 1981. Multiobjective Design of Transportation Networks, Ph.D Thesis, Department of Geography and Environmental Engineering, The John Hopkins University.
2. Current, J. R. & Schilling, D. A. 1989. The covering salesman problem, Transportation Science 23, 208–213.
3. Gendreau, M., Laporte, G. & Semet, F. 1995. The covering tour problem, Operations Research 45, 568–576.
4. Hodgson, M. J., Laporte, G. & Semet, F. 1998. A covering tour model for planning mobile health care facilities in Suhum district Ghana, Journal of Regional Science 38, 621–628.
5. Oppong, J. R. & Hodgson, M. J. 1994. Spatial accessibility to health care facilities in Suhum district, Ghana, The Professional Geographer 46, 199–209.
6. Hachicha, M., Hodgson, M. J., Laporte, G. & Semet, F. 2000. Heuristics for the multi-vehicle covering tour problem, Computers and Operations Research 27, 29–42.

7. Labbé, M. & Laporte, G. 1986. Maximizing user convenience and postal service efficiency in post box location, Belgian Journal of Operations Research Statistics and Computer Science 26, 21–35.

8. Current, J. R. & Schilling, D. A. 1994. The median tour and maximal covering tour problems: formulations and heuristics, European Journal of Operational Research 73, 114–126.

9. Dantzig, G. B. & Ramser, J. H. 1959. The truck dispatching problem, Management Science 6, 1, 80–91.

10. Toth, P. & Vigo, D. 2002. The Vehicle Routing Problem, SIAM Monographs on Discrete Mathematics and Applications, SIAM.

11. Dantzig, G. B., Fulkerson, D. R. & Johnson, S. M. 1954. Solution of a large-scale traveling-salesman problem, Journal of the Operations Research Society of America 2, 4, 393–410.

12. LINDO Systems Inc. LINGO Version 11.0.

13. Microsoft. Microsoft Excel 2007.

14. Solomon, M. 1987. Algorithms for the vehicle routing and scheduling problems with time window constraints, Operations Research 35, 2, 254–265.

15. Solomon VRPTW Benchmark Problems. 2012. http://web.cba.neu.edu/~msolomon/problems.htm, last accessed 9 May 2012.

16. Miller, C. E., Tucker, A. W. & Zemlin, R. A. 1960. Integer programming formulation of traveling salesman problems, Journal of the ACM 7, 4, 326–329.

17. Nemhauser, G. & Wolsey, L. 1988. Integer and Combinatorial Optimization, John Wiley & Sons.

18. Desrochers, M. & Laporte, G. 1991. Improvements and extensions to the Miller-Tucker-Zemlin subtour elimination constraints, Operations Research Letters 10, 1, 27–36.

19. Feige, U. 2005. Rigorous analysis of heuristics for NP-hard problems, In Proceedings of the 16th Annual ACM-SIAM Symposium on Discrete Algorithms, 927–927.

20. Gendreau, M., Hertz, A. & Laporte, G. 1992. New insertion and postoptimization procedures for the traveling salesman problem, Operations Research 40, 1086–1094.

21. Balas, E. & Ho, A. 1980. Set covering algorithms using cutting planes, Heuristics and Subgradient Optimization: A Computational Study, Mathematical Programming 12, 37–60.

22. Bektas, T. 2006. The multiple traveling salesman problem: an overview of formulations and solution procedures, Omega 34, 209–219.

23. Pillac, V., Gendreau, M., Gueret, C. & Medaglia, A. L. 2013. A review of dynamic vehicle routing problems, European Journal of Operations Research 225, 1–11.

3

Near-Optimal Assignment of UAVs to Targets Using a Market-Based Approach

Elad Kivelevitch[1], Kelly Cohen[1] and Manish Kumar[2]
[1]*Department of Aerospace Engineering and Engineering Mechanics, College of Engineering, University of Cincinnati, Cincinnati, OH, USA*
[2]*Department of Mechanical, Industrial & Manufacturing Engineering, University of Toledo, Toledo, OH, USA*

3.1 Introduction

Using collectives of Unmanned Aerial Vehicles (UAVs) in military missions has been suggested by various militaries as a way to increase the tempo of battle, shorten hunter-to-shooter loops, and overall increase the capacity of a modern military to deal with time-critical targets [1]. Yet, to achieve this goal, a major question that needs to be answered is: Given a bank of targets, what is the best way to assign a fleet of UAVs to these targets? This question gets even more complicated when considering various additional operational properties, such as dynamic changes in the battlefield, targets with short lifetimes, need to minimize fuel consumption, and so on.

This question, with its many variants, has been researched by various groups in the past decades [2–5], as described in the literature section below. Indeed, the scope of existing literature is staggering. Still, one can divide the various approaches into two main categories: mathematical approaches and heuristic approaches.

Mathematical approaches have the benefit of providing a guarantee on optimality, or at least how far the solution is from an absolute possible optimal solution. Typically, mathematical approaches use a Mixed Integer Linear Programming (MILP) approach and utilize the advancements in commercial MILP solvers and computer-processing power to

Operations Research for Unmanned Systems, First Edition. Edited by
Jeffrey R. Cares and John Q. Dickmann, Jr.
© 2016 John Wiley & Sons, Ltd. Published 2016 by John Wiley & Sons, Ltd.

tackle problems of growing sizes [2],[3]. Yet, since the underlying formulation of the problem is considered to be non-deterministic polynomial-time hard (NP)-hard, these approaches all suffer from the Curse of Dimensionality [4]. This means that they are limited in the number of UAVs and targets for which they can find an optimal assignment and that they will struggle in finding solutions to dynamic scenarios when the rate of changes is high. This limits the application of mathematical optimization.

Heuristic approaches are varied [6–19]. The main advantage of using heuristics is that a solution can be found relatively quickly, which lends them to be used in dynamic and rapidly changing situations. They can also scale more easily to larger groups of UAVs and bigger target banks. However, there is typically no guarantee on the quality of the solution (how far it is from the optimal solution), and in many cases the quality of the solution may deteriorate with increasing the scale of the problem.

It should be noted that both mathematical and heuristic solutions are usually tailored to a particular type of problem formulation. Thus, there is a difference in the algorithm used if the goal is to reduce the time required to perform the overall mission or the goal is to reduce overall fuel consumption. Similarly, in some cases it is imperative to handle all the targets while in others the limited resources can only allow handling top-priority targets. From the operational point of view, there is a need for flexibility in the definition of the mission, but the underlying algorithms that exist in the contemporary literature to solve these formulations are in many cases vastly different.

As a result, the main requirements from a solution methodology to assign UAVs to targets are:

1. *Quality:* the approach shall provide near-optimal results, when compared to the mathematical approaches and better results than other heuristic approaches.
2. *Real-Time:* the approach shall reach a solution in a time that is still relevant to the ongoing mission.
3. *Scalability:* the approach shall scale to large number of UAVs and large numbers of targets in polynomial time and without loss of quality.
4. *Versatility:* the approach shall be versatile to various optimization goals.
5. *Adaptability:* the approach shall adapt to dynamic changes during the mission, for example, addition or removal of UAVs from the fleet and addition or removal of targets from the target bank.

In this chapter, we describe a solution based on economic markets, in which each vehicle is represented by a software agent [20]. These agents perform several trading activities that mimic the way a rational individual is acting in an economic market, with the egotistical goal of increasing the personal utility that the individual can obtain from trades. By shaping the way the utility is defined for the agents, their trading will lead to an emergent behavior that will obtain near-optimal division of labor.

The chapter is organized in the following way: first, we formulate the operational problem in mathematical terms (an MILP formulation is given in the appendix to the chapter). Existing literature is given in Section 3.3 and our market-based approach is defined in Section 3.4. Results for various formulations, representing different optimization goals, are given in Section 3.5, followed by conclusions.

3.2 Problem Formulation

In this section, we formulate the problem in operational terms. An MILP formulation is given in the appendix to this chapter.

3.2.1 Inputs

We start with a definition of the inputs: there are m UAVs, each UAV starts its motion from some initial location, called a *depot*, whose coordinates are $(x_k^0, y_k^0), k = 1,\ldots,m$. The depots are known at the beginning of the scenario and are assumed to be important for the overall mission and it is assumed that at the end of the tour each UAV will return to its respective depot.

There are n known target locations, denoted by $(x_i, y_i), i = 1,\ldots,n$. The i-th target is given a benefit function, $b_i(t)$, which represents the priority of this target to the overall mission, as a function of time. A map showing a distribution of three depots and eight targets is depicted in Figure 3.1a. The time-varying benefit of each target is depicted in Figure 3.1b, denoting a possibly short-lived target, whose priority may be high at first but reduces after some time.

3.2.2 Various Objective Functions

The distance between every depot and target location is calculated using the Euclidean norm:

$$d_{ki}^0 = d_{ik}^0 = \sqrt{\left(x_k^0 - x_i\right)^2 + \left(y_k^0 - y_i\right)^2}, \forall k = 1,\ldots,m, i = 1,\ldots,n \tag{3.1}$$

Similarly, the distance between every two target locations is calculated as:

$$d_{ij} = d_{ji} = \sqrt{\left(x_j - x_i\right)^2 + \left(y_j - y_i\right)^2}, \forall i, j = 1,\ldots,n \tag{3.2}$$

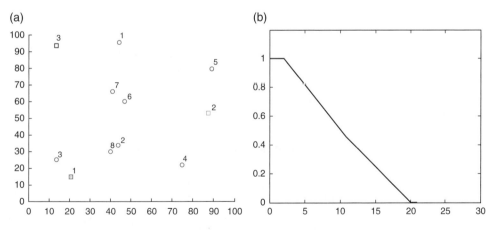

Figure 3.1 (a) A map with three depots, denoted by squares and eight targets, denoted by circles. (b) A time-varying target benefit value, where the benefit decreases over time to account for limited target lifetime.

The cost of travel of the k-th UAV from its depot to a list of targets and back to the depot is then given by:

$$J_k = d^0_{p_1 k} + \sum_{j=2}^{n_k} d_{p_{j-1} p_j} + d^0_{k p_{nk}} \tag{3.3}$$

where we assume that the tour of the k-th UAV is organized as:

$$\{ \text{Depot} \rightarrow \text{Target } p_1 \rightarrow \text{Target } p_2 \rightarrow \ldots \rightarrow \text{Target } p_{n_k} \rightarrow \text{Depot} \}.$$

The overall benefit accrued by the UAV by completing the tour via all the targets assigned to it is:

$$B_k = \sum_{j=1}^{n_k} b_{p_j} \left(t_{p_j} \right) \tag{3.4}$$

Using the notations above, there are three possible objectives to achieve:

1. *Minimizing Fuel Consumption:* assuming fuel consumption is linear with distance and all UAVs consume fuel equally, this will be achieved by reducing the total distance traveled by all the UAVs. Note that the goal is to visit every target. We call this case **Min-Sum** and it is essentially the same as the classical *Multi-Depots Vehicle Routing Problem* (MDVRP).
2. *Minimizing Time:* similar to the above, the goal is to visit every target, but to do it in such a way so that the UAVs minimize the longest time it takes to return to their depots. Assuming equal speeds, this means minimizing the longest tour. We call this case **Min-Max** and it is the same as the *Min-Max Multi-Depots Vehicle Routing Problem* (MMMDVRP).
3. *Maximizing Profit:* in this case, the goal is to maximize the total benefit accrued by the UAVs while minimizing the cost of travel to the targets. Note that there is no need to visit all the targets, and only targets with high enough benefit (priority) value will be visited. Obviously, if all the targets have a high enough time-invariant benefit, all of them will be visited and so maximizing the profit is achieved by reducing the cost of travel, which makes it similar to the first case. If the benefit is high enough but time-varying, maximizing the profit is done by getting to the targets as quickly as possible, which is similar to minimizing time. However, this case is more general than the above because it allows us to ignore some of the lower priority targets and focus on the higher priority targets. We call this case **Max-Pro** and it is the same as the *MDVRP with Varying Profits*.

Using the Eqs. (3.3) and (3.4) above, these cases are defined as:

$$\text{Min} - \text{Sum} : J^*_{MinSum} = \min \left(\sum_{k=1}^{m} J_k \right) \tag{3.5}$$

$$\text{Min} - \text{Max} : J^*_{MinMax} = \min J_{k^+}$$
$$\text{where } k^+ = \arg \max_{k \in \{1,\ldots,m\}} J_k \tag{3.6}$$

$$\text{Max} - \text{Pro}: J^*_{MaxPro} = \min\left(\sum_{k=1}^{m} (J_k - B_k) \right) \tag{3.7}$$

The Eqs. (3.5) and (3.6) above are the same as defined by Shima *et al.* [19]. Equation (3.7) is an extension of a similar equation by Feillet *et al.* to the multi-depots and time-varying case [3]. The *Min-Sum* and *Min-Max* cases are optimized under the constraints that each UAV has a single tour and each target must be visited exactly once. The *Max-Pro* case is optimized under the constraint of a single tour per UAV, and assuming that a target benefit can be collected at most once, that is, no repeat visits of the same target. Please refer to the appendix for a more rigorous MILP formulation.

As described by various works, including Shima *et al.* [19], the number of all possible solutions, if all were to be enumerated, scales by:

$$N_a = m^n n! \tag{3.8}$$

This makes the problem an NP-hard problem, like any other variant of the Vehicle Routing Problem (VRP). For example, a relatively small-sized problem consisting of 5 UAVs and 30 targets will have 2.47×10^{53} different permutations. However, out of this huge number of permutations, the vast majority of them are obviously inefficient. Thus, there is a need for an algorithm that can perform a quick search for good solutions.

3.2.3   *Outputs*

The result of an algorithm that assigns UAVs to targets is a list of UAV tours. The tours define the sequence of target IDs that are assigned to each UAV, that is, the UAV's part of the overall mission. Target IDs can then be matched with target locations and the list of target locations can be the basis for a flight plan for the UAV. Obviously, each UAV will have its own internal format for flight plans and how to execute them, but this is beyond the scope of this chapter.

3.3   **Literature**

As many researchers do, we formulate the problem of assigning UAVs to targets as variants of the MDVRP [15–19]. This section covers existing literature relevant to this problem, as well as a review of techniques that are based on the multi-agent system (MAS) where the agents act in a simulated economic market. It should be noted that the MDVRP and other closely related problems are among the best researched problems in the world of engineering and applied mathematics, and so this section will cover only the works that are most pertinent to our formulation and solution.

3.3.1   *Solutions to the MDVRP Variants*

We begin with the famous and very well-researched Traveling Salesman Problem (TSP). The TSP is defined simply in the following way: given a set of cities, find the shortest path that guarantees that each city will be visited exactly once and that the salesman returns back to the city of origin. A fascinating account of the TSP and various solutions for it can be found in [2]. The best guaranteed optimal solution to the TSP was found using Dynamic Programming and has a non-polynomial complexity, thus the TSP is considered to be an NP-hard problem [4]. Nonetheless, mathematical formulations were devised using MILP, culminating in the best-known TSP solver called Concorde [21]. A near-optimal heuristic solution based on the Lin–Kernighan algorithm is considered to be a close second among the TSP solvers [2], [14].

The VRP is an extension of the TSP in the sense that there exists more than a single vehicle and the vehicles originate and return to a depot, that is, not an arbitrarily chosen city along the tour. There are a variety of additional formulations that are beyond the scope of this chapter, for example, time-windows, pickup and delivery, capacitated vehicles, and so on.

In the VRP, there are two possible objectives: minimizing the total distance traveled by all the vehicles and minimizing the longest tour by any vehicle (*Min-Max*) [11]. A variation of the VRP (and the closely related Multiple Traveling Salesmen Problem or MTSP) with profits was also suggested and a review of existing literature exists in [3].

The MDVRP is a further extension of the VRP, in which the vehicles may originate from and return to several depots, not just a single one. Again, there could be several objectives: minimizing total traveled distance (*Min-Sum*), minimizing the longest tour (*Min-Max*), and maximizing profit (*Max-Pro*). Since the MDVRP is an extension of the VRP, which is an extension of the TSP, the problem is known to be NP-hard. There are both mathematical and heuristic solutions to these three cases.

In the *Min-Sum* case, an excellent review of formulations and solution procedures is given by Bektas in [5], and for more details of the various solution procedures we refer the interested reader to this reference. In summary, there are several mathematical formulations of the MTSP and VRP problems, which are based on MILP, and can yield solutions with guarantee on optimality. Additionally, several types of heuristics have been proposed to solve the MTSP/ VRP, which include an extension to the Lin–Kernighan heuristic [22], Genetic Algorithms [23], [24], Simulated Annealing [25], Tabu Search [26], and Artificial Neural Networks [27–29].

While it is applicable to many scenarios in which a timely execution of the mission is impor-tant [6], the *Min-Max* case is somewhat less researched than the *Min-Sum* case, but is still well researched. Two mathematical formulations and solutions to this problem were proposed in [30]. A variety of heuristic solutions were suggested, among which a well-known heuristic was proposed by Carlsson *et al.* in [7], an Ant Colony Optimization (ACO) [31], Fuzzy Logic clus-tering [32], [33], and a hybrid of Genetic Algorithms with Tabu Search [34], [35]. To generalize, these methods divide the set of targets equitably between the UAVs by dividing the area that each UAV will take and optimizing each UAV tour through the targets in that area.

The TSP, MTSP, and VRP with profits are generally less researched than other variants of the problem. This is particularly true in the Multiple-Depots case and especially if the benefits to visit the targets are time-varying. There are three variations on the VRP with profits: the first one, which is the one we are solving, has an objective of maximizing the profit (*Max-Pro*) that a collective of UAVs can achieve by visiting a set of targets, where profit is total accumulated benefits minus total travel distance. The second, called the "Orienteering Problem" variation puts a constraint on the maximum total travel distance and the UAVs try to collect as much benefit without violating this constraint. The third option puts a constraint on the minimum benefit to be collected and tries to minimize the total distance traveled to collect this benefit, and is called "Prize Collecting TSP." The most commonly researched of these three is the Orienteering Problem [8], [9], [36–38].

A mathematical formulation for the *Max-Pro* case with constant profits for a single vehicle was developed by Feillet *et al.* [3] and extended to the Multiple-Depots case by Kivelevitch *et al.* in [39]. Attempts at mathematical formulation of the problem with time-varying profits were proposed, but they have to assume a particular dependence between the benefit and the time, for example, linear decrease [40–42]. Moreover, in these problems there is usually a single depot.

To summarize this part of the literature review, there are two main gaps that can be seen.

First, the three different objectives that we defined are treated differently by researchers, and each solution is tailored to a particular case, thus a solution that optimizes the *Min-Max* case will not work well in the other two cases. From the operational point of view, this puts a limitation on the flexibility that a military force needs to decide when to use each objective. However, the military may want one day to operate the UAV collective in an attempt to minimize fuel consumption (*Min-Sum*), while on another day the rate of target attacks has to increase (*Min-Max*), and on the next day the force is restricted to engage only high priority targets (*Max-Pro*). Therefore, this flexibility is important.

Second, there is a gap in available methods for the various cases. While the *Min-Sum* is very well researched, the other two cases are less researched, and the heuristic solutions are rather untested as compared to mathematical results. This leaves an unknown optimality gap, that is, how good the heuristic solutions are relative to mathematical optimization methods.

The work described in this chapter is aimed at bridging these gaps by offering a general-purpose and tried approach based on economic markets. The next subsection explains why we chose this option.

3.3.2 Market-Based Techniques

This subsection gives a short introduction to the concept of market-based optimization, but in order to do so, we first have to define the concept of software agents and their uses.

A *software agent*, or simply an *agent*, is defined by Wooldridge [20] in the following way: "An agent is a computer system that is situated in some *environment*, and that is capable of autonomous action in this environment in order to meet its design objectives." In other words, the software agent has the ability to *perceive* the environment around it, via sensing or communication, to *reason* about the situation and how to obtain its goals within it, and to *act* upon that reasoning. Similar to *objects* in object-oriented programming, an agent is an encapsulated entity, but it has the ability to manipulate its surroundings [20].

It is common to build systems that involve several agents, creating an MAS. The agents in an MAS may communicate among themselves, either explicitly or implicitly by changing the environment, for example, ants laying pheromone signals for each other. The agents can either cooperate or compete, or sometimes do both. MASs are particularly useful when there is a need to simulate a complex behavior that emerges from simple, goal-seeking agents. This paradigm lends itself to various applications, including ACO, Flocks and Swarms, and so on. In our solution, we use an MAS to simulate the actions of rational and egotistical agents that act to maximize their gain in an economic setting.

Like many of the algorithms we discussed in the previous subsection, dealing with resource allocation problems was traditionally done in a centralized way [43]. For these solutions to perform perfect optimization, the central node performing the optimization is required to have perfect information about the system it is trying to optimize [44]. This requires that all the resources in the system report to the central node about their capabilities and status, accept the decisions made by the central node, and report to it if deviations from the plan occur. The more resources there are in the system, and the more task requirements there are, the more complex this central processing becomes. In addition, for a complex system, a single system-wide metric to be used by the central node in the optimization process is rarely easy to define [44]. Furthermore, this problem may be dynamic in the sense that new tasks may arrive and depart the system, or new resources may be added to the system or may be taken from it during operation.

In many cases, such a complex problem is intractable to solve in a centralized mechanism [44].

As a result, decentralized resource allocation processes have been proposed. In the extreme case of a decentralized resource allocation process, each resource tries to optimize a metric, called a utility function that defines its own set of goals. The resources then try to reach a consensus that will allow each of them to maximize its utility given a set of constraints. One type of a decentralized solution is a solution based on the model of economic markets.

In such a model, the tasks are represented as buyer agents and the resources are represented as seller agents [45]. As discussed above, agents encapsulate the preferences and modus operandi of the entity they represent, and communicate only an abstract notion of their preferences, using utility functions [45].

The motivation for using economic models such as the one just described comes from the hypothesis of the efficient economic market, in which many agents act egotistically in an attempt to maximize their own benefits. As we are aware, economic markets represent some of the most complex systems known to science, yet for thousands of years economic markets were used to "get things done," reach an equilibrium between opposing goals, and react to various changes in the system [46]. In part, this is based on the egotistic reasoning of the agents in the market: a transaction between two agents will only occur if at least one of them gains something from it and the other at least does not lose. This has the potential to quickly and efficiently lead to a Pareto-optimal solution [47].

Thus, market-based solutions (MBSs) have been suggested to solve a large variety of resource allocation and optimization problems [43], [44], [46–51]. In particular, some of these works suggest the use of MBSs to problems similar to the ones we propose, but they have two main limitations: (i) they only attempt to solve one variant of the problem and not all three and (ii) they only use one type of economic transaction and ignore the other. In other words, they either use a "peer-to-peer" trade or they use "bid-auction," but never both. As we show in our approach, using both types of transactions leads to a faster convergence on near-optimal solutions.

3.4   The Market-Based Solution

We are now ready to describe the MBS used for solving the three cases defined above: Min-Sum, Min-Max, and Max-Pro. We do so in three steps: first, we describe the basic MBS, without hierarchy, and the actions that each software agent can take. Next, we describe an extension of the MBS, using hierarchical clustering of targets, which vastly improves the overall performance of the MBS in the *Min-Max* case. Then, we describe the small modifications needed to solve the

Algorithm 3.1 The Basic Market

Require: m agents, n tasks, r clustering ratio.

    Optional: Previous Assignments, Clusters

1: if No Previous Assignment then   ▷ Previous assignments may come from higher hierarchies run

2:   for all Tasks do   ▷ Initialize using nearest neighbor partitioning of the area

3:     Assign task to nearest agent

4:   end for

5: else   ▷ Previous run assigned clusters to agents

6:   for all Clusters do

7:     for all Tasks in Cluster do

8:       task.assignment ← cluster.assignment

9:     end for

10:   end for

11: end if

12: while $Iteration \le Num\ Of\ Iterations$ do

13:   Step 1: Market Auction

14:   Step 2: Agent-to-Agent Trade

15:   Step 3: Agents Switch

16:   Step 4: Agents Relinquish Tasks

17: end while

    Return: Best Agents, Assignments, Clusters, Costs, Runtime

Max-Pro case. It should be noted, though, that the final result, with the hierarchy and modifications, is used to solve all three cases. In other words, the MBS we use for solving all three cases is a hierarchical market with modifications, which allows the flexibility we discussed above.

3.4.1 The Basic Market Solution

We start the description of the MBS by defining the basic structure of the market and the actions each agent in the market is able to take. In this solution, the UAVs are represented by software agents that bid and trade for *tasks* in a simulated economic market. The tasks are software agents that represent targets and encapsulate all the information regarding the target, in particular its location and benefit vs. time.

The MBS begins with a fast nearest-neighbor assignment of tasks to agents. The purpose of this assignment is to initialize the market to a reasonable solution, instead of a randomized one, but we have shown in [52] that this initial assignment is still far from the best market solution. Following the initial assignment, the MBS is an iterative process, and in each iteration agents perform a set of actions, see Algorithm 3.1.

Step 1: Market Auction: At the beginning of each iteration, the agents are informed of available tasks, and each agent computes the additional cost it will incur if that task is assigned to it and bids that cost. The cost is calculated by inserting the task to an existing tour in a way that will increase the existing tour through tasks, which have already been assigned to the agent, by the smallest distance, and if no other tasks are assigned, the cost is the travel from the depot to the task and back. The agent with the lowest bid is selected to take the task. This is a classic bid-auction mechanism, which was used in many previous works, for example, [53].

Step 2: Agent-to-Agent Trade: One agent selects randomly a second agent. The first agent queries the list of tasks from the second agent, and calculates the cost of travel to these tasks. If the first agent can service a task on the list of tasks from the second agent with a lesser cost than the second agent, that task is assigned to the first agent. This action is similar to subcontracting an agent. This peer-to-peer mechanism was suggested by Karmani *et al.* in [54], but they did not use the auction process we use in Step 1.

Step 3: Agents Switch: So far the first two steps were greedy actions that may result in the solution reaching local minima. The third step is another peer-to-peer process, but it is different than the second action, as it allows the solution to escape local minima. This step is also the only step in which we differentiate between a *Min-Max* objective and the other options. In the *Min-Max* case, two agents are randomly chosen and their tours are examined. If the two tours form polygons that intersect each other, the agents exchange tasks in an attempt to resolve that intersection. It should be noted, though, that in some cases this may not result in a better solution overall, but it is a way to improve solutions that reach a local minimum. In the *Min-Sum* and *Max-Pro* cases, a quick way to resolve local minima was found to be by simply letting one agent take over all the tasks of another agent. We randomly choose two agents and do that. Following each action, each agent invokes a single TSP solver. We can use any TSP solver in the solution, and we use Concorde [21] for large TSP instances, Lin–Kernighan [14] for medium-sized problems, and convex hull with nearest-neighbor insertion (CHNNI) [55] for small problems. These solvers were chosen to get the best runtime performance while still maintaining acceptable quality of the TSP solutions. The fastest solver is the CHNNI, but the quality of its solutions deteriorates when the number of tasks in a tour increases beyond 50 tasks, so it is used only for TSP tours of up to that number. Lin–Kernighan is faster than Concorde, but more limited in problem sizes, so we use Concorde just for the largest TSP tours, with more than 500 tasks per tour.

Step 4: Agents Relinquish Tasks: Finally, agents release randomly chosen tasks in preparation of the next market iteration. The probability to release a task is proportional to the additional cost incurred by that task. By releasing tasks this way, we allow for the possibility of escaping local minima as well as exploitation of best gradient. The market stops when the number of iterations since the last improvement is greater than a parameter. An improvement is defined as either: (i) a lower cost (depending on the case) or (ii) same cost but smaller sum of all tours. The latter is necessary for the improvement of the tours of other agents, which may open the door to further improvement of the longest tour in following market iterations.

3.4.2   The Hierarchical Market

3.4.2.1   Motivation and Rationale

The MBS described above works well for problems that have a relatively low ratio of targets to UAVs. When this ratio become higher, and especially when the total number of targets increases, it becomes more difficult for the MBS to find a good assignment of targets to the UAVs.

One way to improve this is by obtaining a quick and good partitioning of the targets, as done by Carlsson [4] and Narasimha [2], [20]. Then, the MMMDVRP would be simplified to solving a set of TSP problems (preferably, in parallel), by invoking TSP solvers for each sub-region. Yet, it is important to understand that partitioning the problem like this inherently adds artificial constraints on the global optimization – the boundaries between the sub-regions. At best, if the partitioning is perfect, these constraints will not increase the best cost, but if it is imperfect it may increase the global cost.

To the best of our knowledge, there is no quick way to perform this partitioning perfectly. The rationale behind the hierarchical market presented in this work is to recursively refine the partitioning of the problem in a way that does not add rigid artificial constraints to the problem. The way we do this is by gradually clustering tasks into larger and larger sets of tasks, while purposefully ignoring the fine resolution of the problem by representing each area by a single task at the center of the area. It should be noted that the depots are not part of this clustering and remain the same depots at each clustering level. A possible future development could be to cluster depots as well.

To illustrate, suppose the targets truly represent real towns scattered uniformly across the USA. A rigid partitioning will divide the country along state lines, or some major topographical features like riversheds, but some states (or riversheds) are bigger than others and this will result in inequitable partitioning. The result will be that some tours are much longer than others and will result in an increased min-max cost.

Algorithm 3.2   The Hierarchical Market

Require: m agents, $n^{(0)}$ tasks, r

```
1: i ← 0
2: while n^(i) > m * r do                     ▷ Recursive process
3:     i ← i + 1
4:     n^(i) ← n^(i-1)/r                       ▷ Rounded value using Matlab round
5:     Cluster tasks to n^(i) clusters using k-means clustering
6:     tasks ← clusters                        ▷ Each cluster becomes a single task, located at its centroid, in
                                                 the next hierarchy
7: end while
   ▷ At this point, the regular market can run. In the highest hierarchy, there is no
   previous assignment. The market will initialize with a default nearest neighbor
   assignment. Once there is a previous assignment in one hierarchy, the next assignment
   will begin by taking the tasks of each cluster and assigning them to the agent that was
   assigned to the cluster in the higher hierarchy.
8: while i > 0 do                             ▷ Until the solution runs on the level of single tasks.
9:     Invoke Market (Algorithm 2) to assign tasks to agents.
10:    i ← i -1
11: end while
```

Our solution works differently. At first, we group towns into small areas, for example, we group Cincinnati, Mason, and Dayton in Ohio to one small area called SW Ohio, represented as a task near Monroe, Ohio. The second step is to group small areas into larger areas, for example, taking the SW Ohio area from before and grouping it with the Central Ohio area of Columbus, Central Kentucky area of Louisville and Lexington, and SE Indiana and we call this area Ohio-Kentucky-Indiana (OKI) and represent it as a task in the Cincinnati area. We continue by grouping the larger areas into regions, say the whole Ohio Valley. We do this until we get to a point in which the number of such regions is small enough to allow us rapid assignment of large regions to depots.

In essence, this is a fast way to partition the area between the depots. It is possibly a good baseline, but not necessarily a perfect partitioning. From this point on, the idea is to refine the partitioning by gradually reintroducing the finer resolution of the assignment.

Getting back to the aforementioned example, suppose we initially assign the Ohio Valley region to a depot in Columbus, Ohio, and the Mid-South region to a depot in Memphis, Tennessee, which means that all the cities within these regions are initially assigned to either Columbus or Memphis. However, when considering the level of smaller areas, we discover that if we reassign the southernmost area of the Ohio Valley region, which is represented by a single task at this level, to the depot in Memphis, the resulting min-max cost will improve, so we reassign that area to Memphis. This reassignment is done using the market algorithm (Algorithm 3.1) until the solution settles. Then, we refine it again by running the market on the sub-areas, and this process continues until an assignment at the level of single tasks is done. And, while each time the market runs on a finer resolution there is the potential of many new reassignments, the underlying baseline solution achieved at the coarser resolution level prevents many of these permutations since reassignments are done mainly near the partitioning lines.

For example, consider Figure 3.2, that shows the assignment of 5 UAVs to 200 targets. The Figure 3.2a shows the initial assignment of very large clusters to UAVs, and at this level there are only 2–3 clusters per UAV, each representing between 12 and 25 targets. It can already be seen that UAV 5 is assigned to the northwest corner of the map, UAV 2 to the northeast, UAV 4 to the southeast, UAV 3 to the southwest, and UAV 1 is assigned to the middle of the map.

Once this initial assignment is complete, the next level of assignment, shown in Figure 3.2b, is of smaller clusters, but since it's based on the initial assignment, the geographic partitioning remains the same. This continues until the last level of clustering, which at this point is 2–3 targets per cluster, is shown in Figure 3.2c. The geographic partitioning still remains essentially the same, but the finer level of clustering allows for a more balanced division of labor, as shown by the text in the top right of each sub-figure. Finally, Figure 3.2d shows the final assignment of individual targets to UAVs, which still keeps the same initial geographic partitioning. It can be noted that the longest distance, traveled by UAV 5 in this assignment, is only 10 units, or about 4%, longer than the shortest, traveled by UAV 3 here. This is a very equitable division indeed, which is required from any *Min-Max* solution.

Thus, we achieved our goal of reducing the number of permutations by partitioning the problem, but at the same time allowing reassignments from one partition to another and having no constraints on the optimization.

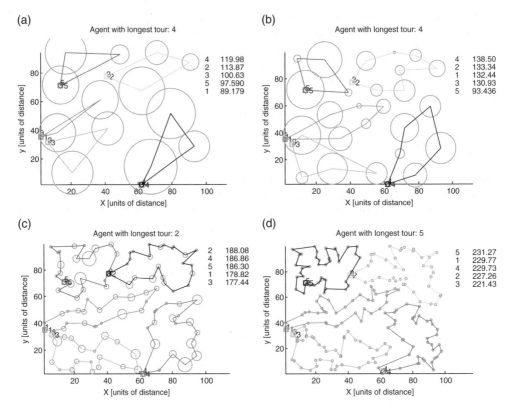

Figure 3.2 Solution development. (a) Initial solution with coarse clustering. (b) Assignment of finer clusters. (c) Assignment of small clusters. (d) Final solution and assignment of individual targets. The circle radii are proportional to the number of cities in the cluster.

3.4.2.2 Algorithm Details

The mechanism is based on a recursive clustering of tasks into larger bundles of tasks, where each bundle is represented by a new task at the centroid of this bundle, then trading on these bundles. We define $r > 1$ as the clustering ratio, which, on average, represents the number of tasks per cluster.

Reasonable values for clustering ratio were found to be between 2 and 5, where a clustering ratio of 2 usually improves the quality of the solution, but requires more time, and higher clustering ratios generally reduce computation time but increase the variability in the solution quality.

The algorithm is given in Algorithm 3.2. With every recursion of the algorithm, the number of tasks the market has to solve for will decrease by a factor equal to the clustering ratio, ending with a small number of tasks to handle in the most aggregated case. The market can very quickly find a solution to these tasks, which represent a sub-region containing possibly many tasks, and then each of these aggregate tasks is broken down to allow the agents to deal with smaller bundles of tasks. This continues until the market works on single tasks instead of bundles.

Obviously, the parameter r is a design parameter that can be used to tweak the number of tasks bundled together at each recursion, and, consequently, the number of recursions during the run.

A large clustering ratio will require a smaller number of recursions, which may result in a shorter runtime, but with the caveat that from one recursion to the next, the market has a higher degree of freedom in solving for a larger number of bundles. This may cause the next recursion to run longer, as the market explores a larger solution space.

On the other hand, a large clustering ratio also means that the partitioning is less rigid, and therefore, at least in theory, there is a better chance of finding a better solution. However, since the market is limited in the number of iterations it is allowed to run, finding a better solution is also affected by searching in the appropriate region of the search space. Thus, there is no right answer for how large r should be.

Empirically, it was found that for more complicated problems, which involve a higher ratio of tasks per agent, and a higher number of tasks as a whole, it is best to keep r smaller, while for simple cases, r can be allowed to be larger.

3.4.3 Adaptations for the Max-Pro Case

The *Max-Pro* case differs from the other two cases in that there is no requirement to assign all the tasks to an agent. In the operational sense, this means that a target is either too low priority or that it disappears too fast, that is, if the UAV cannot find the target in time it will not obtain the benefit of shooting it.

To accommodate this new possibility, we make slight adaptations to the hierarchical market: we introduce the *Null-Agent*. In this method, clustering of tasks accounts also for benefits, agents bidding for a task deduct the task benefit from their cost, and tour generation has to account for possible tours that are not TSP-optimal.

The most notable adaptation is the *Null-Agent*. This agent is added to the collective of m agents representing the UAVs and acts as a "garbage collector." Its behavior is very simple: it bids zero for all the tasks, wins the tasks that no other agent desires, and returns these tasks back to the market when all the agents relinquish their tasks. Thus, tasks are always assigned to an agent, either one that represents a UAV or the *Null-Agent*. This allows us to run the hierarchical market as defined above.

The rest of the modifications are minor: when clustering tasks in the hierarchical market, their benefit is collected. This is important to make sure that a cluster of tasks is visited even if each target alone is not of high priority. The bid that each agent gives in Step 1: Market Auction takes into account the task benefit. In this way, only the profitable tasks get a bid higher than 0, which, combined with the *Null-Agent*, guarantees that only profitable tasks are assigned to agents.

Finally, if the task benefits are time-varying, it may be more beneficial to visit tasks in a sequence that does not minimize tour length, but rather that maximizes profit. An example of such a tour is given in Figure 3.2, where it is better to first visit all three targets closer to the depot before moving on to the other targets. The TSP solver used is augmented by performing k-opt changes on its results.

3.4.4 Summary

In this section, we described the hierarchical market, which solves the three cases we defined. It should be noted that, since the same algorithm is used, it gives us the flexibility to solve all three cases, which is unique relative to other solutions described in the literature. Next, we present the quality of the results of this algorithm when solving each one of these cases.

3.5   Results

In this section, we summarize the main results of running the MBS for the three cases defined above. Due to the stochastic nature of the algorithm, the MBS was run for different randomly generated scenarios and each scenario was solved by several runs of the algorithm.

A *scenario* refers to a single set of depots and targets (both locations and benefits). The locations of both depots and targets were drawn uniformly in a 100×100 square. A *run* refers to a single attempt to solve a scenario by the MBS.

The results of the MBS are compared with the results of other methods using two metrics: the time required to run the algorithms and the quality of the results, measured by the optimality gap:

$$\epsilon = \frac{J_{MBS} - J^*}{J^*} \tag{3.9}$$

Here, J^* is the optimal result found by a mathematical optimization formulation and J_{MBS} is the result obtained by the MBS. If a mathematical optimization method cannot be used, we compare various heuristic methods and define J^* to be the best result obtained by any of the heuristic methods. The gaps of the other heuristic methods is found similarly relative to J^*.

The optimal results are calculated based on the MILP formulation described in the appendix to this chapter. The problem matrices are generated in MATLAB® and solved using the well-known and widely used commercial mathematical optimization solver IBM CPLEX® [56]. On the other hand, the MBS is written fully in MATLAB.

CPLEX is an optimized software tool that runs executable code on all CPU cores, while MATLAB codes are run by interpreting the code in real time and only on a single core. The differences between the modes of execution of the two codes mean that CPLEX has an inherent advantage at obtaining results faster. Based on available comparisons between computer languages, we anticipate that an implementation of the MBS in an executable language running in parallel on CPU cores would improve execution times of the MBS by at least one order of magnitude [57]. As we will see next, this is not a problem at the moment.

To reduce the dependency of the runtime results on the actual hardware, we run both codes on the same machine and compare the ratio of runtimes:

$$\tau = \frac{t_{MBS}}{t_{CPLEX}} \tag{3.10}$$

When comparing other heuristic algorithms, which are written in MATLAB as well, we will just compare runtime directly.

All the runs are done on a PC running Windows operating system. The specific hardware and operating system varies between the cases, but will be specified in each case.

3.5.1 Optimizing for Fuel-Consumption (Min-Sum)

To consider the performance of the MBS in the case of optimizing for fuel consumption, we run 10 scenarios of 20 UAVs and 100 targets. The theoretical number of permutations for this case is 1.183×10^{288}.

These scenarios were solved using CPLEX to obtain the mathematical optimization results as a baseline for comparison. All the runs were done on a laptop with Intel iCore5, 2.5 GHz, processor, and 6 GB RAM. The runtime for the MILP solution obtained by CPLEX ranged from 68 to 392 s, which shows the strength of this commercial optimization suite.

Figure 3.3 depicts the comparison of the MBS performance to the mathematical optimization. From the cost point of view, the median obtained by the MBS is about 1.7% higher than the binary programming cost and the worst case is 4.8% higher. In 99.7% of the runs, the runtime for the MBS is shorter than the CPLEX runtime. The median normalized runtime is 0.125, that is, the MBS is about eight times faster than the CPLEX runtime. These results are particularly impressive because the market-based is written in MATLAB and was not optimized and compiled like CPLEX – a commercially used solver.

From these results it is concluded that the MBS is comparable to the binary programming in terms of computational cost, and significantly reduces the runtime relative to binary programming. Since (as discussed in Section 3.4) the MBS scales better than the mathematical optimization solutions, the performance benefit of MBS only increases for larger cases and since the mathematical optimization is considered to be the state-of-the-art for this problem, having near optimal results provided by MBS demonstrates the power of this method.

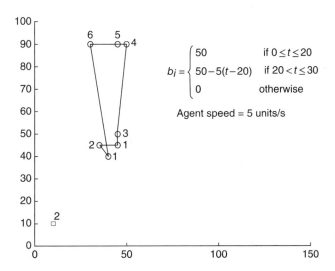

Figure 3.3 An example of a maximum profit tour that is not a minimum cost tour. Graph axes represent units of distance.

3.5.2   Optimizing for Time (Min-Max)

In most scenarios, the *Min-Sum* case results in a very inequitable distribution of targets among the UAVs: a single UAV takes most of the targets and the rest of the UAVs pick up very few targets if at all. This will result in a very long time to perform all the missions, due to the heavy load on one of the UAVs.

In time-critical missions, we want to minimize the time to perform the entire mission. This can be done by minimizing the longest tour (and time) required by each UAV. This means that the load of visiting all the targets needs to be divided among the UAVs as equally as possible. To achieve that, we use a *Min-Max* formulation. The results shown for this case are divided into two parts. The first part, similar to the *Min-Sum* results shown above, compares the MBS with the results of mathematical analysis. However, this can only be done for very small scenarios. Thus, an additional analysis is done to compare the results of the MBS with state-of-the-art heuristic algorithms, which include Carlsson's [7], ACO by Narasimha *et al.* [31] and Fuzzy Clustering (FCL) by Ernest and Cohen [32], [33].

Figure 3.4 depicts the comparison of the MBS to the results obtained by mathematical optimization in three problem sizes: 3 UAVs and 12 targets, 6 UAVs and 12 targets, and 8 UAVs and 12 targets. These were found to be the largest sizes that could be solved reliably by CPLEX before running into memory issues on a PC with 6 GB RAM, and show how limited mathematical optimization is in this kind of optimization. In each problem size, we used 20 scenarios and 100 runs for a total of 2000 MBS runs per size, 6000 runs overall.

As seen in Figure 3.4, the MBS obtained a perfect score (zero gap) in about 80% of the cases: 81.25% of the 3 UAVs scenarios, 78.63% of the runs in the 6 UAVs scenarios, 79.75% of the 8 UAVs scenarios. Ninety percent of the runs ended with an optimality gap of about 3% or less. The worst run in each class was 18% for 3 and 6 UAVs, and 22% for 8 UAVs.

Figure 3.4b depicts the normalized runtime, and it is clear that all values are significantly less than 1, in other words the MBS is much faster than mathematical optimization. The worst case in each class is 0.19 for 8 UAVs, 0.15 for 6 UAVs, and 0.02 for 3 UAVs. These represent

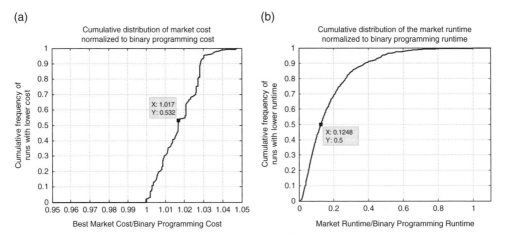

Figure 3.4   Comparison of the MBS to mathematical optimization of the *Min-Sum* case. (a) Optimality gap. (b) Relative runtime.

an improvement in runtime by a factor of about 5, 6, and 50, respectively. The median runtime ratio for each class is 0.00367 for 8 UAVs, 0.00119 for 6 UAVs, and 0.00077 for 3 UAVs; again, these represent an improvement by a factor of 272, 840, and 1299, respectively. So, the MBS runs at least one order of magnitude faster, and in half the cases, two or three orders of magnitude faster, even for such small problems, and even while considering that the mathematical optimization problem is run by a commercial mathematical optimization tool. This allows running the MBS several times to obtain a perfect solution in less time than it takes to run the mathematical optimization.

Yet, these problems are "toy-problems," in the sense that they don't represent any real-life situations. Unfortunately, larger problems could not be solved mathematically without running into memory issues, so we compare the results of the MBS to other heuristic solutions, where none of these solutions (including the MBS) can guarantee an optimal solution.

Table 3.1 gives the average optimality gap of the MBS to the other methods. In all the cases, the optimality gap is obtained by comparing the result of running all the various methods for each scenario to obtain the best-known result. In all the cases, the MBS obtained the best result. It should be noted that in all these cases there are several UAVs originating from the same depot.

The results show that not only does the MBS obtain the optimal result for each case; it is also consistent in these results. The average MBS optimality gap is between 1.7 and 5.2% from the best result. The ACO is second in performance and improves over Carlsson, but its performance is still much worse than the MBS, with optimality gaps ranging from 7.4 to 37.9%. Carlsson's algorithm obtains average optimality gaps ranging from 24.4 to 73.1%. This is significant, because such high gap values will require much longer time to complete time-critical mission calculations

Table 3.2 lists the time required for each method to obtain a solution. It is clear the fastest method is Carlsson's, which runs an order of magnitude faster than the MBS, while the ACO is the slowest and requires another order of magnitude to run. Yet, despite the clear difference between Carlsson and the MBS, the optimality gap between them sheds doubt on the applicability of Carlsson's method in real-world scenarios.

For large-scale problems that include tens of UAVs and a thousand targets or more, we compare the MBS with Carlsson's algorithm and a new method still in development, FCL solution, described by Ernest and Cohen [32], [33]. The ACO is not used because of its limited scalability to larger problems.

Table 3.1 Comparison of the average optimality gap obtained by the market-based solution (MBS) to Carlsson and ant colony optimization (ACO) for the *Min-Max* case.

Case	# of depots	# of UAVs	# of targets	Carlsson	ACO	MBS
1	3	6	80	0.396	0.187	0.017
2	3	9	140	0.244	0.24	0.027
3	4	8	80	0.438	0.379	0.032
4	4	12	140	0.453	0.205	0.019
5	5	12	140	0.291	0.074	0.017
6	5	15	140	0.731	0.156	0.052

Table 3.2 Comparison of the average runtime in seconds required by the market-based solution (MBS) to Carlsson and ant colony optimization (ACO) for the *Min-Max* case.

Case	# of depots	# of UAVs	# of targets	Carlsson	ACO	MBS
1	3	6	80	1.78	245.95	40.07
2	3	9	140	4.42	602.49	47.35
3	4	8	80	2.14	225.75	20.31
4	4	12	140	1.76	484.53	60.28
5	5	12	140	4.10	432.60	51.00
6	5	15	140	5.58	443.86	55.98

Table 3.3 Comparison of the market-based solution (MBS) to Carlsson and fuzzy clustering (FCL) for the *Min-Max* case with 4 depots, 16 UAVs, and 1000 targets.

Metric	Carlsson	FCL	MBS
Best gap	0.1672	0.1015	0
Average gap	0.2550	0.1022	0.0212
Average time (s)	35.2	16.225	1419.4

To accommodate the current requirements of the FCL solution, a problem with 4 depots, 16 UAVs, and 1000 targets was generated. The depots are located at [250,250], [250,750], [750,250], and [750,750]. There are 4 UAVs per depot and 1000 targets randomly located in a 1000×1000 region. This kind of setting is favorable for the FCL, which still has limitations in coping with random depot locations.

Table 3.3 lists the results of the three algorithms for this case. For this case the best-known solution was obtained by the MBS to be 1603.2 units. Once again, the cost of the MBS is the best out of the three algorithms both for best run and average. For the best runs, the optimality gaps of the other methods are 0.1672 and 0.1015 for Carlsson and FCL, respectively. The average costs of each algorithm have a gap of 0.255, 0.1022, and 0.0212 for Carlsson, FCL, and MBS, respectively. This shows that the results of the MBS are consistent even for larger problems.

While the MBS achieves better quality of solutions overall and on average, its runtime is longer than the other methods by two orders of magnitude. Yet this deficiency can be easily mitigated, since much of the time is spent on running the TSP algorithms per each UAV and these calculations can (and will) be done in parallel when implemented in an operational system. This can reduce the computation time by a factor similar to the number of vehicles (16 in this case).

Figure 3.5 depicts a single run of a large-scale problem with 10 UAVs and 5000 targets. The objective of running this case was to visually examine the scalability of the MBS in terms of both runtime and optimality. The tour lengths of all UAVs are listed in Figure 3.5b and show that the four longest tours are less than 1% of each other. This means that the potential for further improvement of this solution is limited. The space is almost perfectly partitioned between the tours with very few intersections, which is another indication of a high-quality *Min-Max* solution. This case was run on a high-end desktop PC with second generation i7,

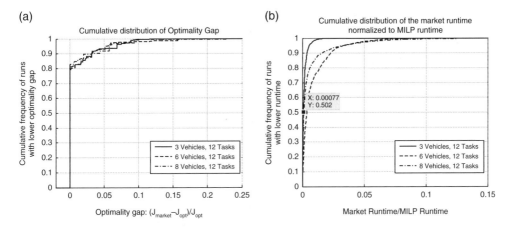

Figure 3.5 Comparison of the MBS to mathematical optimization of the *Min-Max* case. (a) Optimality gap. (b) Relative runtime.

3.4 GHz CPU with 16 GB of RAM. Runtime of this case was less than 12 000 s, or 3 h and 20 min, for a problem of roughly 4.23×10^{21325} permutations.

A similar run was done for a scenario with 100 UAVs and 5000 targets and took 76 477 s (21:14:37 h) to run on the same PC. The behavior was similar, with less than 0.3% difference in the cost of the top 20 UAVs. This problem has 4.23×10^{26325} different solutions.

Overall, the results shown for the *Min-Max* put the MBS as the leading method in terms of solution optimality and fast enough to be used in large-scale problems. Our results show that it is a suitable choice as an overall optimizer for an entire fleet of UAVs.

3.5.3   Optimizing for Prioritized Targets (Max-Pro)

In this section we move on to the case in which not all targets have to be assigned to UAVs, and the decision whether to assign a UAV to a target depends on how important the target is and how far it is from other targets. We formulate this problem as a maximum profit (*Max-Pro*) MDVRP. The benefit (priority) of a target can be time-invariant or time-varying, which leads to three possible cases:

1. *Time-Invariant and Equal Target Benefits:* This case is most similar to the *Min-Sum* case, especially if the benefit is high enough to guarantee that all the targets will be assigned. Nonetheless, as the benefit associated with a single target is reduced, more targets will become less beneficial and an optimal solution will not include them in UAV tours.
2. *Time-Invariant with Different Target Benefits:* This case most resembles a real-world operational scenario in which the targets are not time critical but are of different importance to the overall mission. High priority targets will need to be assigned to the UAVs, medium priority targets will be assigned only if they are close enough to the tours, and low priority targets will most likely be ignored unless they happen to be very close to UAV tours.
3. *Time-Varying Target Benefits:* This case is the closest resemblance to a real-world operational scenario with time-critical targets. A target is assigned a certain initial benefit value,

which may remain constant for a short while, but then the benefit deteriorates as time progresses and diminishes to zero. For example, this could be a case of a highly maneuverable target that is spotted once but has to be serviced immediately afterwards or it might disappear again.

The first two cases (time-invariant benefit values) can readily be solved by mathematical formulations while the third case is, to the best of our knowledge, not solved by any algorithm.

To test the case of time-invariant and equal target benefits we randomly generated 20 scenarios that consist of 10 UAVs and 100 targets uniformly distributed in a 100×100 area. On average, there is a single target in every 10×10 units cell, or the average distance between targets is about 14 units. Three different benefit values were used: 22.5 which is high enough to guarantee that most, but not all, targets will be assigned to UAVs; 12.5, which will assign relatively few targets to UAVs; and 17.5 as a middle ground.

Each scenario is solved by the mathematical optimization and then run 100 times by the MBS for a total of 2000 runs per scenario.

Again, we compare the MBS run in MATLAB to a mathematical solution found by CPLEX, both on a laptop with Intel iCore5, 2.5 GHz, processor, and 6 GB RAM. The performance metrics are once more optimality gap and relative runtime.

The results are depicted in Figure 3.6. As expected, the hardest case to solve by the MBS is the case with lowest benefit values, which leave many targets unassigned. Still, even in this case, the median value has an optimality gap of 1.38% and a worst case of less than 10%. In the best case, about 10% of the MBS runs reached a perfect assignment (zero gap). The results for the other two benefit values are even better.

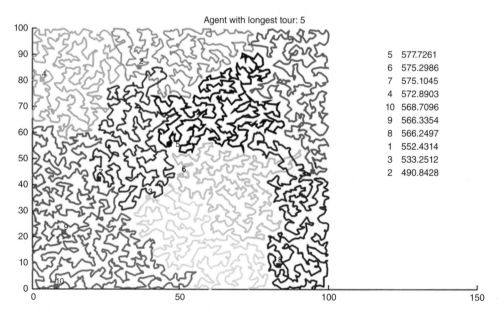

Figure 3.6 A *Min-Max* solution for 10 UAVs and 5000 targets. Runtime was about 12 000 s on a computer with i7, 3.4 GHz CPU with 16 GB RAM. (*For color detail, please see color plate section.*)

In terms of relative runtime, the MBS performs about 5–10 times faster than the mathematical optimization in all three benefit values, with a worst case run a little more than two times faster.

The time-invariant with different benefit values is examined in the following ways: 20 scenarios of 10 UAVs and 100 targets each. The UAVs and targets are randomly placed in a 100×100 units cell with uniform distribution. The targets are randomly assigned three possible benefit values: high (17.5), medium (12.5), and low (7.5), to simulate high priority, medium priority, and low priority targets, respectively. Each scenario is first solved by CPLEX and then solved 100 times by the MBS for a total of 2000 MBS runs.

Figure 3.7 depicts the results of this comparison. The median optimality gap is about 0.7%, with worst run of slightly more than 3.1%. The best result of a perfect zero gap is obtained in about 10% of the runs. In terms of runtime, the median is 0.05305, representing an improvement factor of 18.85 in runtime over CPLEX. Worst relative runtime was 0.3205, about three times faster than the mathematical optimization.

Based on these results, we once more see that the MBS is comparable to mathematical optimization in terms of quality but is much faster. In particular, the runtime is impressive when considering that the comparison is between the MBS implemented in MATLAB and mathematical optimization run on a commercial optimization tool.

We conclude this section with an example of an MBS run with time-varying target benefits, as depicted in Figure 3.8. There are 5 UAVs and 30 targets uniformly distributed in a 100×100 area. All the targets are assigned a maximum benefit value of 50 units, which will means that they are all very high priority values. Nevertheless, these targets are time critical: the benefit value rapidly deteriorates from 50 units at $t = 20$ to 0 at $t = 30$ s. Thus, even though all the targets are high priority, not all of them are assigned to UAVs, as can be seen in the bottom-left corner of the map – the null-agent (marked as agent 6 in the Figure 3.9) takes one target that cannot be visited in time by any agent.

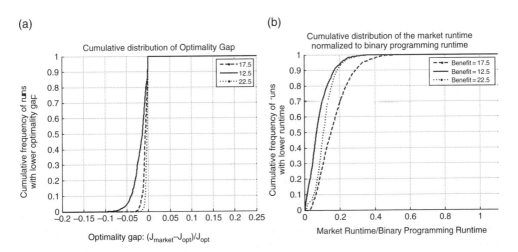

Figure 3.7 Comparison of the MBS to mathematical optimization in the *Max-Pro* case for time-invariant and equal benefit values. (a) Optimality gap. (b) Relative runtime.

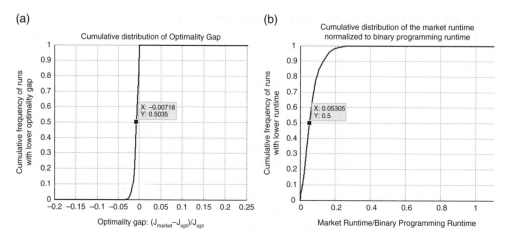

Figure 3.8 Comparison of the MBS to mathematical optimization in the *Max-Pro* with time-invariant and three benefit values case. (a) Optimality gap. (b) Relative runtime.

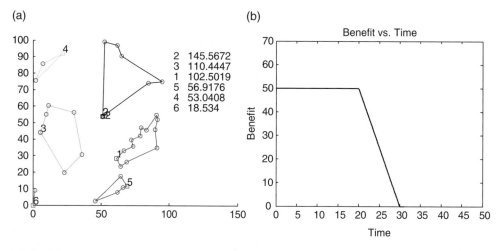

Figure 3.9 A scenario depicting *Max-Pro* with time-varying target benefits. (a) The solution found by MBS. (b) The values of the target benefits vs. time.

Qualitatively, this problem is similar to a *Min-Max* problem in the sense that the assignments attempt to reach as many targets as long as there is a benefit to do so. Thus, the UAVs divide the work between them.

Unfortunately, while the MBS is capable of solving these problems, we have not found any other solution in either the mathematical or heuristic optimization literature. Thus, the only comparison we were able to do was for extremely small cases vs. a direct enumeration of all possible solutions. The latter is very limited in the size of problems it can solve and while the MBS achieves near optimal results, this comparison is too trivial.

3.6 Recommendations for Implementation

In this section, we offer some recommendations for implementation of such a system to create an operational system. The MBS requires two things: a situational picture and communication. The situation picture includes a list of available resources (UAVs) and a list of known targets. These two lists are constantly updated based on data gathered by the UAV package at the area of interest, by intelligence gathered by sources external to the UAV package, by health monitoring data from the UAV, and by tasking orders from central command and control centers, that may shift UAVs to and from the package, based on higher level decision making.

Understandably, any implementation recommendations are very general in nature and should be further adapted to the organization in question and the way this organization operates. Some organizations are more centralized while others are more decentralized in nature, thus the implementation in a real-time operational system will have to address these constraints.

Figure 3.10 depicts a general-use case: a UAV package is deployed at some remote location, connected via satellite link to a Command, Control, Communications, Computing, Intelligence, Surveillance, and Reconnaissance (C4ISR) system that controls them. The C4ISR system is assumed to be connected to some organization-wide Command, Control, and Intelligence (C2I) center, which uses other information sources to generate a higher level situation picture.

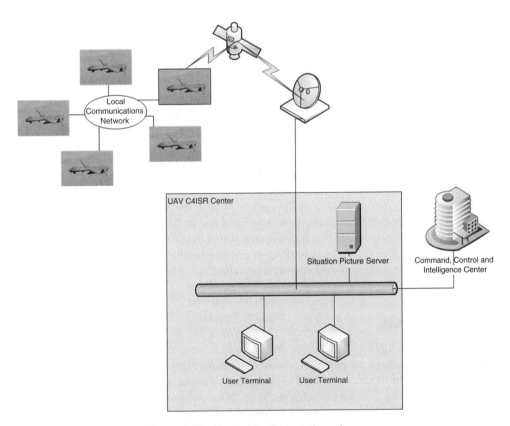

Figure 3.10   General implementation scheme.

In a completely centralized organization, the UAV C4ISR center serves as a center for all the data that come from the UAV package via satellite processes this information, fuses it with the higher level situation picture, and based on this centralized situation picture, the tasking algorithm is run. The benefit of such a scheme is the ability to create a uniform situational awareness based on the information that comes from all the sources, thus the algorithm can run on the most complete and accurate situation picture. In addition, human operators may intervene and improve the tasking algorithm when presented to them. The limitations are the constant need to maintain a satellite link between the UAV package and the C4ISR center, which makes the satellite link a single point of failure for the entire system, and the delay from gathering information by the UAVs in the package to acting on it.

In a completely decentralized organization, the data gathered by the UAV package, using the sensors onboard these UAVs, are shared via a local communications network. The UAVs can act independently to fuse the sensor data, based on available protocols for data merging and fusion, and each UAV will obtain its own situation picture, which may be different from the others. In this scheme, the UAVs in the package will run the MBS onboard and will use communication to share information about targets and trade known targets between the UAVs. The main advantages of such a scheme are the relative independence from a satellite link, there is no single point of failure, and data is processed and acted upon with as little latency as possible, because the center is not part of the processing. The caveats, of course, are the lack of a uniform situational picture, and the constant need to maintain a well-connected local network and consensus between the UAVs. Some of these limitations were addressed by algorithms like the Consensus-Based Bundle Algorithm (CBBA) [53].

A hybrid centralized-decentralized solution can probably improve on the aforementioned solutions. In this scheme, a more high-level situation picture is periodically generated at the UAV C4ISR center, based on all available information from the centralized C2I center and the data gathered by the UAV package. The UAV C4ISR center runs the MBS on this centralized picture and creates a tasking baseline, which is communicated to the UAV package. While the UAV package is operating, local decisions are being made based on data gathered onsite, the UAVs make local adaptations, until a new baseline is communicated from the ground or until the changes from the baseline are deemed too great and the UAV package requests the UAV C4ISR center for a new baseline. This way, satellite link is used only periodically and since the UAV package is autonomous enough to act without it, the link does not become a single point of failure. On the other hand, new developments at the central level or at the UAV package can be addressed in real time in a central way, based on a uniform situation picture.

Naturally, these suggestions are just descriptions of possible implementations of the MBS in a real-time system, and require further consideration.

3.7   Conclusions

The problem of assigning a collective of UAVs to a bank of targets is one of the most fundamental problems in operations research of the impact that UAVs can make in the battle-ground of the future. There are many problem formulations, which differ in the objective function that the collectives should achieve, and each formulation comes with a set of different mathematical and heuristic solutions. Still, from the operational point of view, a commander of a large fleet of UAVs requires the flexibility to plan for various objectives. The optimization must scale well for a large number of UAVs and sizeable target banks, and while it doesn't have to guarantee optimality, the quality of the optimization should be comparable to mathematical optimization methods that do not scale very well.

In this chapter, we proposed a solution based on economic markets, which are known to be able quickly to find near-optimal solutions. We represent each UAV as an agent and each target as a task and, using a set of economic behaviors, we let the agents compete for the tasks and reach assignments in a quick, efficient, and effective way. Our method allows the flexibility of various objective functions to be used, it scales well for very large problems, and can be further improved by parallel processing and implementation in a compiled language.

While a full appreciation of the benefits of using collectives of UAVs may remain elusive, we believe that this work can serve as a benchmark for future operational research in the field of UAVs in military and civilian applications. This work shows that there are methods to assign large groups of UAV to targets and that there is evidence for claims of increasing returns and for how these returns can be obtained.

Appendix 3.A A Mixed Integer Linear Programming (MILP) Formulation

Here we formulate the three cases as an MILP problem. Let $N = \{1,2,\dots,n\}$ be the set of indices for target locations and $D = \{n+1, n+2, \dots, n+m\}$ be the set of indices for depot locations. Note that if a certain depot serves as the initial location of more than a single UAV, the index of that depot will be repeated as many times as the number of UAVs in the depot. Let $N_0 = N \cup D = \{1, 2, \dots, n, n+1, n+2, \dots, n+m\}$ be the set of all indices of both targets and depots.

Let d_{ij} be the Euclidean distance between any two targets $i, j \in N$ and let d_{ki}^0 and d_{ik}^0 be the Euclidean distance between depot $k \in D$ and target $i \in N$. Let b_i be the time-invariant benefit associated with target i. We assume that the problem is symmetric so that:

$$d_{ij} = d_{ji} \ \forall i, j \in N \tag{A1}$$

$$d_{ki}^0 = d_{ik}^0 \forall i \in N, k \in D \tag{A2}$$

Furthermore, we assume that the problem satisfies the triangle inequality, so that:

$$d_{ij} + d_{jl} \geq d_{il} \ \forall i, j, l \in N_0 \tag{A3}$$

Let x_{ijk} be a binary variable equal to one if UAV k moves from location i to location j along its tour and zero otherwise, where $k \in D$ and $i, j \in N_0$. We assume that $x_{iik} = 0$ for all combinations of i and k to prevent a UAV from moving from one location to the same location on its tour.

Based on the definitions above, the tour length of a single UAV is:

$$L_k = \sum_{j=1}^{n} x_{kjk} d_{kj}^0 + \sum_{i=1}^{n}\sum_{j=1}^{n} x_{ijk} d_{ij} + \sum_{i=1}^{n} x_{ikk} d_{ik}^0 \tag{A4}$$

The first term in Eq. (A4) represents the cost of travel from the depot to the first target in the tour, the second term sums all the distances along the tour from one target to the next, and the last term represents the cost to return to the depot from the last target in the tour.

As defined in Section 3.1, there are three main objective cases: *Min-Sum*, *Min-Max*, and *Max-Pro*. We formulate them here:

$$\textbf{\textit{Min - Sum}} \quad Minimize \sum_{k=1}^{m} L_k \tag{A5a}$$

$$\textbf{\textit{Min - Max}} \quad Minimize \, L_{max} \tag{A5b}$$

$$\textbf{\textit{Max - Pro}} \quad Minimize \left(\sum_{k=1}^{m} L_k - \sum_{i=1}^{n} \left(\sum_{k=1}^{m} x_{kik} + \sum_{j=1}^{n} \sum_{k=1}^{m} x_{jik} \right) b_i \right) \tag{A5c}$$

In short, Eq. (A5a) sets the objective as the sum of all tour lengths, Eq. (A5b) as the maximum tour length by any UAV, and Eq. (A5c) subtracts the benefits associated with completing targets from the total tour length.

The optimization is subject to the following constraints:

$$\sum_{k=1}^{m} x_{kik} + \sum_{j=1}^{n} \sum_{k=1}^{m} x_{jik} \leq 1 \, \forall i \in N \tag{A6}$$

$$x_{kik} + \sum_{j=1}^{N} x_{jik} - \left(x_{ikk} + \sum_{j=1}^{n} x_{ijk} \right) = 0 \, \forall i, j \in N, k \in D \tag{A7}$$

$$\sum_{i=1}^{n} x_{kik} \leq 1 \, \forall k \in D \tag{A8}$$

$$\sum_{i=1}^{n} x_{kik} - \sum_{i=1}^{n} x_{ikk} = 0 \, \forall k \in D \tag{A9}$$

$$L_k \leq L_{max} \tag{A10}$$

3.A.1 Sub-tour Elimination Constraints

The constraint in Eq. (A6) guarantees that each target is assigned at most once. In the *Min-Sum* and *Min-Max* cases, we replace the inequality sign with an equality sign and force the target to be performed *exactly once*. Equation (A7) makes sure that if a target is assigned, the same UAV that is assigned to that target will continue in its tour. Equation (A8) guarantees that a UAV has at most one tour. UAVs may not be assigned to any targets, which is common in the *Min-Sum* and *Max-Pro* cases, where the cost of leaving a depot and traveling to a target precludes many UAVs from being assigned to targets. The constraint in Eq. (A9) guarantees that a UAV tour will always return to the depot. The constraint in Eq. (A10) is used only in the *Min-Max* case. Sub-tour elimination constraints make sure that no subset of targets is assigned to a sub-tour that does not originate or end in a depot. It should be noted that the number of sub-tour constraints is the main driver on the computation time required for solving this mathematical formulation. In particular, in the *Max-Pro* and *Min-Max* cases, targets that are closely clustered together far from a depot tend to form sub-tours that need to be eliminated, especially if the benefit associated with the targets is low.

It may be noted that the formulations for the *Min-Sum* and *Max-Pro* (with constant benefits) cases are essentially binary programming problems. On the other hand, due to inclusion of L_{max} in cost function, the *Min-Max* formulation becomes the MILP problem.

References

1. Department of Defense (DOD). "Unmanned Systems Integrated Roadmap FY 2013-2038", Reference Number: 14-S-0553, published online in 2013 http://www.defense.gov/pubs/DOD-USRM-2013.pdf, retrieved June 3, 2014.
2. Applegate DL, Bixby RE, Chvatal V, Cook WJ. *"The Traveling Salesman Problem: A Computational Study"*, 1 ed., Princeton Series in Applied Mathematics, Princeton University Press, Princeton, NJ, 2006.
3. Feillet D, Pierre D, Michel G. "Traveling salesman problem with profits". *Transportation Science* 2005;**39**(2):188–205.
4. Held M, Karp RM. "A dynamic programming approach to sequencing problems". *Journal of the Society for Industrial and Applied Mathematics* 1962;**10**(1):196–210.
5. Bektas T. "The multiple traveling salesman problem: an overview of formulations and solution procedures". *Omega* 2006;**34**(3):209–219.
6. Campbell AM, Vandenbussche D, Hermann W. "Routing for relief efforts". *Transportation Science* 2008;**42**(2):127–145.
7. Carlsson J, Ge D, Subramaniam A. *"Lectures on Global Optimization"*, illustrated ed., vol. 55 of Fields Institute communications, ch. "Solving the Min-Max Multi-Depot Vehicle Routing Problem", 31–46, American Mathematical Society, Providence, RI, 2009.
8. Chao IM, Golden BL, Wasil EA. "A fast and effective heuristic for the orienteering problem". *European Journal of Operational Research* 1996;**88**(3):475–489.
9. Chao IM, Golden BL, Wasil EA. "Theory and methodology: the team orienteering problem". *European Journal of Operational Research* 1996;**88**(3):464–474.
10. Dorigo M, Maniezzo V, Colorni A. "The ant system: optimization by a colony of cooperating agents". *IEEE Transactions on Systems, Man, and Cybernetics – Part B* 1996;**26**(1):29–41.
11. Golden BL, Laporte G, Taillard ED. "An adaptive memory heuristic for a class of vehicle routing problems with minmax objective". *Computers & Operations Research* 1997;**24**(5):445–452.
12. Kennedy J, Eberhart R. "Particle swarm optimization". *IEEE Int'l Conference on Neural Networks* 1995;**IV**:1942–1948.
13. Kirkpatrick S Jr, Gelatt CD, Vecchi MP. "Optimization by simulated annealing". *Science* 1983;**220**(4598): 671–680.
14. Lin S, Kernighan BW "An effective heuristic algorithm for the traveling salesman problem". *Operations Research* 1973; **21**(2):498–516.
15. Passino K, Polycarpou M, Jacques D, Pachter M, Liu Y, Yang Y, Flint M, Baum M. "Cooperative control for autonomous air vehicles". In Proceedings of the Cooperative Control Workshop, Orlando, FL, 2000.
16. Rasmussen S, Chandler P, Mitchell JW, Schumacher C, Sparks A. "Optimal vs. heuristic assignment of cooperative autonomous unmanned air vehicles". In: Proceedings of the AIAA Guidance, Navigation and Control Conference, Austin, TX, 2003.
17. Schumacher C, Chandler P, Pachter M. "UAV Task Assignment with Timing Constraints", AFRL-VA-WP-TP-2003-315, 2003, United States Air Force Research Laboratory.
18. Schumacher C, Chandler PR, Rasmussen S. "Task allocation for wide area search munitions via network flow optimization". In: Proceedings of the 2001 AIAA Guidance, Navigation, and Control Conference, Montreal, Canada, 2001.
19. Shima T, Rasmussen SJ, Sparks AG, Passino KM. "Multiple task assignments for cooperating uninhabited aerial vehicles using genetic algorithms". *Computers & Operations Research* 2006;**33**(11):3252–3269, Part Special Issue: Operations Research and Data Mining.

20. Wooldridge M. *"An Introduction to Multi-Agent Systems"*, 1 ed., John Wiley & Sons, Chichester, England, 2002.
21. Applegate DL, Bixby RE, Chvatal V, Cook WJ, "Concorde", can be downloaded online at http://www.tsp.gatech.edu/concorde/index.html, 2004.
22. Potvin J, Lapalme G, Rousseau J. "A generalized k-opt exchange procedure for the MTSP". *INFOR* 1989;**27**(4):474–81.
23. Zhang T, Gruver WA, Smith MH. "Team scheduling by genetic search". In: Proceedings of the Second International Conference on Intelligent Processing and Manufacturing of Materials, Honolulu, Hawaii. vol. 2, 1999. p. 839–844.
24. Yu Z, Jinhai L, Guochang G, Rubo Z, Haiyan Y. "An implementation of evolutionary computation for path planning of cooperative mobile robots". In: Proceedings of the Fourth World Congress on Intelligent Control and Automation, Shanghai, China. vol. 3, 2002. p. 1798–1802.
25. Song C, Lee K, Lee WD. "Extended simulated annealing for augmented TSP and multi-salesmen TSP". In: Proceedings of the International Joint Conference on Neural Networks, Portland, Oregon. vol. 3, 2003. p. 2340–2343.
26. Ryan JL, Bailey TG, Moore JT, Carlton WB. "Reactive tabu search in unmanned aerial reconnaissance simulations". In: Proceedings of the 1998 Winter Simulation Conference, Washington, DC. vol. 1, 1998. p. 873–879.
27. Goldstein M. "Self-organizing feature maps for the multiple traveling salesmen problem". In: Proceedings of the IEEE International Conference on Neural Network. San Diego, CA, 1990. p. 258–261.
28. Torki A, Somhon S, Enkawa T. "A competitive neural network algorithm for solving vehicle routing problem". *Computers and Industrial Engineering* 1997;**33**(3–4):473–476.
29. Modares A, Somhom S, Enkawa T. "A self-organizing neural network approach for multiple traveling salesman and vehicle routing problems". *International Transactions in Operational Research* 1999;**6**:591–606.
30. Kivelevitch E, Sharma B, Ernest N, Kumar M, Cohen K. "A hierarchical market solution to the min–max multiple depots vehicle routing problem". *Unmanned Systems* 2014;**2**(1):87–100.
31. Narasimha KV, Kivelevitch E, Sharma B, Kumar M. "An ant colony optimization technique for solving minmax multi-depot vehicle routing problem". *Swarm Evolutionary Computing* 2013;**13**:63–73.
32. Ernest N, Cohen K. "Fuzzy logic clustering of multiple traveling salesman problem for self-crossover based genetic algorithm". In: Proceedings of the IAAA 50th ASM. AIAA, Nashville, TN, January 2012.
33. Ernest N, Cohen K. "Fuzzy clustering based genetic algorithm for the multi-depot polygon visiting dubins multiple traveling salesman problem". In: Proceedings of the 2012 IAAA Infotech@Aerospace. AIAA, Garden Grove, CA, June 2012.
34. Ren C. "Solving min-max vehicle routing problem". *Journal of Software* June 2011;**6**:1851–1856.
35. Ren C., "Heuristic algorithm for min-max vehicle routing problems". *Journal of Computers* April 2012;**7**:923–928.
36. Tang H, Miller-Hooks E. "A tabu search heuristic for the team orienteering problem". *Computers & Operations Research* 2005;**32**(6):1379–1407.
37. Vansteenwegen P, Souriau W, Van Oudheusden D. "The orienteering problem: A survey". *European Journal of Operational Research* 2011;**209**:1–10.
38. Boussier S, Feillet D, Gendreau M. "An exact algorithm for team orienteering problems". *4OR: A Quarterly Journal of Operations Research* 2007;**5**:211–230.
39. Kivelevitch E, Cohen K, Kumar M. "A binary programming solution to the multiple-depot, multiple traveling salesman problem with constant profits". In: Proceedings of 2012 AIAA Infotech@ Aerospace Conference, Garden Grove, CA. 2012.
40. Ekici A, Keskinocak P, Koenig S. "Multi-robot routing with linear decreasing rewards over time". In: IEEE International Conference on Robotics and Automation, Kobe, Japan. 2009. p. 958–963.
41. Erkut E., Zhang J. "The maximum collection problem with time-dependent rewards". *Naval Research Logistics (NRL)* 1996;**43**:749–763.
42. Li, J. "Model and Algorithm for Time-Dependent Team Orienteering Problem", *"Advanced Research on Computer Education, Simulation and Modeling"* (S Lin, X Huang, eds.), Communications in Computer and Information Science, vol. **175**, 1–7, Springer, Berlin, Germany, 2011.
43. Baker AD. *"Market-Based Control: A Paradigm for Distributed Resource Allocation"*, no. 9810222548, ch. "Metaphor or Reality: A Case Study Where Agents Bid with Actual Costs to Schedule a Factory", 184–223, World Scientific Publishing, Singapore, 1996.
44. Ferguson DF, Nickolaou C, Sairamesh J, Yemini Y, *"Market-Based Control: A Paradigm for Distributed Resource Allocation"*, no. 9810222548, ch. "Economic Models for Allocating Resources in Computer Systems', 156–183, World Scientific Publishing, Singapore, 1996.

45. Kuwabara K, Ishida T, Nishibe Y, Suda T. *"Market-Based Control: A Paradigm for Distributed Resource Allocation"*, no. 9810222548, ch. "An Equilibratory Market-Based Approach for Distributed Resource Allocation and Its Application to Communication Network Control", 53–73, World Scientific Publishing, Singapore, 1996.

46. Clearwater SH. *"Market-Based Control: A Paradigm for Distributed Resource Allocation"*, no. 9810222548, ch. Preface, v–xi, World Scientific Publishing, Singapore, 1996.

47. Beinhocker ED. *"The Origin of Wealth"*, 35–36, Harvard Business School Press, Cambridge, MA, 2007.

48. Gagliano RA, Mitchem PA. *"Market-Based Control: A Paradigm for Distributed Resource Allocation"*, no. 9810222548, ch. "Valuation of Network Computing Resources", 28–52, World Scientific Publishing , Singapore, 1996.

49. Harty K, Cheriton D. *"Market-Based Control: A Paradigm for Distributed Resource Allocation"*, no. 9810222548, ch. "A Market Approach to Operating System Memory Allocation", 126–155, World Scientific Publishing, Singapore, 1996.

50. Kulkarni AJ, Tai K, "Probability collectives: A multi-agent approach for solving combinatorial optimization problems". *Applied Soft Computing* 2010;**10**:759–771.

51. Wellman MP. *"Market-Based Control: A Paradigm for Distributed Resource Allocation"*, no. 9810222548, ch. "Market-Oriented Programming: Some Early Lessons", 74–95, World Scientific Publishing, Singapore, 1996.

52. Kivelevitch E, Cohen K, Kumar M. "A market-based solution to the multiple traveling salesmen problem". *Journal of Intelligent & Robotic Systems* October 2013;**72**(1):21–40.

53. Choi H-L, Brunet L, How JP. "Consensus-based decentralized auctions for robust task allocation". *IEEE Transactions on Robotics* August 2009;**25**:912–926.

54. Karmani RK, Latvala T, Agha G. "On scaling multi-agent task reallocation using market-based approach". In: Proceedings of the first IEEE International Conference on Self-Adaptive and Self-Organizing Systems, Boston, MA. July 2007. p. 173–182.

55. Giaccari L. Tspconvhull, MATLAB Central (12, 2008).

56. IBM, Cplex Web, 2012, http://www-01.ibm.com/software/integration/optimization/cplex-optimizer/.

57. Andrew T. "Computation Time Comparison Between Matlab and C++ Using Launch Windows", available online at http://digitalcommons.calpoly.edu/aerosp/78/

58. Somhom S, Modares A, Enkawa T. "Competition-based neural network for the multiple traveling salesmen problem with minmax objective". *Computers and Operations Research* 1999;**26**(4):395–407.

4

Considering Mine Countermeasures Exploratory Operations Conducted by Autonomous Underwater Vehicles*

Bao Nguyen[1], David Hopkin[2] and Handson Yip[3]

[1] Defence Research and Development Canada, Centre for Operational Research and Analysis, Ottawa, ON, Canada
[2] Defence Research and Development Canada–Atlantic, Dartmouth, NS, Canada
[3] Supreme Allied Command Transformation, Staff Element Europe, SHAPE, Mons, Belgium

4.1 Background

The availability of affordable Commercial-Off-The-Shelf (COTS) Autonomous Underwater Vehicles (AUVs) equipped with side-scan sonar is beginning to transform the way detection is conducted in Mine Countermeasures (MCM) operations. These AUVs can achieve higher efficiency and effectiveness than conventional assets, at near zero risk to human life. Cost-effective, lightweight COTS AUVs are also particularly favored by the military community because of their minimal logistical requirements for rapid deployment and handling.

A frequent and important MCM operation is the "MCM exploratory operation" (also known as an "MCM reconnaissance operation"). The aim of an MCM exploratory operation is to determine the presence or absence of mines in an area of operation. It is an essential step before

---

*Contribution originally published in MORS Journal, Vol. 19, No. 2, 2014. Reproduced with permission from Military Operations Research Society.
© Her Majesty the Queen in Right of Canada (2016).

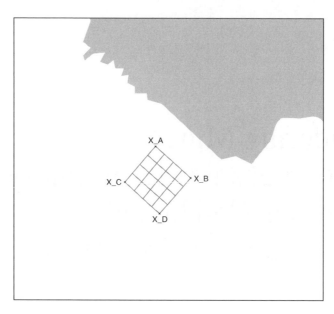

Figure 4.1   Search areas (Gulf of La Spezia, Italy). *Source*: Nguyen, Bao; Hopkin, David; Yip, Handson, "Considering Mine Countermeasure Exploratory Operations Conducted by Autonomous Underwater Vehicles," *MORS Journal*, Vol. 19, No. 2, 2014, John Wiley and Sons, LTD.

Table 4.1   Scenario parameters.

Parameter	Value
Total area: 16 segmented areas	12 km × 12 km
Segmented area	3 km × 3 km
AUV speed	9 knots
AUV endurance	30 h

deploying any high-value asset in potentially mined waters. If a mine is detected, classified, and identified during MCM exploratory operations, then mine clearance operations will follow.

The use of AUVs in MCM exploratory operations is relatively recent; therefore, there is a lack of knowledge on how to employ them efficiently and effectively. In this chapter, we will focus on two key metrics that characterize the efficiency and effectiveness of AUVs in such operations. The first metric is the confidence that mines are or are not present in the search area. The second metric is the time required to achieve that confidence level.

Any search strategy must take into account the current mindset and operating practices of the operator, as well as the technological constraints of today's AUV systems. The preferred way to conduct an exploratory search of a large area is to segment the area into smaller areas that can be searched in a series of short missions. Typically, the operator would plan each AUV mission such that the AUV would bring back data for analysis within a few hours. In this experiment, the trial area has been partitioned into 16 3 km × 3 km square cells, as shown in Figure 4.1. Table 4.1 summarizes the parameters defining the scenario and the capabilities of the *Dorado*, the AUV developed by Defence Research and Development Canada (DRDC)–Atlantic, depicted in Figure 4.2.

Figure 4.2 The *Dorado*. *Source*: Nguyen, Bao; Hopkin, David; Yip, Handson, "Considering Mine Countermeasure Exploratory Operations Conducted by Autonomous Underwater Vehicles," *MORS Journal*, Vol. 19, No. 2, 2014, John Wiley and Sons, LTD.

4.2 Assumptions

In this chapter, we assume that the AUV carries a side-scan sonar. The mine-hunting environment plays an important role in the performance of side-scan sonar, as it does for any MCM sensor. Seabed conditions such as clutter, composition, and topology have the most significant effects on the sensor's detection performance. However, in this first attempt to analyze the effectiveness of AUVs, we assume benign seabed conditions with high-target-strength mines. Such circumstances allow reasonable assumptions to be made about sensor performance. Specifically, the probability of identifying a mine as a mine is very high (approximately 100%) while the probability of identifying a non-mine as a mine is very low (approximately 0%).

The interior of the rectangular frame in Figure 4.3 is the search area, within which solid lines represent the AUV path while circles indicate mines. The arrow in the search area indicates the general direction of the AUV's transit. The vehicle starts at the bottom left, travels from left to right, travels upward, travels from right to left, and then travels upward and repeats this motion.

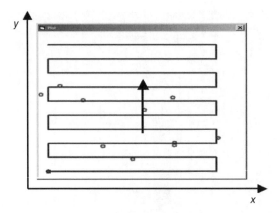

Figure 4.3 Coordinate system. *Source*: Nguyen, Bao; Hopkin, David; Yip, Handson, "Considering Mine Countermeasure Exploratory Operations Conducted by Autonomous Underwater Vehicles," *MORS Journal*, Vol. 19, No. 2, 2014, John Wiley and Sons, LTD.

The AUV path shown in Figure 4.3 is a typical search pattern called the "lawn-mowing pattern" or the "parallel search pattern." This and every search pattern is characterized by its resulting Measures Of Effectiveness (MOEs). These MOEs may include probability of detection, coverage, overlap, and search time. They can be determined using the scenario's geometry and the sonar's Measures Of Performance (MOPs).

4.3 Measures of Performance

To assess the effectiveness of MCM operations, we need to understand the sonar's MOPs: the probability of detection as a function of range, as shown on the left-hand side (LHS) of Figure 4.4, and the probability of detection as a function of aspect angle, as shown on the right-hand side (RHS) of Figure 4.4. (The aspect angle is defined as the angle between the sonar beam and the axis of symmetry of a cylindrical mine.) We assume that the range and angular probabilities are independent.

Probability of detection as a function of range represents a characteristic of the side-scan sonar, while probability of detection as a function of angle represents a characteristic of the mine. These probability curves are modeled using the Johnson distribution. Equation (4.1) [1] is parameterized using the values in Table 4.2. The scale (λ) shown below is a factor compounded to Johnson's distribution such that the maximal value of each curve is equal to 1.

Johnson's curve can be expressed as:

$$f(x) = \begin{cases} \dfrac{\lambda \cdot \alpha_2 \cdot (x_2 - x_1)}{(x - x_1) \cdot (x_2 - x) \cdot \sqrt{2\pi}} e^{-\frac{1}{2}\left(\alpha_1 + \alpha_2 \cdot \ln\left(\frac{x - x_1}{x_2 - x}\right)\right)^2} & x_1 < x < x_2 \\ \\ 0 & \text{otherwise} \end{cases}$$

(4.1)

The range probability curve implies that if a mine lies between the minimum range and the maximum range, then it will be detected with a probability close to 100%. We refer to the

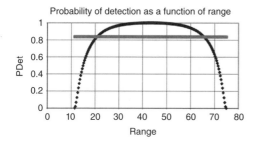
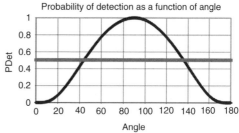

Figure 4.4 Probability of detection as a function of range (LHS) and angle (RHS). *Source*: Nguyen, Bao; Hopkin, David; Yip, Handson, "Considering Mine Countermeasure Exploratory Operations Conducted by Autonomous Underwater Vehicles," *MORS Journal*, Vol. 19, No. 2, 2014, John Wiley and Sons, LTD.

Table 4.2 Johnson parameters defining the range and angle probability curves.

	Johnson's parameters	
Range probability curve	$\alpha_1 = 0, \alpha_2 = 0.75$	$x_1 = 11.5\,\text{m}, x_2 = 75\,\text{m}$
Angular probability curve	$\alpha_1 = 0, \alpha_2 = 1.25$	$x_1 = 0, x_2 = \pi$

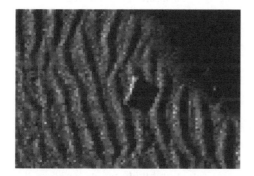

Figure 4.5 Mine observed at 85° (LHS) and at 0° (RHS). *Source*: Nguyen, Bao; Hopkin, David; Yip, Handson, "Considering Mine Countermeasure Exploratory Operations Conducted by Autonomous Underwater Vehicles," *MORS Journal*, Vol. 19, No. 2, 2014, John Wiley and Sons, LTD.

minimum range as the gap in coverage in the sense that if a mine lies between the sonar and its minimum detection range then it will not be detected. The angular probability curve implies that probability of detection of the mine reaches a maximum when the mine's aspect angle is perpendicular to the side-scan sonar beam, and decreases symmetrically as the angle deviates from the perpendicular case.

It is clear from Figure 4.4 that the probability of detection is substantially degraded if the aspect angle differs from $\pi/2$. The impact of this degradation is shown in Figure 4.5. On the LHS of Figure 4.5, a mine is observed at an angle of 85° and would without a doubt be identified by an experienced operator. On the RHS of Figure 4.5, the same mine is observed at an angle of 0°, and is not recognizable at all.

4.4   Preliminary Results

If we assume that the spacing between two consecutive horizontal legs of the lawn-mowing search pattern shown in Figure 4.3 is typically equal to twice the maximum range of detection, then there is no overlap in coverage between two consecutive horizontal legs and the probability of detection obtained by the lawn-mowing search pattern is approximately 40%. We arrive at this figure by noting that the mean range probability of detection is about 80% (LHS of Figure 4.4) while the mean angular probability of detection is about 50% (RHS of Figure 4.4). Thus, the final probability is the product, or 40%.

The usual lawn-mowing pattern poses two problems. First, the MOEs are affected by the gap in detection range. Second, there is a substantial degradation in the probability of detection due to angular effects. The next section presents solutions to these problems.

4.5   Concepts of Operations

4.5.1   Gaps in Coverage

If the AUV conducts an uneven lawn-mowing pattern, as shown in Figure 4.6, then it will eliminate gaps in coverage. This search pattern is designed to provide 100% coverage of the search area, in addition to minimizing overlaps and search time with respect to other mowing paths. With that in mind, we derive that the small spacing between horizontal legs must be equal to the maximum detection range minus the minimum detection range $(r_{max} - r_{min})$, while the large spacing between them must be equal to twice the maximum detection range $(2 \cdot r_{max})$. Note that this solution can be fulfilled only if $r_{max} \geq 3 \cdot r_{min}$, a condition that is met by the parameters shown in Table 4.2.

4.5.2   Aspect Angle Degradation

Degradation due to angular probability of detection, described in the previous section, can be mitigated by fusing more than one observation of a mine at different aspect angles [2–4]. Assume a cylindrical mine is observed from two angles: 30° and 60°. The fusion of these two observations improves the angular probability of detection by 6% points based on the

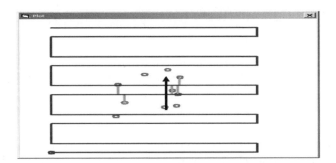

Figure 4.6   Uneven lawn-mowing pattern. *Source*: Nguyen, Bao; Hopkin, David; Yip, Handson, "Considering Mine Countermeasure Exploratory Operations Conducted by Autonomous Underwater Vehicles," *MORS Journal*, Vol. 19, No. 2, 2014, John Wiley and Sons, LTD.

probability curve shown on the RHS of Figure 4.4. We can understand this improvement by observing that the angular probability of detection by the first AUV leg is $P_1 = 77\%$ while the angular probability of detection by the second AUV leg is equal to $P_2 = 24\%$. Combining the results of the two AUV legs, the angular probability of detection becomes $P_{1-2} = 1 - (1 - P_1)(1 - P_2) = 83\%$ – an improvement of 6% points with respect to $P_1 = 77\%$.

4.6   Optimality with Two Different Angular Observations

If the AUV can observe the same mine from two different angles, then it is natural to ask how the two angles should be related to one another such that the angular probability of detection is optimal. We define P_1 to be the mean angular probability of detection generated by one AUV leg while $P_2(\phi)$ to be the mean angular probability of detection generated by two AUV legs, calculated as a function of the angle ϕ between them. $P_2(\phi)$ achieves a maximum when $\phi = \pi / 2$. That is, to maximize the angular probability of detection with two observations at different angles, the difference between the angles must be $\pi/2$.

The proof of this assertion is described below. However, for the purpose of this section, it is convenient to redefine the aspect angle θ by shifting it to the left by $\pi/2$ such that the maximum angular probability of detection occurs at $0\,\text{rad}$.

To obtain P_1 and P_2, we assume that a mine lies between the minimum and maximum range of an AUV leg and two AUV legs, respectively. We compute P_1 by integrating the angular probability of detection, shown on the LHS of Figure 4.4, from $-\pi / 2$ to $+\pi / 2$ and then averaging it through a division by π. We compute P_2 in the same way, except that it is based on two AUV legs whose relative angle is ϕ.

$$P_1 = \frac{1}{\pi} \cdot \int_{-\frac{\pi}{2}}^{\frac{\pi}{2}} d\theta \cdot f(\theta) \tag{4.2}$$

$$P_2(\phi) = 1 - \frac{1}{\pi} \cdot \int_{-\frac{\pi}{2}}^{\frac{\pi}{2}} d\theta \cdot (1 - f(\theta)) \cdot (1 - f(\theta + \phi)) \tag{4.3}$$

here, $f(\theta)$ is the Johnson function whose parameters are defined in Table 4.2. The fact that P_2 reaches a maximum at $\phi = \pi / 2$ is true in general. It is due to the shape and symmetry of the angular probability of detection curve. Observe that the derivative of $P_2(\phi)$ at $\phi = \pi / 2$ is:

$$\frac{d}{d\phi} P_2(\phi) \bigg]_{\phi = \frac{\pi}{2}} = \frac{1}{\pi} \cdot \int_{-\frac{\pi}{2}}^{\frac{\pi}{2}} d\theta \cdot (1 - f(\theta)) \cdot \frac{d}{d\phi} f(\theta + \phi) \bigg]_{\phi = \frac{\pi}{2}} \tag{4.4}$$

Since $f(\theta)$ is an even function of θ, $1 - f(\theta)$ is also an even function of θ. In addition, $\frac{d}{d\phi} f(\theta + \phi) \bigg]_{\phi = \pi / 2}$ is an odd function of θ. Hence, the integrand of Eq. (4.4) is odd at $\phi = \pi / 2$, implying that $\dfrac{d}{d\phi}P_2(\phi)\bigg]_{\phi=\pi/2} = 0$. Therefore, $P_2(\phi)$ achieves an optimum at $\phi = \pi/2$. Appendix 4.A further shows that this is not merely an optimum but in fact a global maximum. Based on the parameters shown in Table 4.2, we get $P_1 = 0.50$ and $P_2(\phi = \pi/2) = 0.88$.

4.7 Optimality with N Different Angular Observations

It turns out that we can generalize the result in the previous subsection. That is, we are able to determine the angles such that the angular probability of detection is optimal when a mine is observed N times from the same range. We show that if $\phi_i = i \cdot \varphi$ where $i = 0 \ldots N-1$ and $\phi = \pi/N$, then the resulting mean angular probability of detection is maximal. For example, if there are three angular observations, then the mean angular probability of detection is maximal when $\phi_0 = 0$; $\phi_1 = \pi/3$; $\phi_2 = 2\pi/3$.

To prove this assertion, we only need to show that $\left(\dfrac{\partial}{\partial\phi_0}P_N(\bar\phi)\right)_{\phi_i=i\cdot\theta} = 0$, since we can always choose a coordinate system that is rotated in such a way that for any i, ϕ_i becomes zero. Note that $\left(\dfrac{\partial}{\partial\phi_0}P_N(\bar\phi)\right)_{\phi_i=i\cdot\theta}$ can be expressed as:

$$\left(\frac{\partial}{\partial\phi_0}P_N(\bar\phi)\right)_{\phi_i=i\cdot\theta} = \int_{-\frac{\pi}{2}}^{\frac{\pi}{2}}\frac{d\theta}{\pi}\cdot g'(\theta)\cdot g(\theta+\varphi)\cdots\cdots g(\theta+(N-1)\cdot\varphi) \qquad (4.5)$$

The above is zero if $g'(\theta)$ is odd and $g(\theta+\varphi)\cdot g(\theta+2\cdot\varphi)\cdots\cdot g(\theta+(N-1)\cdot\varphi)$ is even, as this implies that the integrand is odd, and integrating an odd function from $-\pi/2$ to $\pi/2$ gives zero.

Since $g(\theta)$ is even, we infer that $g'(\theta)$ is odd. Moreover,

$$g(-\theta+\varphi)\cdot g(-\theta+2\cdot\varphi)\cdots\cdot g(-\theta+(N-1)\cdot\varphi)$$
$$= g(\theta-\varphi)\cdot g(\theta-2\cdot\varphi)\cdots\cdot g(\theta-(N-1)\cdot\varphi)$$
$$= g(\theta-\varphi+N\cdot\varphi)\cdot g(\theta-2\cdot\varphi+N\cdot\varphi)\cdots\cdot g(\theta-(N-1)\cdot\varphi+N\cdot\varphi)$$
$$= g(\theta+(N-1)\cdot\varphi)\cdot g(\theta+(N-2)\cdot\varphi)\cdots\cdot g(\theta+\varphi)$$

The first equality is true due to the fact that $g(\theta)$ is even. The second equality is true due to the fact that $g(\theta)$ is periodic with a period equal to $N\cdot\varphi = \pi$. The last equality shows that the product is an even function. Hence the integrand of Eq. (4.5) is odd and thus the integral is zero. In practice, given the angular measure of performance, we can determine the number of optimal looks such that the angular probability of detection meets a desired threshold. There is evidence in the literature such as [5] where an area of interest is searched using a star search pattern to maximize the overall probability of detection. In this context, a star search pattern is a pattern that is composed of straight legs whose angular orientations are offset by $\pi/3$ with respect to one another. On the one hand, employing a multiple look search pattern such as the star search pattern generates overlaps which increases the search time. On the other hand, the overlaps will significantly improve the probability of detecting a mine and thus offer a higher probability of detection to naval undersea warfare commanders. This agrees with [6] where search tactics that involved overlaps improved the probability of detection.

4.8   Modeling and Algorithms

Now that we have found a solution that eliminates gaps in coverage and mitigates degradation in probability of detection due to aspect angle, we shall implement this solution in a Monte Carlo model and a deterministic model. The Monte Carlo model is coded in Visual Basic (VB) while the deterministic model is coded in Mathcad. The two are distinct but equivalent. They allow validation of the MOEs by comparing those from the Monte Carlo simulation to those obtained by the deterministic model.

4.8.1   Monte Carlo Simulation

For each run, the AUV employs the same search path, while the mines' positions and angles are generated using an *a priori* distribution. The simulation then computes the probability of detection using the range probability curve and the angular probability curve. It also fuses data when the same mine is detected by multiple legs and/or multiple AUVs. The program collects the results of each run, performs the statistics, and outputs the MOEs.

We make use of the Chernoff bound [7] to determine the number of Monte Carlo runs (m) required to achieve a given accuracy (ε) and a predetermined confidence level ($1 - \delta$):

$$m \geq \frac{1}{2 \cdot \varepsilon^2} \cdot \ln\left(\frac{2}{\delta}\right)$$

This criterion implies that:

$$P\left(\left|\hat{P} - P_T\right| \leq \varepsilon\right) = 1 - \delta$$

where $\hat{P}$ is the estimated probability collected from the Monte Carlo simulation of m runs while P_T is the true probability. As an example, to meet a 5% accuracy and a 95% ($\delta = 0.05$) confidence level, we need 738 runs.

4.8.2   Deterministic Model

We divide the search area into 200-by-200 grid of cells. For each leg of the AUV's sweep, we look at each cell and determine whether it lies between the AUV's minimum and maximum detection ranges. If it does, then we compute the probability of detection using the probability curves and the probability density of the mines.

For example, if the mines are uniformly random over the entire search area, then the probability density of a cell is equal to the area of the cell divided by the size of the search area. The corresponding contribution of each cell to the total probability of detection is simply equal to ($\Delta \cdot P_g / A$), where A is the size of the search area, Δ is the size of the cell, and P_g is the probability of detection of that cell based on range, aspect angle, and the AUV's search path. The probability P_g can be expressed as $P_g = 1 - \prod_{i=1}^{d}(M_i)$, where d is the number of AUV legs whose coverage contains that cell and $M_i(\leq 1)$ is the probability that the ith leg misses the same cell. In general, as d increases, so does P_g.

In Nguyen *et al.* [8], we determine the probability of detection, based on the Monte Carlo simulation and the deterministic calculation, and demonstrate that our calculations are consistent, because the two models, despite being designed differently, provide the same statistical result. However, this Monte Carlo simulation offers more flexibility. For example, we can use the simulation to model the details of the seabed. We will be enhancing the simulation with this feature and will report the results in a future paper.

4.9   Random Search Formula Adapted to AUVs

Before we present the results of the models described above, we will derive a formula for the probability of detection of an AUV conducting a random search. This formula will allow comparisons of effectiveness between a random search and the novel concepts of operations proposed for AUVs. Since the side-scan sonar MOPs (Figure 4.4) indicate that the detection probability is not perfect in general, and in fact depends on both range and angle, it is clear that we need to modify Koopman's formula [9] to determine the detection probability of an AUV conducting random search operations. Fortunately, this modification is simple, so we will present the derivation below.

We divide the search length L into n search segments, each with length L/n. The maximum detection range r_{max}, the minimum detection range r_{min}, and the length of each segment cover two small rectangles. Based on a uniform density distribution, the probability of detection of a mine lying in this subarea is equal to:

$$p_K(\theta) = \frac{L}{n} \cdot \frac{2 \cdot \int_{r_{min}}^{r_{max}} p(r) \cdot dr}{A} \cdot p(\theta)$$

where $p(r)$ is the detection probability as a function of range (r), as shown on the LHS of Figure 4.4, while $p(\theta)$ is the detection probability as a function of angle (θ), as shown on the RHS of Figure 4.4. (The K stands for Koopman.) In a random search, this angle θ is uniformly random because the mine's orientation is random, which is equivalent to the AUV changing course randomly relative to a mine of fixed orientation. The above result can be further simplified by recalling that the detection probability as a function of range was modeled as a Johnson density distribution scaled by λ_r as defined in Eq. (4.1):

$$\int_{r_{min}}^{r_{max}} p(r) \cdot dr = \int_{r_{min}}^{r_{max}} f(r) \cdot dr = \lambda_r \cdot \int_{r_{min}}^{r_{max}} f_J(r) \cdot dr = \lambda_r$$

The probability that a mine lies in the subarea can now be written as:

$$p_K(\theta) = \frac{(L/n) \cdot 2 \cdot \lambda_r}{A} \cdot p(\theta)$$

The probability that a mine lies outside of this rectangle is simply:

$$q_K(\theta) = 1 - p_K(\theta)$$

The only way that the AUV misses a mine after n independent and random search segments is to miss it in every search segment. As a result, the detection probability due to n search segments is equal to:

$$P_K = 1 - \prod_{i=1}^{n}\left(q_K(\theta_i)\right) = 1 - \prod_{i=1}^{n}\left(1 - \frac{2 \cdot \lambda_r \cdot (L/n)}{A} \cdot p(\theta_i)\right)$$

Expanding the product:

$$P_K = 1 - \left(1 - \frac{L \cdot 2 \cdot \lambda_r}{n \cdot A} \cdot \sum_{i=1}^{n} p(\theta_i) + \cdots\right)$$

$$= 1 - \left(1 - \frac{L \cdot 2 \cdot \lambda_r}{A \cdot \pi} \cdot \frac{\pi}{n} \sum_{i=1}^{n} p(\theta_i) + \cdots\right)$$

$$= 1 - \left(1 - \frac{L \cdot 2 \cdot \lambda_r}{A \cdot \pi} \cdot \int_{0}^{\pi} d\theta \cdot p(\theta) + \cdots\right)$$

The key argument in the above equation is the third equality. The sum in the second equality can be sorted in increasing order of θ_i. As θ is uniformly random, for large n, $\theta_i = i \cdot \pi / n$ after sorting. This sum compounded to π/n is exactly $\int_0^\pi p(\theta) \cdot d\theta$. Again, recalling that $p(\theta)$ was modeled as a Johnson's density distribution scaled by λ_θ, we get $\int_0^\pi p(\theta) \cdot d\theta = \lambda_\theta$. Taking the limit as n tends to infinity yields the following formula [10]:

$$P_K = 1 - \left(1 - \frac{L \cdot 2 \cdot \lambda_r \cdot \lambda_\theta}{A \cdot \pi} + \cdots\right)$$

$$= 1 - \left(1 - \frac{L \cdot 2 \cdot \lambda_r \cdot \lambda_\theta}{n \cdot A \cdot \pi}\right)^n$$

$$= 1 - e^{-\left(\frac{2 \cdot \lambda_r \cdot L}{A} \cdot \frac{\lambda_\theta}{\pi}\right)}$$

$$= 1 - e^{-\left(\frac{2 \cdot \lambda_r \cdot V \cdot t}{A} \cdot \frac{\lambda_\theta}{\pi}\right)}$$

(4.6)

here, P_K is the probability of an AUV detecting a mine while conducting a random search at velocity V for time t. This formula has an attractive interpretation. The detection range R is replaced by an effective range $R_{effective} = \lambda_r$ due to two characteristics of the probability of detection as a function of range. First, there is a minimum detection range; that is, if a mine is observed at a range less than the minimum detection range, then it will not be detected. Second, detection is not perfect for a mine lying between the minimum and maximum ranges. Hence, the original detection range is replaced by an effective detection parameter λ_r. For example, in our scenario the minimum detection range is 11.5 m while the maximum detection range is 75.0 m. This generates an effective detection range λ_r of approximately 53.06 m.

Additionally, the exponential's power includes a factor of λ_θ/π, which represents the mean angular probability of detection. In our scenario, the effective angular range, λ_θ, is approximately $\pi/2$, implying a mean angular detection probability of approximately 50%. This modification is essential, as Koopman's original formula assumes no angular dependence. That is, the probability of detection as a function of angle is 100% in Koopman's original derivation.

In the case of more than one AUV, the same formula can be used with the exception that the power in the exponential term must be multiplied by the number of AUVs. The reason for this is simple: Each AUV conducts a random search, which means that the motions of multiple AUVs are completely independent. Thus, using two AUVs, for instance, is equivalent to having one AUV execute a random search for twice the search time. In general, then, the probability of detection by a identical AUVs can be expressed as:

$$P_K(a) = 1 - e^{-\left(\frac{2 \cdot a \cdot \lambda_r \cdot V \cdot t}{A} \cdot \frac{\lambda_\theta}{\pi}\right)} \tag{4.7}$$

4.10 Mine Countermeasures Exploratory Operations

This experiment is intended to test concepts of MCM exploratory operations. The search area is divided into a total of N subareas or cells. Each cell is a square, has the same size, and contains at most one mine. For simplicity's sake, we assume that there is a uniformly random probability of having n mines in the search area, where n ranges from zero to N; that is, $p(n) = 1/N$. This choice is arbitrary and can be replaced by any other density distribution; it does not affect the derivation below. These n mines are further distributed randomly among the N cells. The experiment's objective is to determine the resources required to establish the presence or absence of mines in the search area.

Resources are measured in terms of the number of cells that are fathomed, denoted by m. When searching m subareas, there is a probability that m_d of these subareas have mines and the remaining $m_a = m - m_d$ have no mines. The probability associated with such an event belongs to a hypergeometric density distribution:

$$P_{m_d,m_a} = \frac{\binom{n}{m_d} \cdot \binom{N-n}{m_a}}{\binom{N}{m = m_d + m_a}} \tag{4.8}$$

There are two possible outcomes from a search in a cell with a mine: detected or not detected. Assuming there is a mine, we will denote the probability of detection with P_d and the probability of non-detection with Q_d, where d stands for detection. There are also two possible outcomes from a search in a cell with no mines: false alarm and no false alarm. Assuming there is no mine, we will denote the probability of a false alarm with P_a and the probability of a no false alarm with Q_a, where a stands for alarm. Combining the detection probability and the false-alarm probability of each subarea with the hypergeometric density distribution, we obtain the probability of finding at least one object that looks like a mine:

$$P_e = \frac{1}{N} \cdot \sum_{n=0}^{N} \left(1 - \sum_{m_d+m_a=m} P_{m_d,m_a} \cdot (Q_d)^{m_d} \cdot (Q_a)^{m_a}\right) \tag{4.9}$$

As this is a first attempt to analyze exploratory operations, we have picked an area where the environment is benign; that is, where $Q_a = 1$. In this scenario, then, the probability of finding at least one mine is equal to:

$$P_e = \frac{1}{N} \cdot \sum_{n=0}^{N} \left(1 - \sum_{m_d + m_a = m} P_{m_d, m_a} \cdot \left(1 - P_d \right)^{m_d} \right) \tag{4.10}$$

4.11   Numerical Results

Figure 4.7 shows the probabilities of detecting at least one mine as a function of the number of searched cells. We pick the same search time, 8 h, for each cell and for each search pattern. We assume that there is at most one mine in each cell.

Figure 4.8(a) shows three different search patterns for one AUV. Figure 4.8(b) shows three different hatching search patterns for two AUVs. The search patterns in Figure 4.8(a) are variants of the lawn-mowing pattern while the search patterns in Figure 4.8(b) are simple extensions of those in Figure 4.8(a).

P2Me (e for exploratory operations) corresponds to two perpendicular, regular lawn-mowing patterns 2M, as shown on the LHS of Figure 4.8(b); P2MUe corresponds to two perpendicular, uneven lawn-mowing patterns 2MU, as shown in the middle of Figure 4.8(b); P2Ze corresponds to two complementary zigzag patterns 2Z, as shown on the RHS of Figure 4.8(b); and P2Ke corresponds to random search patterns 2K conducted by two AUVs. For completeness, Figure 4.8(a) also illustrates patterns conducted by one AUV. The values of the probabilities of detection for the search patterns 2M, 2MU, and so on, are displayed in Table 4.3.

Figure 4.7 shows that P2MUe is the best among the four scenarios. This improvement is most significant for numbers of searched cells less than six $(m \leq 6)$. Such an improvement helps reduce resource and time requirements when conducting MCM operations. For instance, a 2Z search pattern takes at least 48 h (searching six cells) to achieve the same probability of

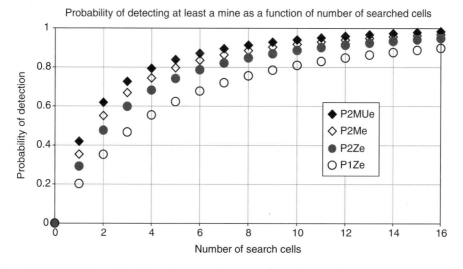

Figure 4.7   Probability of detecting at least one mine vs. number of searched cells. *Source*: Nguyen, Bao; Hopkin, David; Yip, Handson, "Considering Mine Countermeasure Exploratory Operations Conducted by Autonomous Underwater Vehicles," *MORS Journal*, Vol. 19, No. 2, 2014, John Wiley and Sons, LTD.

(a)

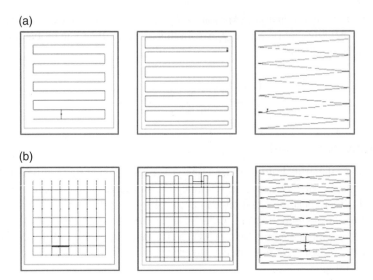

(b)

Figure 4.8 (a) Search with one AUV – even lawn-mowing pattern (LHS), uneven lawn-mowing pattern (middle), and zigzag pattern (RHS). (b) Search with two AUVs – even lawn-mowing patterns (LHS), uneven lawn-mowing patterns (middle), and zigzag patterns (RHS). *Source*: Nguyen, Bao; Hopkin, David; Yip, Handson, "Considering Mine Countermeasure Exploratory Operations Conducted by Autonomous Underwater Vehicles," *MORS Journal*, Vol. 19, No. 2, 2014, John Wiley and Sons, LTD.

Table 4.3 Probability of detection for a search time of 8 h.

Search pattern	2M	2MU	2Z	1Z	2K	1K
Probability of detection	0.67	0.79	0.55	0.39	0.70	0.45

detecting at least one mine as a 2MU pattern that takes 32 h (searching four cells). Equivalently, the probability of detecting at least one mine after searching six cells using the 2MU search pattern is more than 10% higher than that achieved using the 2Z search pattern – a substantial improvement in the confidence of finding at least one mine. This shows the importance of choosing the right search pattern.

4.12 Non-uniform Mine Density Distributions

While the choice is clear in the case of a uniform mine density distribution, as described above, it is less obvious in the case of a non-uniform density distribution. Below, we make use of a derivation to analyze an exponential mine density distribution along the y (vertical) axis. This technique can be used to derive the optimal AUV path in general, assuming an *a priori* mine density distribution in the search area. To simplify the mathematics, we look for the AUV path as a line, rather than as an area, that maximizes the probability of detection. The optimal path is a solution to the Euler equation for $W[y(x)]$ in the search area, which can be expressed as:

$$W\left[y(x)\right] = \int_{\Delta x}^{X-\Delta x} dx \cdot \sqrt{1 + \left(\frac{dy}{dx}\right)^2} \cdot f(x) \cdot g(y) \qquad (4.11)$$

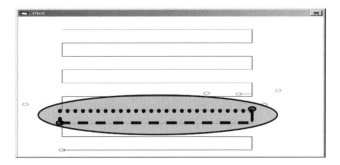

Figure 4.9 Optimal path (dotted) in a sub-box area. *Source*: Nguyen, Bao; Hopkin, David; Yip, Handson, "Considering Mine Countermeasure Exploratory Operations Conducted by Autonomous Underwater Vehicles," *MORS Journal*, Vol. 19, No. 2, 2014, John Wiley and Sons, LTD.

where Δx is the margin on the left and the right of the AUV path. As an example, we assume $f(x)$ is the uniform density distribution along x:

$$f(x) = \begin{cases} 1/X & \text{if } 0 \le x \le X \\ 0 & \text{otherwise} \end{cases} \tag{4.12}$$

while $g(y)$ is the exponential density distribution along y:

$$g(y) = \begin{cases} \dfrac{e^{-|y-Y/2|/b}}{2 \cdot b \cdot \left(1 - e^{-Y/(2 \cdot b)}\right)} & \text{if } 0 \le y \le Y \\ 0 & \text{otherwise} \end{cases} \tag{4.13}$$

As illustrated in Figure 4.9, it is typical in search operations to divide the search area into sub-box areas (such as the shaded rectangular area) when the AUV employs the lawn-mowing pattern. The figure shows the AUV path in the sub-box area within the ellipse. Note that the AUV's starting point is the dot at the lower left and its ending point is the dot at the upper right. We will answer the following question: Given a sub-box area like the one just described, what is the optimal path starting from the lower left dot and ending at the upper right dot?

Substituting the density distributions $f(x)$ and $g(y)$ into Eq. (4.11), and assuming that $y(x)$ is monotonically increasing, we get:

$$
\begin{aligned}
W &\le \int_{\Delta x}^{X-\Delta x} \frac{dx}{X} \cdot \left(1 + \frac{dy}{dx}\right) \cdot \frac{e^{-(Y/2-y)/b}}{2 \cdot b \cdot \left(1 - e^{-Y/(2 \cdot b)}\right)} \\
&\le \int_{\Delta x}^{X-\Delta x} \frac{dx}{X} \cdot \frac{e^{-(Y/2-y)/b}}{2 \cdot b \cdot \left(1 - e^{-Y/(2 \cdot b)}\right)} + \int_{y_s}^{y_e} \frac{dy}{X} \cdot \frac{e^{-(Y/2-y)/b}}{2 \cdot b \cdot \left(1 - e^{-Y/(2 \cdot b)}\right)} \\
&\le P_X + P_Y
\end{aligned}
\tag{4.14}
$$

where y_s is the starting y coordinate of the AUV path and y_e is the ending y coordinate.

$$P_X = \int_{\Delta x}^{X-\Delta x} \frac{dx}{X} \cdot \frac{e^{-(Y/2-y)/b}}{2 \cdot b \cdot \left(1 - e^{-Y/(2 \cdot b)}\right)}; \quad P_Y = \int_{y_s}^{y_e} \frac{dy}{X} \cdot \frac{e^{-(Y/2-y)/b}}{2 \cdot b \cdot \left(1 - e^{-Y/(2 \cdot b)}\right)} \tag{4.15}$$

Examining these expressions for P_X and P_Y closely, we realize that P_X is the weight of the horizontal dotted line shown in Figure 4.9, while P_Y is the weight of the vertical dotted line. Since we have just demonstrated that the weight of any curve is less than the sum $P_X + P_Y$, this shows that the dotted line represents the optimal search path. This also confirms our intuition; that is, to optimize the weight of a search path, we need to let it be as close as possible to the half-width line, as this line corresponds to the highest mine density. Note that we obtain the same solution by solving Euler's equation through the functional expression of W [11].

We can also derive the optimal path for a more complex mine density distribution using the same formalism. For example, we can assume that both $f(x)$ and $g(y)$ are exponential distributions. That is, we replace the above definition of $f(x)$ in Eq. (4.12) by:

$$f(x) = \begin{cases} \dfrac{e^{-|x-X/2|/a}}{2 \cdot a \cdot \left(1 - e^{-X/(2 \cdot a)}\right)} & \text{if } 0 \le x \le X \\ 0 & \text{otherwise} \end{cases}$$

while $g(y)$ remains the same. The solution to the Euler–Lagrange equation is:

$$C \cdot e^{-\frac{1+\alpha^2}{a} \cdot x + \alpha \cdot z} = \left(-\sin(z) + \alpha \cdot \cos(z)\right)$$

$$z = \frac{1}{b} \cdot \left(x - \frac{y}{\alpha}\right) + D; \quad \alpha = \frac{a}{b}$$

where C and D are constants that can be determined at the boundary conditions. We will further analyze this solution in a future paper.

4.13 Conclusion

In this chapter, we have examined a number of search patterns: lawn-mowing, zigzagging, and random searches. The MOEs of the first two patterns were evaluated through the use of a stochastic model and a deterministic one, whose outcomes were verified to be consistent.

Based on the assumed MOPs, we demonstrated that the 2MU search pattern (two AUVs conducting perpendicular, uneven mowing patterns) provides the best probability of detection as a function of search time. This optimality is due to two features. First, the uneven mowing pattern is constructed so that there are no gaps between consecutive legs. Second, two perpendicular search patterns maximize the angular probability of detection, as proved in the main text and discussed in further detail in Appendix 4.A.

We have also derived a novel result for an AUV conducting a random search while carrying a side-scan sonar. Although such an AUV cannot in reality collect data while employing a random search, it was useful to show that the 2MU search pattern achieves better effectiveness than a random search – a clue that 2MU is the right search pattern to use. The impact of search patterns on time and resource requirements for MCM exploratory operations was also determined.

In closing, we wish to note that all of the concepts of operations that this chapter introduces can be implemented and tested in real-life scenarios, which might in fact be its true contribution.

Appendix 4.A   Optimal Observation Angle between Two AUV Legs

In the main text, we have shown that the mean angular probability of detection, $P_2(\phi)$, generated by two AUV legs achieves an optimum when the angle ϕ between the two legs is equal to $\phi = \pi/2$. This appendix further shows that this value is in fact a global maximum. It is true because of the shape and symmetry of the angular probability of detection with respect to $\phi = \pi/2$.

For the purpose of this appendix, it is convenient to redefine the aspect angle θ by shifting it to the left by $\phi = \pi/2$ such that the maximum angular probability of detection occurs at 0. There are many ways to prove that $P_2(\phi = \pi/2)$ is a global maximum. We choose to show that $P_2(\phi = \pi/2) > P_2(\phi = \pi/2 + \varphi)$ for any φ. However, due to the periodicity of $f(\theta)$, it is sufficient to show that this is true for all φ between 0 and $\phi = \pi/2$.

Recall that $P_2(\phi)$ is defined as:

$$P_2(\phi) = 1 - \frac{1}{\pi} \cdot \int_{-\frac{\pi}{2}}^{\frac{\pi}{2}} d\theta \cdot (1 - f(\theta)) \cdot (1 - f(\theta + \phi)) \tag{A1}$$

Hence:

$$P_2\left(\frac{\pi}{2} + \varphi\right) = 1 - \frac{1}{\pi} \cdot \int_{-\frac{\pi}{2}}^{\frac{\pi}{2}} d\theta \cdot (1 - f(\theta)) \cdot \left(1 - f\left(\theta + \frac{\pi}{2} + \varphi\right)\right) \tag{A2}$$

Since $f(\theta)$ is a periodic function with a period of π:

$$P_2\left(\frac{\pi}{2} + \varphi\right) = 1 - \frac{1}{\pi} \cdot \int_{-\frac{\pi}{2}}^{\frac{\pi}{2}} d\theta \cdot \left(1 - f(\theta) - f\left(\theta + \frac{\pi}{2} + \varphi\right) + f(\theta) \cdot f\left(\theta + \frac{\pi}{2} + \varphi\right)\right) \tag{A3}$$

$$= \frac{2 \cdot \lambda_\theta}{\pi} - \frac{1}{\pi} \cdot \int_{-\frac{\pi}{2}}^{\frac{\pi}{2}} d\theta \cdot f(\theta) \cdot f\left(\theta + \frac{\pi}{2} + \varphi\right)$$

where λ_θ is the scale compounded to Johnson's curve as defined in the main text. Thus, showing that $P_2(\phi = \pi/2)$ is a maximum is equivalent to showing that $\int_{-\frac{\pi}{2}}^{\frac{\pi}{2}} d\theta \cdot f(\theta) \cdot f\left(\theta + \frac{\pi}{2}\right)$ is a minimum, in the sense that:

$$\int_{-\frac{\pi}{2}}^{\frac{\pi}{2}} d\theta \cdot f(\theta) \cdot f\left(\theta + \frac{\pi}{2} + \varphi\right) \geq \int_{-\frac{\pi}{2}}^{\frac{\pi}{2}} d\theta \cdot f(\theta) \cdot f\left(\theta + \frac{\pi}{2}\right) \tag{A4}$$

Rewrite the LHS as follows:

$$\int_{-\frac{\pi}{2}}^{\frac{\pi}{2}} d\theta \cdot f(\theta) \cdot f\left(\theta + \frac{\pi}{2} + \varphi\right) = \int_{-\frac{\pi}{2}}^{0} d\theta \cdot f(\theta) \cdot f\left(\theta + \frac{\pi}{2} + \varphi\right) + \int_{0}^{\frac{\pi}{2}} d\theta \cdot f(\theta) \cdot f\left(\theta + \frac{\pi}{2} + \varphi\right)$$

$$= \int_{0}^{\frac{\pi}{2}} d\theta \cdot \left[f(\theta) \cdot f\left(\theta - \frac{\pi}{2} - \varphi\right) + f(\theta) \cdot f\left(\theta + \frac{\pi}{2} + \varphi\right) \right] \tag{A5}$$

Similarly, rewrite the RHS as:

$$\int_{-\frac{\pi}{2}}^{\frac{\pi}{2}} d\theta \cdot f(\theta) \cdot f\left(\theta + \frac{\pi}{2}\right) = \int_{-\frac{\pi}{2}}^{0} d\theta \cdot f(\theta) \cdot f\left(\theta + \frac{\pi}{2}\right) + \int_{0}^{\frac{\pi}{2}} d\theta \cdot f(\theta) \cdot f\left(\theta + \frac{\pi}{2}\right)$$

$$= \int_{0}^{\frac{\pi}{2}} d\theta \cdot \left[f(\theta) \cdot f\left(\theta + \frac{\pi}{2}\right) + f(\theta) \cdot f\left(\theta - \frac{\pi}{2}\right) \right] \tag{A6}$$

Subtract the RHS from the LHS:

$$A + B + C =$$

$$\int_{0}^{\frac{\pi}{2}} d\theta \cdot f(\theta) \cdot \left(f\left(\theta - \frac{\pi}{2} - \varphi\right) - f\left(\theta - \frac{\pi}{2}\right) \right) + \int_{0}^{\frac{\pi}{2}} d\theta \cdot f(\theta) \cdot \left(f\left(\theta + \frac{\pi}{2} + \varphi\right) - f\left(\theta + \frac{\pi}{2}\right) \right) \tag{A7}$$

where:

$$A = \int_{\varphi}^{\frac{\pi}{2}} d\theta \cdot f(\theta) \cdot \left(f\left(\theta - \frac{\pi}{2} - \varphi\right) - f\left(\theta - \frac{\pi}{2}\right) \right) + \int_{0}^{\frac{\pi}{2} - \varphi} d\theta \cdot f(\theta) \cdot \left(f\left(\theta + \frac{\pi}{2} + \varphi\right) - f\left(\theta + \frac{\pi}{2}\right) \right) \tag{A8}$$

$$B = \int_{0}^{\varphi} d\theta \cdot f(\theta) \cdot \left(f\left(\theta - \frac{\pi}{2} - \varphi\right) - f\left(\theta - \frac{\pi}{2}\right) \right) \tag{A9}$$

$$C = \int_{\frac{\pi}{2}-\varphi}^{\frac{\pi}{2}} d\theta \cdot f(\theta) \cdot \left(f\left(\theta + \frac{\pi}{2} + \varphi\right) - f\left(\theta + \frac{\pi}{2}\right) \right) \tag{A10}$$

To prove that $P_2(\phi = \pi/2)$ is a global maximum, we will show that $A,\ B,\ C \geq 0$. Note that we can remove the parameter λ_θ in the proof, as it is only a constant multiplied to Johnson's density distributions.

Proof for $A \geq 0$. Make a change of variable $\theta' = \theta + \varphi$ in the second integral, and use the knowledge that $f(\theta)$ is a periodic function with a period of π:

$$A = \int_{\varphi}^{\frac{\pi}{2}} d\theta \cdot f(\theta) \cdot \left(f\left(\theta - \frac{\pi}{2} - \varphi\right) - f\left(\theta - \frac{\pi}{2}\right) \right) + \int_{0}^{\frac{\pi}{2}-\varphi} d\theta \cdot f(\theta) \cdot \left(f\left(\theta + \frac{\pi}{2} + \varphi\right) - f\left(\theta + \frac{\pi}{2}\right) \right)$$

$$= \int_{\varphi}^{\frac{\pi}{2}} d\theta \cdot f(\theta) \cdot \left(f\left(\theta - \frac{\pi}{2} - \varphi\right) - f\left(\theta - \frac{\pi}{2}\right) \right) + \int_{\varphi}^{\frac{\pi}{2}} d\theta \cdot f(\theta - \varphi) \cdot \left(f\left(\theta + \frac{\pi}{2}\right) - f\left(\theta + \frac{\pi}{2} - \varphi\right) \right)$$

$$= \int_{\varphi}^{\frac{\pi}{2}} d\theta \cdot f(\theta) \cdot \left(f\left(\theta + \frac{\pi}{2} - \varphi\right) - f\left(\theta + \frac{\pi}{2}\right) \right) + \int_{\varphi}^{\frac{\pi}{2}} d\theta \cdot f(\theta - \varphi) \cdot \left(f\left(\theta + \frac{\pi}{2}\right) - f\left(\theta + \frac{\pi}{2} - \varphi\right) \right)$$

Rewrite the integrand as a product:

$$A = \int_{\varphi}^{\frac{\pi}{2}} d\theta \cdot \left(f(\theta - \varphi) - f(\theta) \right) \cdot \left(f\left(\theta + \frac{\pi}{2}\right) - f\left(\theta + \frac{\pi}{2} - \varphi\right) \right) \tag{A11}$$

Since $f(\theta)$ is monotonous for θ between 0 and $\pi/2$, the first term of the integrand is non-negative; that is, $f(\theta - \varphi) - f(\theta) \geq 0$. Similarly, the second term of the integrand is also non-negative; that is, $f(\theta + \pi/2) - f(\theta + \pi/2 - \varphi) \geq 0$. Hence, the integrand of A is non-negative and thus A itself is non-negative.

Proof for $B \geq 0$. Rewrite B as a sum of two integrals:

$$B = \int_{0}^{\frac{\varphi}{2}} d\theta \cdot f(\theta) \cdot \left(f\left(\theta - \frac{\pi}{2} - \varphi\right) - f\left(\theta - \frac{\pi}{2}\right) \right) + \int_{\frac{\varphi}{2}}^{\varphi} d\theta \cdot f(\theta) \cdot \left(f\left(\theta - \frac{\pi}{2} - \varphi\right) - f\left(\theta - \frac{\pi}{2}\right) \right)$$

Make a change of variable $\theta' = \varphi - \theta$ in the second integral:

$$B = \int_{0}^{\frac{\varphi}{2}} d\theta \cdot f(\theta) \cdot \left(f\left(\theta - \frac{\pi}{2} - \varphi\right) - f\left(\theta - \frac{\pi}{2}\right) \right) + \int_{0}^{\frac{\varphi}{2}} d\theta \cdot f(\theta - \varphi) \cdot \left(f\left(\theta - \frac{\pi}{2}\right) - f\left(\theta - \frac{\pi}{2} - \varphi\right) \right)$$

where we have used the even parity of $f(\theta)$ and its period of π in the second integral. Rewrite the sum of the two integrands as a product:

$$B = \int_0^{\frac{\varphi}{2}} d\theta \cdot \left(f(\theta) - f(\theta - \varphi) \right) \cdot \left(f\left(\theta - \frac{\pi}{2} - \varphi \right) - f\left(\theta - \frac{\pi}{2} \right) \right) \tag{A12}$$

For $\theta \in [0, \varphi/2]$, the first term of the integrand is non-negative; that is, $f(\theta) - f(\theta - \varphi) \geq 0$. Similarly, the second term of the integrand is non-negative; that is, $f(\theta - \pi/2 - \varphi) - f(\theta - \pi/2) \geq 0$. Hence, the integrand of B is non-negative and thus B itself is non-negative.

Proof for $C \geq 0$. The proof for $C \geq 0$ is identical to that for B.

Conclusion. Thus, we have shown that A, B, $C \geq 0$. Hence, $A + B + C \geq 0$. This confirms that $P_2(\phi = \pi/2)$ is a global maximum.

Appendix 4.B   Probabilities of Detection

In this Appendix, we provide the inputs that were used to determine the probabilities of detection in Table 4.3. The search area is a 3 km × 3 km square. The AUV speed is equal to 9 knots. The probabilities of detection as a function of range and as a function of angle are shown in Figure 4.4.

Search pattern 2M (even lawn mowing):

• Spacing between consecutive and parallel legs is equal to 100 m.
• Offset from the boundary of the search area is equal to 100 m.
• Number of long and parallel legs is equal to 29 in each direction.
• Turning time is equal to 1.25 min for each rotation of 90°.
• Total time (search time and turning time) is equal to 6 h and 7 min.
• Probability of detection is equal to 0.62.

The above spacing for search pattern 2M is chosen such that the total time is approximately 6 h and the coverage is essentially 100% if we do not account for gaps in the detection performance.

Search pattern 2MU (uneven lawn mowing):

• Small spacing between consecutive and parallel legs is equal to 63.5 m.
• Large spacing between consecutive and parallel legs is equal to 150.0 m.
• Offset from the boundary of the search area is equal to 63.5 m.
• Number of long and parallel legs is equal to 28 in each direction.
• Turning time is equal to 1.25 min for each rotation of 90°.
• Total time (search time and turning time) is equal to 6 h and 2 min.
• Probability of detection is equal to 0.75.

As described in the main text, the 2MU large spacing is equal to $2 \bullet r_{max} = 150$ m and the small spacing is equal to $r_{max} - r_{min} = 63.5$ m.

Search pattern 2Z (zigzag):

• Zigzag angle between the horizontal axis and an AUV leg is equal to 2.0454°.
• Number of legs is equal to 28 in each direction.

- Turning time is equal to 2.5 min for each u-turn.
- Total time (search time and turning time) is equal to 6 h and 5 min.
- Probability of detection is equal to 0.55.

Search pattern 1Z (zigzag):

- Zigzag angle between the horizontal axis and an AUV leg is equal to 2.0454°.
- Number of legs is equal to 28.
- Turning time is equal to 2.5 min for each u-turn.
- Total time (search time and turning time) is equal to 6 h and 5 min.
- Probability of detection is equal to 0.40.

Search pattern 2K and 1K (Koopman's random search):

- Eq. (4.7): $P_K\left(a\right) = 1 - e^{-\left(\frac{2 \cdot a \cdot \lambda_r \cdot V \cdot t}{A} \cdot \frac{\lambda_\theta}{\pi}\right)}$
- $\lambda_r = 53.057$ m.
- $\lambda_\theta = 1.575$ rad.
- $A = 9$ km^2.
- $V = 9$ knots.
- $t = 6$ h.
- $\alpha = 2$ for 2K and $\alpha = 1$ for 1K
- Probability of detection is equal to 0.69 for 2K and 0.45 for 1K.

References

1. Law, A.M. and W.D. Kelton, *Simulation Modeling and Analysis*, *3rd Edition*, McGraw-Hill, Boston MA, U.S.A., 2000, pp. 314–315.
2. Nguyen, B. and D. Hopkin, "Modeling Autonomous Underwater Vehicle (AUV) Operations in Mine Hunting," Conference Proceedings of IEEE Oceans 2005 Europe, Brest, France, 20–23 June 2005.
3. Nguyen, B. and D. Hopkin, "Concepts of Operations for the Side-Scan Sonar Autonomous Underwater Vehicles Developed at DRDC–Atlantic," Technical Memorandum TM 2005-213, October 2005.
4. Zerr, B., E. Bovio, and B. Stage, "Automatic Mine Classification Approach Based on AUV Manoeuvrability and the COTS Side Scan Sonar," Conference Proceedings of Autonomous Underwater Vehicle and Ocean Modelling Networks: GOAT2 2000, 2000, pp. 315–322.
5. Bays, M.J., A. Shende, D.J. Stilwell, and S.A. Redfield, "A Solution to the Multiple Aspect Coverage Problem," International Conference on Robotics and Automation, 2011.
6. Hill, R.R., R.G. Carl, and L.E. Champagne, "Using Agent-Based Simulation To Empirically Examine Search Theory Using a Historical Case Study," Journal of Simulation 1 (1), 2006, pp. 29–38.
7. Vidyasagar, M. "Statistical Learning Theory and Randomised Algorithms for Control," IEEE Control System Magazine 18 (6), 1998, pp. 69–85.
8. Nguyen, B., D. Hopkin, and H. Yip, "Autonomous Underwater Vehicles: A Transformation in Mine Counter-Measure Operations," Defense & Security Analysis 24 (3), 2008, pp. 247–266.
9. Koopman, B.O. *Search and Screening: General Principles with Historical Applications*, Military Operations Research Society, Alexandria VA, U.S.A., 1999, pp. 71–74.
10. Zwillinger, D. (editor), *CRC Standard Mathematical Tables and Formulae, 30th Edition*, CRC Press, Boca Raton FL, U.S.A., 1996, p. 333.
11. Gel'fand, I.M. and S.V. Fomin, *Calculus of Variations, 13th Edition*, Prentice-Hall, Englewood Cliffs NJ, U.S.A., 1963, p. 19.

5

Optical Search by Unmanned Aerial Vehicles: Fauna Detection Case Study

Raquel Prieto Molina[1], Elena María Méndez Caballero[1],
Juan José Vales Bravo[1], Isabel Pino Serrato[1], Irene Rosa Carpintero Salvo[1],
Laura Granado Ruiz[1], Gregoria Montoya Manzano[1],
Fernando Giménez de Azcárate[2], Francisco Cáceres Clavero[2],
and José Manuel Moreira Madueño[2]

[1] *Environmental Information Network of Andalusia, Environment and Water Agency, Johan Gutenberg, Seville, Spain*
[2] *Regional Ministry of Environment and Spatial Planning, Avenida Manuel Siurot, Seville, Spain*

5.1 Introduction

Many Mediterranean forests are still used for traditional hunting. To appropriately manage use of public lands for hunting and to regulate the hunting capacity within these territories, decision makers require accurate information about game populations, the conservation status of the forests, and conditions of environmental development. Classical census-taking methods typically involve estimation of game populations through itinerary sampling, counting at feeders, and other time- and labor-consuming techniques. Recently, the use of remote sensing from airborne sensors on Unmanned Aerial Vehicles (UAVs) has been identified as a supplement or alternative to these methods. In particular, thermal infrared detection methods [1–4] have been used to identify animal species and estimate the number of individuals in these populations over significant territorial areas in large hunting reserves.

This chapter discusses the use of unmanned vehicles for sensing tasks and describes the approaches and considerations for effective search using census taking as an example. It explores the technological solutions to what historically were tasks conducted by humans,

Operations Research for Unmanned Systems, First Edition. Edited by
Jeffrey R. Cares and John Q. Dickmann, Jr.
© 2016 John Wiley & Sons, Ltd. Published 2016 by John Wiley & Sons, Ltd.

namely identification of a field of view, *in situ* flight planning to prevent gaps in coverage, and methods to ensure consistency throughout a task. Without a human in the vehicle these tasks, which in manned aviation are relegated to the routine, become very difficult to achieve optimally. Important conclusions have been drawn, aimed mainly at improving observation tasks (capture of images and supplementary data) and post-processing. This work was carried out by the Environmental Information Network of Andalusia (REDIAM).

5.2   Search Planning for Unmanned Sensing Operations

Search planning for UAV operations occurs in four distinct phases: Preliminary Analysis, Flight Geometry Control, Image Preparation, and the Analysis and Identification of Individuals.

The first phase consists of analysis of all information available prior to the flight (camera specifications, search area size, likely flight paths, image-processing requirements, etc.) that determine which specific flight dynamics must be executed for the effective acquisition and subsequent processing of images.

In the second phase, an analysis of imaging scaling and altitude is conducted (what level of detail is appropriate and what flight level should be maintained for this detail). In manned air search, the human eye provides a remarkably adaptive viewer that allows for relatively easy in-flight adjustment to plans as the need arises. For many aerial photographic tasks, particularly those conducted with low-cost UAVs, it makes little sense to invest in an extraordinarily expensive camera that mimics the adaptability of the human eye. Also, adaptable cameras return images with different resolutions, which cause havoc with post-flight image processing. So planners start the information analysis with the camera, which is usually relatively limited in resolution and field of view. The first task in information analysis is to determine the *scale* of the images that are expected during the flight. Each camera has a *focal length*, which is the distance in millimeters from the lens to the photographic receptor (either film or a digital device), which is in the *focal plane*. The lens captures light and projects it onto the receptor in the focal plane. Each point of light or, in the case of digital photography, each *pixel* in the focal plane has a corresponding point on the earth's surface. The relationship of the distance from the lens to the focal plane and the lens to the ground determines what is called the photographic scale, defined by the following equation:

$$Scale = \frac{Focal\ Length}{Altitude\ Above\ Ground\ Level\,(AGL)} \tag{5.1}$$

The scale relates the distance between two points in the focal plane with two points on the ground. For example, if a camera's focal length is 153 mm and Above Ground Level (AGL) is 10 000 ft, then the scale is 1 : 19 922 (or 1 in. on the photo is 19 922 in. on the ground, or 1660 ft). Obviously, such a camera–altitude combination would not work for detecting game. Closely related to scale is *Ground Sample Distance* (GSD), which in digital aerial photography is the distance between pixel centers measured on the ground. Figure 5.1 shows the relationship between GSD and scale for a UAV flight and Figure 5.2 shows the relationship between GSD and altitude. Flight planners use these two relationships to determine the correct scale, altitude, and pixel size for a particular search flight.

Proper planning (consistent with the GSD, pixel size on the terrain) allows for an adequate photographic coverage, that is, an adequate relationship between the number of flight paths/photographs and the area to be covered. Since this flight was carried out for cataloging purposes,

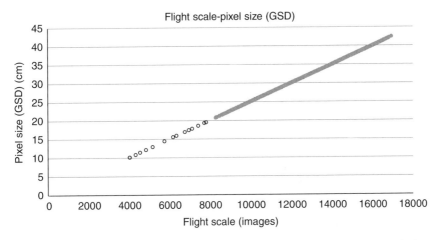

Figure 5.1   Ratio between the flight scale and pixel size (GSD) for each image. Open circles are the UAV positions from take-off until reaching the planned height.

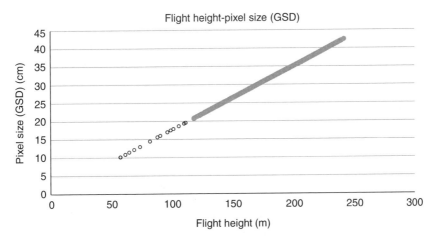

Figure 5.2   Ratio between flight height and pixel size (GSD) for each image. Open circles are the UAV positions from take-off until reaching the planned height.

there are stringent cartographic technical requirements of image orientation (triangulation) and ortho-rectification, which facilitate the creation of mosaics. These include synchronization of positional and angular location with image capturing, selection of an appropriate Global Navigation Satellite System (GNSS) positioning system, and establishment of land surface coverage tolerances to account for gaps, overlap, blind spots due to flight maneuvers, and so on. Depending on the products used for the analysis (images themselves or post-flight automatic image processing) planners must select from the following alternatives:

- *Approximated Geo-referencing with a Constant GSD:* If image frames are geo-referenced without a highly accurate geo-referencing process, the size of the images will be constant for the entire flight, but the objects contained in those images might be re-sized substantially depending on the actual scale of each frame.

- *Rigorous Geo-referencing:* When image frames are geo-referenced by a GSD, according to the scale of each frame, the above problems are attenuated, but distortion problems arising from the conical capture perspective of a camera and the effect passing over undulating terrain at constant height still persist.
- *Ortho-rectification:* When each image frame is ortho-rectified and submitted to a mosaic process the above-mentioned problems (those derived from the approximated geo-referencing with constant scale, conical capture perspective, topography, etc.) are mitigated and a consistent product is achieved in which the objects (animals) have similar size and correct positioning.

Considering the previously mentioned aspects, a thorough control of the flight's geometry is carried out based on the Projection Center positions obtained for each frame. Control checks are performed for flight height, image scaling, pixel size, longitudinal overlapping between images, and transversal overlapping between flight paths. This is usually accomplished using computer-based planning tools.

If each image were to be studied independently by visual interpretation or by digital processing, this would likely cause duplicity of counts due to overlap between frames and between flight paths. Additionally, digital processing would be required for each and every digital image file. Post-processing can be optimized by working on a mosaic of images so that any classification process is applied only once. The third phase in planning is to determine the mosaic software process most appropriate to the analysis. The fourth phase is the selection of software and processes to conduct an automated interpretation of the photomosaic.

The remaining sections of this chapter discuss how these four phases were applied to auto-mating game preserve census taking with a UAV platform.

5.2.1 Preliminary Flight Analysis

For this study, we examined information previously obtained from an unmanned flight over one part of a large gaming estate, Las Navas near Seville, carried out by researchers from ELIMCO, Inc., in collaboration with the Environment and Water Agency of Andalusia, Spain. The data were collected on January 11, 2012, during the early morning when the temperature differences between the animals and soil are greatest. Approximately 725 ha (almost 1800 acres) were surveyed with quick image acquisition to observe and count wildlife without any of the technical challenges of mapping procedures. Since it is very important to account for previous experience, specific knowledge of the species studied and the behavior of game in this habitat were obtained by an on-the-field observation campaign. This was conducted simultaneous to the flight to determine the most probable locations of species, but certainly could have been completed well in advance of the flight.

Two sensors were available for the acquisition of images during the flight: one camera to acquire thermal images (for identifying individual animals) and another camera to acquire visible images (to support photo-interpretation and image-processing tasks). The thermal camera was a Miricle Camera LVDS (Low Voltage Digital Signaling) 307 K from Thermoteknix Systems Ltd. (array size 640×480, spectral response 8–12 µm, and pixel pitch 25 µm). The thermal camera was onboard an unmanned "E-300 Viewer" flight platform (Figure 5.3).

Figure 5.3   "E-300 viewer" platform moments before take-off.

5.2.2   Flight Geometry Control

Recall that the scale of the images related to the flying height above the ground is directly linked to the terrain detail, that is, the GSD (see Figures 5.1 and 5.2). These relationships were used to establish an appropriate nominal GSD for the elements we wanted to capture (animals), which in this case could be defined as between 15 and 20 cm. The scale of the images can vary substantially depending on the flight paths over uneven terrain, and as a result so might the size of the animals in the captured images. But by fixing the GSD at an appropriate range, the flight itself can be conducted at a constant height, which is preferred, since attempting to conform to rapid terrain elevation changes is not only difficult, but is itself another source of distortion and error. An important aspect that must be taken into account is the altimetric reference. The available GPS (Global Positioning System) coordinates are according to the World Geodesic System 1984 reference database (WGS84) and the flight height is available according to the ellipsoidal reference system. So it is essential to obtain the orthometric height (above sea level) in order to compare it with the global Digital Elevation Model (DEM) height (from satellite radar data). Figure 5.4 shows that the flight was conducted at 600 m above sea level, but the DEM height analysis was conducted to see if the actual ground level altitude in the game preserve kept the GSD within the 15–20 cm range. The DEM database, however, has less accuracy than the Spanish National Plan of Aerial Orthophotography (PNOA) database, introducing a significant element of uncertainty in the scale of the images (directly related to the sensor–land distance).

Employing onboard avionics, the six degrees of flight control (latitude, longitude, altitude, roll, pitch, and yaw) positioning and estimated sensor orientation angle to the image projection center are assigned at the moment that the image is taken. XYZ locations were obtained by the GNSS system (GPS) and the inertial data (roll, pitch, yaw) by an inertial system (Inertial Measurement Unit (IMU)) during the flight. This allows for the positioning and orientation to be available in great detail and accuracy at the moment of capturing the image, resulting in easier geo-referencing of the images as well as for the external orientation by aerial triangulation.

Considering the previously mentioned aspects, a thorough check of the flight's geometry was carried out based on the Projection Center positions obtained for each frame.

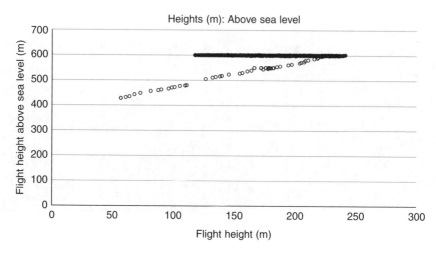

Figure 5.4 Ratio between projection center heights above sea level and projection center heights above the ground. Points below 600 m (open circles) represent the UAV positions from take-off until reaching the planned height for image capturing.

Control checks were performed for flight height, image scaling, pixel size, longitudinal overlapping between images, and transversal overlapping between flight paths. This was conducted with the tools developed by the REDIAM for the Quality Control of National and Regional Photogrammetric Flights Program (supporting PNOA and other projects).

Changes due to greater details in the GSD employed can cause problems related to longitudinal or transversal overlaps while lower levels of detail in the GSD employed will result in details being lost. The results of these tests show several problems mainly related to the size of the GSD (Figure 5.5) and overlaps (Figure 5.6), mainly between flight paths. In Figure 5.5 we can observe problem areas (in red) in lower elevations in which the GSD is out of the 15–20 cm range. In Figure 5.6 we can observe problem areas (in red) where the surface ground cover, primarily in the pilot area – the area selected for initial mosaic and post-processing method testing – where animals might remain unregistered. For tasks primarily focused on orthophotographic restitution or generation, the degree of overlapping should range from 60% between frames (longitudinal overlap) and 30% between flight paths (transversal overlap). This was not achievable for this flight, which caused additional challenges in the mosaic and post-processing activities.

5.2.3 Images and Mosaics

The flight produced 2500 raw images with a size of 640×480 pixels. These images were captured along 22 flight paths (following approximately parallel trajectories). To improve efficiency, the mosaic process and identification of individuals were first tested on a pilot area; after successful testing the methods were extended to the whole study area. The pilot process was first tested on one image and then on a pilot mosaic composed of 153 images, in order to later carry out the analysis on the whole area covered by the flight. The images were obtained as a working format, by transforming them from "raw data" to TIFF 16-bit unsigned format.

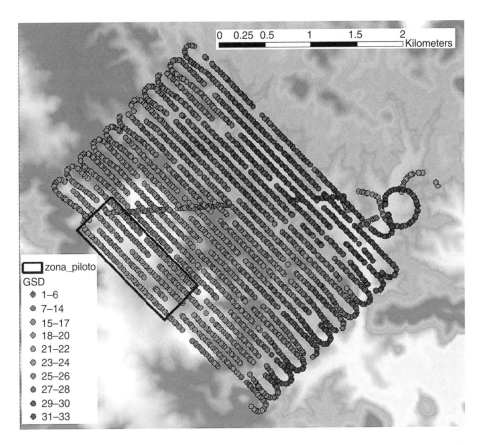

Figure 5.5   Results of the pixel size (cm) analysis control. Location of the pilot area marked in blue. (*For color detail, please see color plate section.*)

Each TIFF image is then associated with a World File (Tagged image File World (TFW)) and the specific parameters of the image (image orientation, resolution depending on its scale, etc.) in a common data file for Geographic Information System (GIS) processing. This allows for the images to be aligned geographically as a mosaic (first for 153 images and subsequently for the entire flight of 2500 images). The generation of the World File for image geo-referencing was also carried out with REDIAM tools. The images mosaic was carried out with typical tools, ERDAS IMAGINE™ (for individual flight paths) and ArcGIS™ (for the complete mosaic). Figure 5.7 shows the mosaic for the pilot area.

The use of these images is affected by the limitations of the conical perspective, to which are added those of the geo-referencing process applied (non-systematic errors reaching even tens of meters). However, when geo-referencing images, if the scale of each image is taken into account and adjusted for, registered elements on those images will have comparable sizes. This is particularly important in the identification of animals (which must be of a similar size no matter in which image they appear). Comparing the photomosaic with a more accurate cartographic reference, such as the PNOA orthophotography (Figure 5.8), the geometric limitations discussed during the image geo-referencing (and the distortions caused by the conical perspective of frames) can be observed.

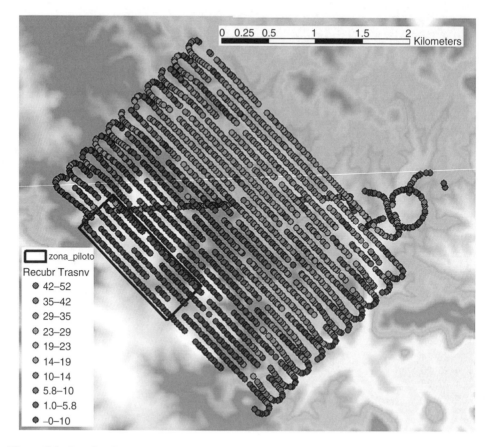

Figure 5.6 Results of the overlap between flight paths analysis. (*For color detail, please see color plate section.*)

5.2.4   Digital Analysis and Identification of Elements

The process was optimized by working on a mosaic of images so that any classification process would have to be applied only once. While technical problems prevented the availability of visible images, thermal images were collected and made available (which is why Figures 5.7 and 5.8 show the mosaics projected onto PNOA orthophotography). That limitation notwithstanding, the availability of individual images and of the photomosaic (in the pilot area), this phase consisted of digital processing for identifying potential individuals of game animals, deer in the gaming estate.

Initially the work was carried out on individual images and later on the pilot photomosaic to study the feasibility of identifying animals and verify that the results obtained were similar in both cases. The process was later extrapolated to the whole flight (mosaic consisting of 2500 images). In all cases the work was carried out on the actual values of the pixels of the images, and none of the work considered any of the physical parameters, for example, soil temperature, animal temperature, and so on. Given that the images were captured during the first hours of the day, the analysis was based on the differences found between the elements detected. To this purpose, different treatments were employed using ENVI™ software, where

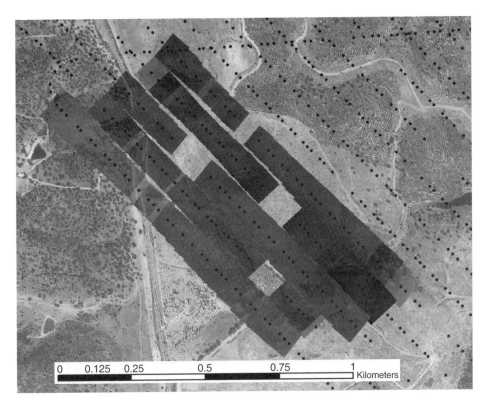

Figure 5.7   Pilot area mosaic (153 images from five flight paths, on PNOA orthophotograph). (*For color detail, please see color plate section.*)

Figure 5.8   Positional differences and geometric limitations between one of the geo-referenced images (red elements) versus PNOA orthophotograph (green). (*For color detail, please see color plate section.*)

a texture filter was finally selected based on the fact that the images contained high amounts of different spatial frequencies, with areas that accounted for a high level of brightness variation. After the analysis of the results, a parallel, supervised classification was carried out on the variance band as well as on the contrast band, identifying elements (interest areas) where animals are evidently observed (as well as other elements, such as image edges and noise). A sample analysis was carried out on two classes ("animals" and "others") and three classes ("animals," "image edges," and "others"). Similar tryouts are carried out on the homogeneity band, without any confirming results. The best results were obtained with variance band classification and the use of two classes ("animals" and "others").

Some good results were also obtained by classifying through a decision tree, discriminating classes by thresholds, working directly with two classes, again, "animals" and "others." Once the thresholds of the elements to be identified were determined, the process was applied directly to mosaics from different flights, as long as the flights were carried out under the same conditions: technical flight details, sensors employed, time and dates of flights, species to identify, and so on. Given the propagation of errors inherent in the level of uncertainty that these conditions may contain, this alternative might be universally effective in practice.

Figures 5.9 and 5.10 show in detail the results in locating potential gaming individuals on the pilot mosaic. These elements can be duplicated (registered in consecutive frames), either because of the overlap between the images or because the animals might have moved (less probable given the airspeed of the UAV). The analysis carried out on the whole flight mosaic identified different areas that account for the presence of potential individuals.

After the identification phase, post-classification processes are carried out to improve the results obtained, filtering the results in order to eliminate information that is of no interest (such as image borders, artifacts, and other elements). This produces an image with the location of all the elements of interest, which will then be vectorized (that is, digitally outlined) and

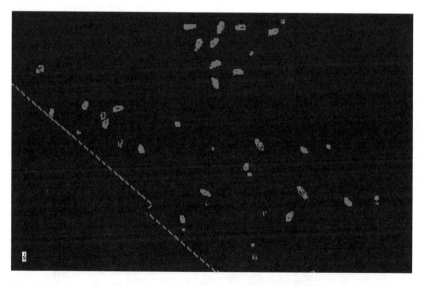

Figure 5.9 Imaging results from the pilot mosaic. Elements (animals) detected through digital processing and classification. (*For color detail, please see color plate section.*)

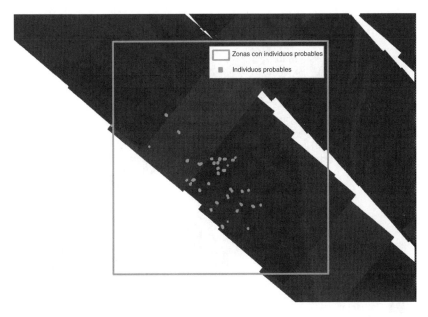

Figure 5.10 Results on the pilot area. Potential individuals detected on the thermal images mosaic and the PNOA orthophotograph. (*For color detail, please see color plate section.*)

refined. This phase includes interpretation of the image to discard non-probable elements from the entire 2500-image sample. This is where direct observations from the field bring valuable information, but is unfortunately restricted to very few and specific cases and areas. On the other hand, the use of other sensors aboard the UAV platforms could provide additional information, such as cross-referencing the presence of groups of individuals. This means that images can be taken by different thermal and visible cameras so that the data obtained can bring a large amount of information to the interpretation of the results. These classification processes have even been able to detect potential individuals that were hidden under trees, something a visual image inspection would not have been able to do. Another interesting line of research could be the monitoring of individuals who carry mini GPS receptors to verify on-the-field data or to study the behavior of a species specifically to provide better census taking by UAVs.

5.3 Results

In processing the digital images from this UAV flight, a number of elements which potentially might be animals were detected. Focusing on the pilot area, Figure 5.11 shows the elements selected as probable gaming individuals in green and those in red considered as non-probable. The behavior of the species and the simultaneous verification on the field and by image capturing confirm the presence of herds of deer. The non-probable elements are distributed almost systematically throughout many frames and therefore may be considered as errors due to various factors as discussed above. In an analysis of the whole area covered by the flight (complete 2500-image mosaic), different areas have been located where elements were detected, some probable and some non-probable. The size of individuals, behavior of the

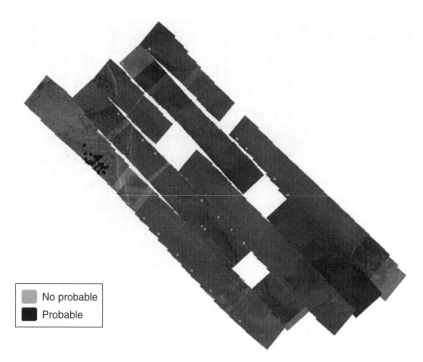

Figure 5.11   Results on the pilot mosaic of locating probable (in green) and non-probable (in red) individuals in the study area (refined vectorization). (*For color detail, please see color plate section.*)

species (for example, herding habits), and interpretation of elements through PNOA ortho-photographs therefore allow estimating probable and non-probable areas the populations within the gaming reserve (see Figure 5.12).

5.4   Conclusions

The following conclusions summarize areas in which observation tasks (capture of images and supplementary data) and post-processing might be improved. There were systematic errors involved in the air search phase of this study that post-mission analysis brought to light; for the most part, their resolution is within the current state of technology.

For example, one main inconvenience found during this flight was the lack of compact integration and synchronization between data acquisition system components (thermal camera, GPS, IMU, etc.), which on occasion causes the distance between image captures to be abnormally uneven. The analysis confirmed that there were offsets of 20–40 m between captured images and their assigned positions, non-systematic errors that substantially affect their estimated geo-referencing. Another related aspect to consider is that there is no correct information concerning the plane's deviation from its planned flight due to crosswinds and other factors. Both of these issues can impact the images of different flight paths since the directions between consecutive paths are opposite to each other.

As another example, the flight plan was carried out according to a global DEM from satellite radar data. A more appropriate approach would be preparing a flight plan with a more accurate

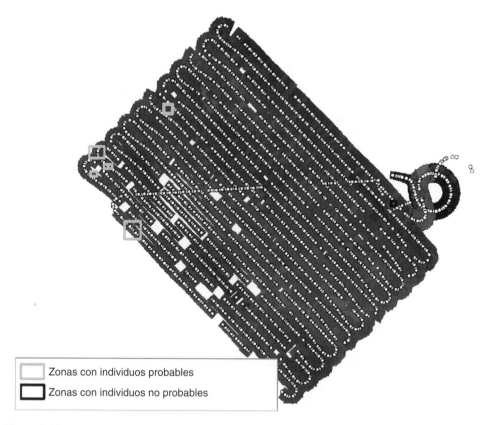

Figure 5.12 Location of areas with probable presence of animals after the classification, processing, and interpretation of the images. (*For color detail, please see color plate section.*)

DEM (PNOA, for instance) and thereby defining similar scale flight paths (coherent with the elements to be mapped). Cartographically it might make more sense if the base reference were the orography (topology) of the reference area, but considering the fact that the elements of interest are mobile, it can be presumed that a systematic coverage of the area by parallel flight paths can provide better results if there are no sudden orographical changes.

UAVs can be considered as good platforms for the observation and estimation of gaming populations. Yet there are certain limitations and specific technical requirements that must be met in order to obtain proper results, and these must be contrasted with field visits carried out simultaneously with the acquisition of images or other elements of interests. In addition, conducting these flights on specific dates and times is very important to ensure that the greatest thermal differences are obtained between individuals and the rest of the elements detected (mainly soil). For this purpose, the flight carried out for this study is an excellent example of meeting these requirements.

The proper planning of the flight, by selecting a reasonable GSD according to the size of the elements to be detected (in this case gaming wildlife) and optimizing the GSD to the total surface to be covered, is very important. In addition, generating an orthophotograph from UAV images provides better representations with similar sizes for captured territorial elements (animals). The photomosaic method employed in this study remains a useful approximation, but the proper procedure is to work on an orthophotograph to eliminate ambiguities and errors that appear due to the duplicity of areas, concealments, covering problems, and gaps.

It must be noted that during the time employed in carrying out a UAV flight (or similar) some animals may move all over the territory, introducing an element of uncertainty in the estimation. In cases of extremely mobile animals, the speed of the flight should be increased so that the capture time for the whole area is low. In the same area, different species of animals might coexist, so it is important to be knowledgeable on the behavior as well as on the approximate and potential distribution of each of them to avoid serious estimation mistakes. In cases where different species might be present, digital analysis of images could be studied as an option for discerning between species (always followed up by an on-the-field verification). In a deeper analysis, both methodologies could be compared: classic detection and counting of animals compared with employing UAVs, analyzing the pros and cons of both methodologies (costs, field work, information processing, error in estimating the individuals, etc.). Of course, both could be complementary, and the possibility of installing GNSS micro-receptors in individuals could improve the results of both approaches. The greater the background knowledge concerning the area to be studied, the better the results obtained.

Acknowledgments

We would like to thank ELIMCO SISTEMAS for the flight and the initial information: images and acquired data, as well as the R&D+i department of the Environment and Water Agency for its interest in developing UAV applications as support for environmental management.

References

1. Kissell, R.E., Tappe, P.A. & Gregory, S.K. 2004. Assessment of population estimators using aerial thermal infrarred Videography data. Lakeside Farms, Wingmead Farms Inc., and the Arkansas Game and Fish Commission.
2. Naugle, D.E., Jenks, J.A. & Kernohan, B.J. 1996. Use of thermal infrared sensing to estimate density of white-tailed deer. Wildlife Society Bulletin, 24 (1), 37–43.
3. Wilde, R.H. & Trotter, C.M. 1999. Detection of Himalayan thar using a thermal infrared camera. Arkansas Game and Fish Commission.
4. Zarco, P.J. 2012. Creación de Algoritmo para búsqueda automática de avutardas usando imágenes adquiridas en los vuelos realizados en 2011 y 2012. Informe de prestación de servicio. Consejería de Medio Ambiente y Ordenación del Territorio, European Project.

6

A Flight Time Approximation Model for Unmanned Aerial Vehicles: Estimating the Effects of Path Variations and Wind*

Matthew J. Henchey[1], Rajan Batta[2], Mark Karwan[2] and Agamemnon Crassidis[3]

[1] Herren Associates, Washington, DC, USA
[2] Department of Industrial and Systems Engineering, University at Buffalo (State University of New York), NY, USA
[3] The Kate Gleason College of Engineering, Rochester Institute of Technology, Rochester, NY, USA

Nomenclature

Lat_a	=	the latitude centered at waypoint a (decimal degrees)
$Long_a$	=	the longitude centered at waypoint a (decimal degrees)
$psi0$	=	the assumed heading at the initial waypoint (degrees)
$psiF$	=	the desired heading at the final waypoint (degrees)
x_{ab}	=	horizontal Cartesian coordinate distance between points a and b (feet)
y_{ab}	=	vertical Cartesian coordinate distance between points a and b (feet)
$feetToDegrees$	=	2.742980561538962e–06 (constant)
$Radius$	=	the required radius of the surveillance task (feet)

*Contribution originally published in MORS Journal, Vol. 19, No. 1, 2014. Reproduced with permission from Military Operations Research Society.

Operations Research for Unmanned Systems, First Edition. Edited by Jeffrey R. Cares and John Q. Dickmann, Jr.
© 2016 John Wiley & Sons, Ltd. Published 2016 by John Wiley & Sons, Ltd.

$Angle_{entry}$	=	the angle from the center location to the entry location (degrees)
$Long_{entry}$	=	the longitude of the entry point on the surveillance task (decimal degrees)
Lat_{entry}	=	the latitude of the entry point on the surveillance task (decimal degrees)
$Duration$	=	the required duration of the loitering task (seconds)
$Circumfer$-$ence$	=	the circumference of the circular loitering path (feet)
$Airspeed$	=	the specified airspeed of the resource (ft/s)
$Distance$	=	the distance traveled while on the circular loitering path (feet)
θ	=	the angle of change between the entry point and exit point (degrees)
$Angle_{entry}$	=	the angle from the center location to the entry location (degrees)
$Long_{exit}$	=	the longitude of the exit point on the circular loitering path (decimal degrees)
Lat_{exit}	=	the latitude of the exit point on the circular loitering path (decimal degrees)
$Angle_{exit}$	=	the heading at the exit point on the circular loitering path (degrees)
u	=	the speed of the head and tail winds (ft/s)
v	=	the speed of the crosswinds (ft/s)
$Airspeed_{UAV}$	=	the airspeed of the UAV (ft/s)

6.1 Introduction

Accurate distances and travel times are important to any vehicle routing problem because they influence the cost of a mission or delivery schedule. Missing time windows for task completion or delivery can result in much greater costs or even mission failure. When considering a vehicle routing problem with time windows (VRPTW) in which unmanned aerial vehicles (UAVs) must be scheduled, accurate flight times are required to determine mission plans that are feasible for a single UAV or even a group of UAVs.

Creating a model that can calculate accurate flight time approximations required consideration of two situations. The first was solving a VRPTW knowing the task locations and the availability of the UAVs to determine a mission plan that minimizes the cost or maximizes mission effectiveness. The second situation was a dynamic resource management problem that alters the plan during a mission with minimal cost for making changes as new tasks become available. In both situations, the algorithms benefit by the use of a parameter $(f_{r,i,j,k})$, representing flight time, assigned to a resource r, and a set of three points: i, j, and k. The flight time $(f_{r,i,j,k})$ calculated is from point j to k, knowing that the resource r came from point i. In our formulation $r \in R$ where R is the set of resources available and $i \in I$, $j \in J$, and $k \in K$ where I, J, and K are the sets of waypoints which must be visited and will range in cardinality based on the size of the mission space. Therefore, the total number of flight time approximations $(|I| \times |J| \times |K|)$ will likely become very large for many missions. It is also important for these vehicle routing problems to consider constraints of UAVs' fuel capacity and range to achieve mission goals.

The goal of this study was to create a flight time approximation model capable of producing a large set of estimated flight times across combinations of possible waypoints in a real-time scenario. This model, in combination with the practical tasks a UAV may be required to perform and the wind effects on the UAV, must limit its computational burden despite the obvious combinatorial implications of increasing the number of intermediate waypoints. The main contribution resulting from our work is a near real-time method for approximating flight times in a more accurate representation than the typical straight-line assumption currently used in real-time UAV mission planning. Realistic surveillance tasks and the effects of wind on fuel burn rates within the flight time approximation model also provide a further contribution to UAV mission planning.

6.2   Problem Statement

This research considers idealized flight kinematics of a UAV to provide more accurate flight time approximations for current mission planning and dynamic resource management models [1, 2]. In urban environments, Weinstein and Schumacher [3] claim using rectilinear distances is appropriate as UAVs must follow the grid layout of a city. They also suggest using Euclidean distance is reasonable if buildings are not too tall. When utilizing larger UAVs outside of a city, neither rectilinear nor Euclidean distances guarantee accurate flight time approximations due to the minimum turning radii of UAVs. This directly affects the shortest path the UAV is able to use. Solving the routing problems requires accurate flight time approximations and fuel consumption rates, which are critical for mission success.

The full-nonlinear simulation of the Pioneer aircraft constructed by Grymin and Crassidis [4] is not a viable option for real-time scenarios as it lacks the necessary computational performance to estimate flight times quickly. Using the Dubins set developed by Shkel and Lumelsky [5], Grymin and Crassidis [4] were able to find the shortest path between two points given a vehicle with a minimum turning radius much faster than the simulation model. Assuming the use of accurate UAV specifications, the Dubins set produces the shortest path that the UAV can follow. Using this as the basis enables us to develop a model that is capable of calculating a large number of flight time approximations quickly and efficiently.

Using a combination of three waypoints, the i, j, k triplet, allows us to approximate flight time ($f_{r,i,j,k}$) assuming constant altitude and airspeed in a 2D planar field. Including surveillance tasks and sensors with a specified range allows the model to produce approximations that account for realistic mission scenarios. We assume the sensor is a gimbaled camera and video recording device, which allows rotation about two axes in order to observe while maneuvering throughout the UAV's flight path. Surveillance tasks include a radius of sight, loitering, and radius of avoidance task. These tasks incorporate a circular area or flight path centered on the specified waypoint with the loitering task having longer observation duration. Given a wind speed and heading acting over the mission space, the model approximates the fuel consumption rate of the UAV while on its flight path between waypoints j and k. These specifications outline the developed model, which can generate flight time and fuel consumption approximations for a large set of possible waypoint combinations in a real-time scenario.

6.3   Literature Review

In this section, we present the current state of the art for approximation methods. Next, the modeling and importance of including surveillance tasks in previous work is established. Finally, we demonstrate the reasoning and methods of adding wind effects into approximation models.

6.3.1   Flight Time Approximation Models

In 1957, L.E. Dubins proved the theory that provides the framework for determining the set of curves that connect two points in a 2D space. Theorem I states that every planar $R - geodesic$, a path composed of arcs of a circle of radius R and straight lines, is necessarily a continuously differentiable curve. The curve can either be (i) an arc of a circle of radius R, a line segment, and then another arc of a circle of radius R, (ii) three arcs of circles of radius R, or (iii) a sub-path of (i) or (ii) [6]. Determining the set of these curves that connect two points provides the shortest path within the set. In the journal article "Classification of the Dubins Set," Shkel and Lumelsky [5] were able to further Dubins work and develop a method to limit the number of calculations needed in order to determine the minimum distance path for a Dubins problem, which they have called the Dubins set. The Dubins problem generally consists of an object moving from some initial point to a final point with specified initial and final orientation angles. The Dubins set is a collection of calculations that produces the shortest path without having to calculate all possible paths of circle-line-circle or circle-circle-circle combinations.

In "Simplified Model Development and Trajectory Determination for a UAV using the Dubins Set," Grymin and Crassidis [4] compare a Dubins aircraft model developed from the use of a full-nonlinear simulation model and the Dubins curve path model. Developed for a specific UAV, however, the Dubins aircraft model is limited and uses a simulation model to develop the flight path. Because the goal of this research revolves around creating and dynamically changing a mission plan for multiple UAVs, a more robust approach is needed that can approximate flight times for multiple types of UAVs. The Dubins curve path model uses the method described in [5], the Dubins set, to determine the shortest path distance. Using these distances provides the flight time to a given waypoint based on distance divided by speed. The ability of a UAV to follow a predetermined path created from the Dubins set was also shown by Grymin and Crassidis [4] and Jeyaraman *et al.* [7]. This research is only interested in the flight time approximation and does not require information on the entire flight path. We determined the Dubins set was a good approximation method for calculating the distance of the shortest path, the Dubins curve path model being the basis of our model.

6.3.2 Additional Task Types to Consider

Research using limited sensor range, areas of avoidance, and loitering to detect possible targets with no a priori knowledge of the exact location of possible threats is extensive. Enright and Frazzoli [8] studied one problem in which a team of UAVs must visit a set of targets stochastically generated over time and within a range that permits sensor detection of a target located at the point of interest. Sujit *et al.* [9] also used the knowledge of limited sensor ranges to develop a negotiation scheme for the task allocation of multiple UAVs. In their study, the UAVs must search for targets in an unknown region and then attack the detected targets.

Mufalli *et al.* [1] and Weinstein and Schumacher [3] developed planning models for UAVs that consider the loitering task as duration of time and the flight paths as Euclidean. Schumacher *et al.* [10] used a loiter ability and staged departure ability in their mixed integer linear programming (MILP) model to account for the need of the air vehicles to delay their arrival to a target to complete a task given the timing and sequencing constraints. Schouwenaars *et al.* [11] specify a zone where the UAV can loiter and wait for task assignments that is outside of threat zones. They specify points on the loitering path in which the UAV can enter and leave in order to determine flight paths to and from those points on the loitering path. This provided a basis for our implementation of many of the surveillance tasks that we included. Schouwenaars *et al.* [11] give an example of the loitering area as well as no fly zones.

The radius of avoidance task is a unique combination of the radius of sight and loitering tasks, as we do not allow the UAV to enter into the radius of avoidance area but rather to only fly along the perimeter of the area. This takes advantage of the sensor range by allowing the UAV to carry out the task without having to enter an area of very high risk. Myers *et al.* [12] and Schouwenaars *et al.* [11] use no-fly zones and obstacle avoidance but assume that the waypoints are outside of the no-fly zones. They provide no way to take advantage of the sensors range in order to complete tasks for waypoints that may actually be within a no-fly zone.

Modeling these additional tasks allows the flight time approximation model to accurately reflect necessary and practical tasks that one would expect to find in a real UAV mission planning problem. Our work uses the sensor range in order to complete surveillance task rather than for detection of targets by allowing the UAV to observe at a specified distance. Furthermore, we include flight kinematics and durations in modeling surveillance tasks rather than the assumption of Euclidean distances and delays.

6.3.3   Wind Effects

Including wind effects into the flight time approximation model is a very important factor to consider, because wind causes changes in airspeed, fuel consumption, and heading of the UAV. Ceccarelli *et al.* [13] used the MultiUAV2 simulation [14] developed by the Air Force Research Laboratory in the development of a path planner for micro unmanned aerial vehicles (MAVs) in the presence of wind. This simulation demonstrated how the presence of wind could cause the MAV to fly at a trajectory that is actually turning into the wind to maintain its flight path. It also exposed instances where the flight paths vary because of the MAV's inability to maintain either its airspeed or heading. McNeely *et al.* [15] have shown how modifications to the Dubins vehicle account for the presence of a time-varying wind. This would also require the use of a system of equations and knowledge about the actual path of the UAV flight between two waypoints. As shown with the simulation models from [4], these simulations tend to have computation times do not provide real-time solutions. Instead, these simulations provide data that can be helpful in trying to approximate the effect of wind on the fuel burn rate of the UAV given the heading of the UAV and constant wind speed.

From our literature review and study of various types of flight time approximation models, we decided that a simplified approach was needed that accounts for the effects of wind on a UAV. The requirements of this problem constrain the flight time approximation model to certain level of complexity that allows for calculating a large number of flight time approximations in near real time. Studying the full-nonlinear simulation developed by Grymin and Crassidis [4] further, we established the effects of wind on fuel burn rates using a simple regression model.

6.4   Flight Time Approximation Model Development

Generating approximations that are both accurate and quickly computed required some important assumptions. First, the UAV will maintain a constant airspeed and altitude throughout the duration of its flight in a 2D planar path. Second, a "Flat-Earth" approximation will be sufficient in converting longitude and latitude into feet as the waypoints will not be far enough away to create a large error. The final waypoint k for each combination will be used as the anchor point when doing the conversions. The third and final assumption is that the initial heading at the initial waypoint j is given by the heading of a straight line from the previous waypoint i to the initial waypoint j. This assumption is made because the desired headings are made equivalent to the heading of the straight line from the initial waypoint j to the final waypoint k. This provides the basis for the assumptions of the i, j, k triplets and the flight time approximation model. We present any further assumptions needed in the sections in which they are required. Finally, our base model considers the task being completed as a "snapshot" task. We define the task as one that allows the UAV to satisfy waypoint requirements by flying directly over the waypoint with no extended duration, such as taking a picture as part of a surveillance task.

6.4.1   Required Mathematical Calculations

The implementation of the initial approximation models requires a MATLAB program that reads in waypoint location information and determines the flight time for every possible triplet combination. The Dubins aircraft model required the initial heading, initial waypoint, final waypoint, and desired heading at the final waypoint. Using the assumptions described previously, the approximation model calculates the initial heading and desired heading and converts the longitude and latitude into Cartesian coordinate distances in feet. The following shows the formulation that determines the Cartesian coordinate distances and the initial and final headings. The triplets are given as the previous i, initial j, and final k waypoints.

$$x_{ij} = \left(Long_j - Long_i \right) / \left(feetToDegrees \times \left(1 / cos \left(Lat_j \times \left(180 / \pi \right) \right) \right) \right) \tag{6.1}$$

$$y_{ij} = \left(Long_j - Long_i \right) / \left(feetToDegrees \right) \tag{6.2}$$

$$psi0 = mod \left(atan2 \left(x_{ij}, y_{ij} \right), 2 \times \pi \right) \times \left(180 / \pi \right) \tag{6.3}$$

$$x_{jk} = \left(Long_k - Long_j \right) / \left(feetToDegrees \times \left(1 / cos \left(Lat_k \times \left(180 / \pi \right) \right) \right) \right) \tag{6.4}$$

$$y_{jk} = \left(Long_k - Long_j \right) / \left(feetToDegrees \right) \tag{6.5}$$

$$psiF = mod \left(atan2 \left(y_{jk}, x_{jk} \right), 2 \times \pi \right) \times \left(180 / \pi \right) \tag{6.6}$$

Following from our assumption, the heading at the initial waypoint is set to the desired heading from the previous triplet, where i would have been the initial waypoint and j would have been the final waypoint. Both of these models read in files containing all waypoint numbers (or other identifiers such as a name), making longitudes and latitudes sufficient to calculate all the possible combinations for any mission by creating an input file. In cases where there are no previous waypoints i, such as when the UAV is leaving from a base at a known heading, the models are also able to have the initial heading input specified rather than calculated. We now describe how each model is implemented to calculate flight time approximations given a unique combination of previous i, initial j, and final k waypoints for a given resource r.

6.4.2 Model Comparisons

The Dubins aircraft model must run a Simulink simulation in order to determine the approximate flight time. The Dubins curve path model follows a series of logical and mathematical equations to determine the approximate flight times. Considering this difference is important when comparing the Dubins aircraft model to the Dubins curve path model. Figure 6.1

(a)

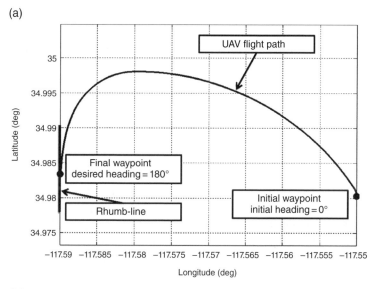

(b)

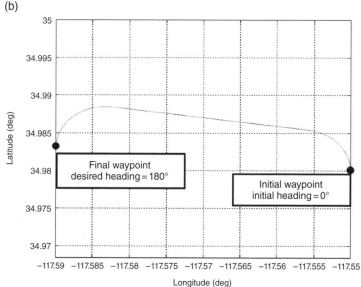

Figure 6.1 (a) Path from Dubins aircraft model and (b) path from Dubins curve path model. *Source*: Henchey, Matthew J.; Batta, Rajan; Karwan, Mark; Crassidis, Agamemnon, "A Flight Time Approximation Model for Unmanned Aerial Vehicles: Estimating the Effects of Path Variations and Wind," *MORS Journal*, Vol. 19, No. 1, 2014, John Wiley and Sons, LTD.

demonstrates the paths produced by each approach is much different, with the Dubins curve path model being shorter than the Dubins aircraft model using rhumb line navigation.

The two main points of comparison between the Dubins aircraft model and the Dubins curve path model are the difference in flight paths and time required to approximate flight time. The significant difference in the path is a result Dubins aircraft model using rhumb line navigation. Usually navigation systems would approximate a great circle by a series of rhumb lines, where a great circle is the shortest route between two points on Earth and a rhumb line is a course with constant bearing. Defining rhumb line navigation as steering along a pre-defined rhumb line, the Dubins aircraft model attempts to maneuver the UAV onto a rhumb line to the final waypoint and desired heading as quickly as possible. Additionally the Dubins aircraft model requires greater runtime as the simulation actually creates the flight path in small step sizes until it reaches the waypoint or stops if reaching the waypoint was not possible. Generally, the simulation has a runtime of approximately 0.5–1 s for each path. This is dependent upon the distance between the initial waypoint and the final waypoint, the simulation runtime increasing as the distance increases.

Prior to modeling the surveillance tasks and fuel consumption rates, the Dubins aircraft model required 543 s to compute 720 combinations, whereas the Dubins curve path model required only 0.53 s. These were run on a laptop operating on Microsoft Windows 7 64-bit with an Intel Core2 Duo CPU T9300 2.50GHz processor and 4.0GB of RAM, with the models running in MATLAB 2009a. For these reasons the Dubins aircraft model was clearly not a possible choice for the basis of the flight time approximation model. Reprogramming the Dubins curve path model into Java to improve the code structure and extensibility of the model further limited the computation time. The MATLAB version required 55 s, whereas the Java version only required 2 s to compute 308 700 approximations.

6.4.3   Encountered Problems and Solutions

In certain cases, the Dubins set did not provide a shortest path due to the calculations returning imaginary numbers. Solving this problem required searching through the Dubins set for the shortest path that did not result in imaginary numbers. Such cases resulted from two waypoints being located close together with respect to the minimum turning radius of the UAV, explained by Shkel and Lumelsky [5]. After applying this solution and running many test cases no other instances occurred where the Dubins set with modifications was not able to produce a shortest path.

When there are a very large number of combinations, it is possible to limit these combinations in order to improve upon the runtime of the model. For example, one could consider UAV type specific tasks rather than computing all combinations for each UAV. Given a 10-waypoint example there would be $(10 \times 9 \times 8) = 720$ combinations for a single UAV. If five of the waypoints required one UAV type and the other five required another UAV type there would only be $2 \times (5 \times 4 \times 3) = 120$ combinations to consider. Considering the time windows and possible durations of tasks and eliminate those combinations that are not possible due to time constraints is another possible solution to this problem. Although limiting the number of combinations may be useful in certain scenarios, implementing the Dubins curve path model to calculate a large number of combinations still provides a method of generating flight time approximations in a very small amount of time.

6.5   Additional Task Types

Up to this point, we have presented the implementation for a "snapshot" task in the Dubins curve path model. To make the flight time approximation model practical, incorporating a radius of sight task, circular loitering path task, and a radius of avoidance task was necessary. These tasks, along with idealized flight kinematics, produce accurate approximations for mission planning models.

6.5.1   *Radius of Sight Task*

The radius of sight task requires the UAV to enter a circular zone specified by a given radius rather than fly directly over the specified waypoint. Once the resource has entered this circular zone, it is able to complete its task and can continue on to the next waypoint specified in its mission plan. To complete the task we consider a minimum observance time in which the UAV must be within a specified range of the waypoint. Selecting an appropriately sized circular zone defined by the radius of sight and minimum observance time, we can guarantee that the UAV will remain within range of the waypoint given the airspeed of the UAV and the duration of the minimum observance time no matter where it enters the circular zone. This is done by creating a smaller circular zone than the one created by the radius of sight alone that has a radius equal to the radius of sight minus (minimum observance time × UAV airspeed). Figure 6.2 demonstrates this concept, showing that the shortest time between the two zones is equal to the minimum observance time. To eliminate confusion we assume the radius of sight specified in our input file reflects both the range of the gimbaled camera and video device and the minimum observance time in the remainder of this section.

The radius of sight task also uses the i, j, k combination, assuming the flight time from j to k knowing that the resource was previously at i. Modeling the radius of sight task requires the location of entry into the radius of site area as well as the heading at that location for each waypoint. Figure 6.2 demonstrates a possible i, j, k combination with a radius of sight task at j. The example shows a simple solution to the problem. Upon further investigation this solution did not provide a practical approximation of flight time as it forces the UAV to enter at the intersection

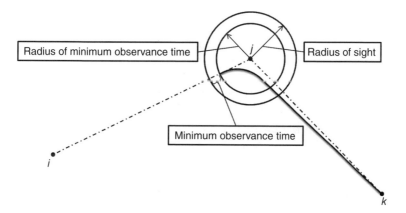

Figure 6.2   An i, j, k example of the radius of sight task. *Source*: Henchey, Matthew J.; Batta, Rajan; Karwan, Mark; Crassidis, Agamemnon, "A Flight Time Approximation Model for Unmanned Aerial Vehicles: Estimating the Effects of Path Variations and Wind," *MORS Journal*, Vol. 19, No. 1, 2014, John Wiley and Sons, LTD.

of the heading from i to j and the edge of the radius of sight area. This may lower the amount of time saved by allowing the resource to enter at some other point and some other heading.

Only the current i, j, k combination is known to the flight time approximation model. It is not able to choose the optimal entry point and heading for the radius of sight task that minimizes flight time. Therefore, we decided to determine the expected flight time rather than forcing the resource to enter at one specific point and heading. When j is a radius of sight task we assume that the resource will enter the radius of sight area with an equal probability at any point on the half of the circle that is closest to i, with multiple headings possible. This allows us to calculate the expected flight time, as well as worst case and best case scenarios of flight times to be obtained based on all the possible flight times calculated for any given i, j, k combination.

Segmenting the radius of sight area into eight equal parts allows the model to calculate the expected flight time. Depending on the location of i relative to j, four of these eight points are used as entry points into the radius of sight. Although we could use all eight points on the perimeter, the four best points provide a compromise between accuracy and computational complexity. In the loitering task section, we show how using four points instead of eight points requires much less computation. Figure 6.3 shows a scenario in which i is located in the sixth segment relative to j. The four points of entry are at the intersection of the upper and lower limit of the sixth segment and the radius of sight area, as well as the adjacent segment intersections.

The following set of equations demonstrates how we determine the actual longitude and latitude point for each of the four best entry points. These equations calculate the location based on the angle and radius from the center of the radius of sight task.

$$Long_{entry} = Long_i + Radius \times feetToDegrees \times \left(1 / cos\left(Lat_i\right)\right) \times sin\left(Angle_{entry}\right) \quad (6.7)$$

$$Lat_{entry} = Lat_i + Radius \times feetToDegrees \times cos\left(Angle_{entry}\right) \quad (6.8)$$

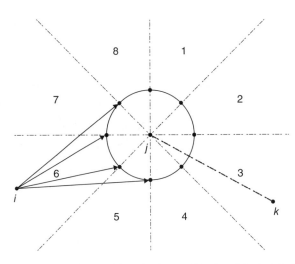

Figure 6.3 Segmentation of the radius of sight task. *Source*: Henchey, Matthew J.; Batta, Rajan; Karwan, Mark; Crassidis, Agamemnon, "A Flight Time Approximation Model for Unmanned Aerial Vehicles: Estimating the Effects of Path Variations and Wind," *MORS Journal*, Vol. 19, No. 1, 2014, John Wiley and Sons, LTD.

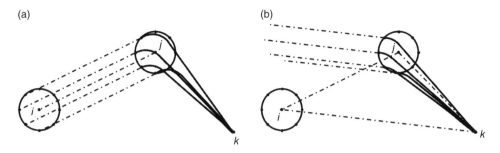

Figure 6.4 The radius of sight task heading sets. (a) shows the headings that are parallel to the heading from i to j and (b) shows the headings that are parallel the heading from i to k for each of the four entry points. *Source*: Henchey, Matthew J.; Batta, Rajan; Karwan, Mark; Crassidis, Agamemnon, "A Flight Time Approximation Model for Unmanned Aerial Vehicles: Estimating the Effects of Path Variations and Wind," *MORS Journal*, Vol. 19, No. 1, 2014, John Wiley and Sons, LTD.

We consider two sets of points and headings for the radius of sight task. One assumes the UAV enters the radius of sight parallel to the heading from waypoint i to j and another that is parallel to the heading from waypoint i to k. Figure 6.4 demonstrates an example of the headings described above, where the left figure (a) shows the headings that are parallel to the heading from i to j and the right figure (b) shows the headings that are parallel the heading from i to k for each of the four entry points.

In certain cases the UAV will be able to turn toward the final waypoint k before reaching the initial waypoint j. In other cases it may stay on a heading parallel to the straight-line heading from i to j. Most likely, operators will use a heading in between these two headings when creating the flight plan. Since we do not know which of these points and headings will be best or always possible, these sets enable the flight time approximation model to return a mean approximation that is representative of the radius of sight task.

6.5.2 Loitering Task

We define the loitering task as one that requires the UAV to remain in position around a specified waypoint in order to accomplish its mission. This may include gathering information using surveillance sensors or simply waiting for some duration of time. Although many patterns exist for loitering tasks, we model a circular loitering path task for our approximation model. Modeling this task requires the radius of the circular loitering path, as well as the entry point location and heading and the loitering duration. From this information, we can obtain the exit point location and heading required by the flight time approximation from waypoint j to k where j is a loiter task. Figure 6.5 demonstrates this concept.

In Figure 6.5 the angle θ represents the change in angle between the entry point and the exit point on the circular loitering path. The following set of equations demonstrate how θ is obtained as well as the exit point location and heading in Java.

$$Distance = Duration \times Airspeed \tag{6.9}$$

$$Circumference = 2 \times \pi \times Radius \tag{6.10}$$

$$\theta = Math.IEEEremainder\left(Distance, Circumference\right) \times Radius \tag{6.11}$$

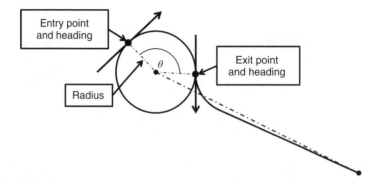

Figure 6.5   The circular loitering path task. *Source*: Henchey, Matthew J.; Batta, Rajan; Karwan, Mark; Crassidis, Agamemnon, "A Flight Time Approximation Model for Unmanned Aerial Vehicles: Estimating the Effects of Path Variations and Wind," *MORS Journal*, Vol. 19, No. 1, 2014, John Wiley and Sons, LTD.

$$Long_{exit} = Long_i + Radius \times feetToDegrees \times \left(1 / cos\left(Lat_i\right)\right) \times sin\left(Angle_{entry} + \theta\right) \quad (6.12)$$

$$Lat_{exit} = Lat_i + Radius \times feetToDegrees \times cos\left(Angle_{entry} + \theta\right) \quad\quad\quad (6.13)$$

$$Heading_{exit} = Angle_{entry} + \theta \pm 90 \quad\quad\quad\quad\quad\quad\quad\quad\quad\quad\quad (6.14)$$

In Eq. (6.14) the ± is dependent on the resource traveling clockwise or counterclockwise around the circular loitering path. We add 90° when traveling clockwise and subtract 90° when traveling counterclockwise. The modeling of the circular loitering path task uses the above equations throughout this section.

To ensure the implementation accurately represented the loitering task and was practical in terms of computation times, the loitering task had important requirements. The loitering task needed to have a defined entry and exit point as well as a duration for which the resource must loiter around the waypoint center of the task. In the initial attempt of modeling the loitering task, we found the exit point from a single specified entry point and duration. After realizing this was not practical given the lack of knowledge of UAV location from using i, j, k combinations, we modeled eight equally likely points to define locations where the UAV can enter and exit the loitering path. Although neither of the original two approaches satisfied the requirements of the task, they provided a basis for the final implementation. Obtaining an approximated flight time for the i, j, k combination when either i, j, or k were loitering tasks was the goal of the final implementation of the circular loitering path task. The durations of the tasks are not included in the flight time as they are inputs from the mission planning and dynamic resource management models that will implement this flight time approximation model. For example, if both j and k were loitering tasks, the approximated flight time would be from the exit point at j to the entry point at k.

Entry into the circular loitering affects the flight time to get into the loitering task but also the point at which the resource will exit the loitering path. In the original two approaches, the model either forced the resource to enter at a single point or assigned an equal likelihood of entering any of the eight points around the circular loitering path. Neither of these two approaches fully represented what would actually happen in a practical scenario. One forces the resource to enter at a single point, which can be suboptimal, and the other does not take the previous location of the resource or the duration of the loitering task into account.

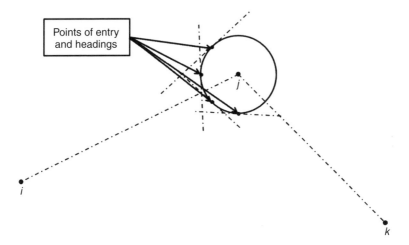

Figure 6.6 Choosing four equally likely points out of eight possible points. *Source*: Henchey, Matthew J.; Batta, Rajan; Karwan, Mark; Crassidis, Agamemnon, "A Flight Time Approximation Model for Unmanned Aerial Vehicles: Estimating the Effects of Path Variations and Wind," *MORS Journal*, Vol. 19, No. 1, 2014, John Wiley and Sons, LTD.

To combine the two approaches we define the eight points of entry as described in the second approach as the possible entry points of the circular loitering path task, which will be dependent upon the previous location of the resource as in the first approach. Of the eight possible entry points, we choose the four best entry points as in the radius of sight task. If the heading to the center waypoint is greater than the heading to the entry point a clockwise heading is assigned otherwise a counterclockwise heading is assigned as demonstrated in Figure 6.6.

Choosing the four best entry points and headings limits the number of calculations, as now there are only four possible combinations to consider at each loitering task rather than 16 when using eight entry and exit points. If j and k are both loitering tasks there will now be (four points at $j \times 1$ heading at each point) $\times$ (four points at $k \times 1$ heading at each point) or 16 possible combinations rather than 256 possible combinations if all eight points were considered. This allows the model to calculate a more realistic mean as a resource is not equally likely to enter all eight points in a practical situation. At the same time, this approach also allows for faster overall computation time as it greatly limits the number of calculations that need to be completed.

This approach uses the equations presented in this section to determine the set of four exit points that are associated with the set of four entry points. In the Java program, we use a collection of waypoints to store the four entry points and then pass these four entry points in as an argument to a method contained in the circular loitering path task class definition. The equations determine the number of rotations the UAV completes within the specified duration and from an entry point the location of the exit point can be determined based on the angle θ. Running a for-loop that contains the set of equations determines all four exit points and headings. Figure 6.7 demonstrates the four entry and exit points on the loitering path task at j and the possible Dubins curve paths from loitering task j to the four best possible entry points into the loitering path task at k. This is done for each of the four exit points at j so in total there are 16 possible paths considered, providing a minimum, maximum, and mean flight time for the loitering task.

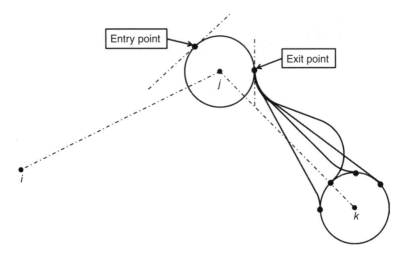

Figure 6.7 Determining the exit points based on entry point location and duration. *Source*: Henchey, Matthew J.; Batta, Rajan; Karwan, Mark; Crassidis, Agamemnon, "A Flight Time Approximation Model for Unmanned Aerial Vehicles: Estimating the Effects of Path Variations and Wind," *MORS Journal*, Vol. 19, No. 1, 2014, John Wiley and Sons, LTD.

This task also enabled us to implement a radius of avoidance task. We considered the radius of avoidance task as a special case of the loitering task in which a specified radius allows the UAV to avoid a certain waypoint but still make some observation with equipped sensors. Assigning an arbitrarily short minimum observance time, the UAV follows the loitering path for a short time to complete its task before moving to the next waypoint. The three additional tasks discussed enable the flight time approximation to include realistic task types to calculate flight time approximations for mission planning and dynamic resource management models.

6.6 Adding Wind Effects

We studied three independent variables for the purposes of including wind effects into our flight time approximation model. These were the head and tail wind speeds acting on the UAV, the crosswind speeds acting on the UAV and the airspeed of the UAV itself. Using the full non-linear simulation of the Pioneer UAV from [4] to study the wind effects, these results are only valid for the Pioneer UAV specifications. The simulation allowed us to place the UAV in a true airspeed hold at three different levels. To ensure that the only influences on the data collected were wind and the UAV, we placed everything else in the simulation in a constant hold and set the UAV to follow a straight path for 200 s. Setting the wind to attain the full force of the specified speed over the first 10 s of the simulation allowed the UAV to react and maintain its true airspeed. Running the simulation at the different levels of the independent variables allowed us to collect data on the average fuel burn rate throughout the entire simulation, which is the dependent variable.

The factor levels used as inputs to the full nonlinear simulation to collect data and build the fuel burn rate model are shown in Table 6.1. Crassidis and Crassidis [16] advised the maximum wind speed magnitude of 10 ft/s (6.82 mph) based on the abilities of the UAV that the full nonlinear simulation represented. All Pioneer UAV aircraft are subject to this

Table 6.1 Average fuel burn rate factor levels.

Independent variable	Minimum level	Maximum level	Step size
Head/tail wind speed (u)	−10 ft/s	10 ft/s	1 ft/s
	(−6.82 mph)	(6.82 mph)	(0.682 mph)
Crosswind speed (v)	−10 ft/s	10 ft/s	1 ft/s
	(−6.82 mph)	(6.82 mph)	(0.682 mph)
UAV airspeed ($Airspeed_{UAV}$)	137.86 ft/s	167.86 ft/s	10 ft/s
	(93.99 mph)	(114.45 mph)	(6.82 mph)

maximum wind speed as a constraint. We set the maximum UAV airspeed to the actual maximum airspeed of the UAV represented by the full nonlinear simulation of the Pioneer UAV. We then used a step size of 10 ft/s (6.82 mph) to determine three additional UAV velocities that were greater than the minimum airspeed needed for the UAV to maintain its altitude. The development of additional full nonlinear simulations for different UAVs or knowledge obtained from UAV testing would be necessary to further these results and implement additional fuel burn rate equations for different UAVs. We believe the methodology of determining the fuel burn rate equations and the implementation would be valid for higher wind speeds given the use of different UAVs.

Although the crosswinds were initially a part of the study to determine a fuel burn rate model, upon further investigation and consultation with [16], the complexity of the crosswinds surpassed the complexity of our model. Therefore, the fuel burn rate model is limited to using the UAV airspeed and the head and tail wind speed. We discuss crosswinds again in the discussion of future research. For now, we assume those piloting the UAV or the UAV's automated systems will be able to handle crosswinds. This could be by making the appropriate maneuvers to turn into or against the crosswinds or make the decision that the crosswinds are too strong for the UAV to handle. As the flight time approximation model does not require knowledge of the entire flight path, it is impossible to determine the effects of crosswinds along the entire flight path without obtaining this knowledge. Doing so would likely turn this model into a simulation model that is unable to quickly compute flight time approximations for a large number of i, j, k combinations as described in the problem statement.

Using the full nonlinear simulation and the data collected, we determined linear regression equations to fit the data generated, one for each of the UAV airspeeds with the head and tail wind being a predictor. Using MINITAB, the regressions all had R-squared values greater than 99%, showing the simple linear equation is able to account for almost all of the variance in the data. The following equations are the fuel burn rate equations for each of the UAV velocities studied and the units are gallons per hour. These equations follow what one would expect to find as the head winds (negative u values) increase the fuel burn rate while tail winds (positive u values) decrease the fuel burn rate.

$$FuelBurnRate_{167.86\,ft/sec} = 3.03107 - 0.0146952u \tag{6.15}$$

$$FuelBurnRate_{157.86\,ft/sec} = 2.90315 - 0.0133689u \tag{6.16}$$

$$FuelBurnRate_{147.86\,ft/sec} = 2.79312 - 0.0117382u \tag{6.17}$$

$$FuelBurnRate_{137.86\,ft/sec} = 2.70043 - 0.0100805u \tag{6.18}$$

6.6.1   Implementing the Fuel Burn Rate Model

Implementing the fuel burn rate model into the flight time approximation model requires more knowledge about the actual flight paths the model is predicting. Although one assumption implemented simply uses the straight-line distance from j to k in the i, j, k combination this may not produce an average fuel burn rate that represents the true flight path of the UAV. In order to obtain a better approximation the flight path is broken up into three linear segments representing the three curves of the Dubins curve path as shown in Figure 6.8.

Using the flight time approximation model, we know the distances of the curve or straight line as well as the headings at the beginning and end of the curves. From this, we calculate the points and headings in a similar manner to the loitering task in which we determined exit points and headings based on the entry point and the distance traveled on the loitering circle. Once this has been done the wind must be broken down into its components, u and v, in order to determine the wind acting along each of the three segments. The example in Figure 6.9 in which the UAV is flying at a 45° heading and the wind is acting at a 270° heading shows how we determine the components.

The following equations calculate the u and v components of the wind that are acting on the UAV. Using this for each of the headings of the three segments, the fuel burn rate model determines the approximate fuel burn rate over the segments. A weighted average then determines the average fuel burn rate over the entire path based on the lengths of each of the three segments.

$$u = speed_{wind} \times cos\left(heading_{wind} - heading_{UAV}\right) \tag{6.19}$$

$$v = speed_{wind} \times sin\left(heading_{wind} - heading_{UAV}\right) \tag{6.20}$$

Now, every time a flight time approximation is computed in the Java program, an instance of the segmented Dubins curve path is created for the i, j, k combination. At that point, the model uses the method that computes the fuel burn rate approximation of the path. This allows the mission planning models to plan missions effective, based on time and endurance of the UAV.

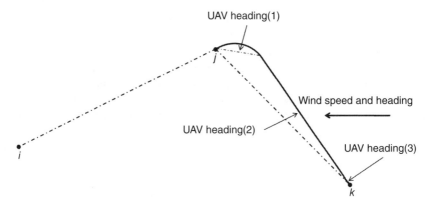

Figure 6.8 Breaking the flight path from j to k into linear segments. *Source*: Henchey, Matthew J.; Batta, Rajan; Karwan, Mark; Crassidis, Agamemnon, "A Flight Time Approximation Model for Unmanned Aerial Vehicles: Estimating the Effects of Path Variations and Wind," *MORS Journal*, Vol. 19, No. 1, 2014, John Wiley and Sons, LTD.

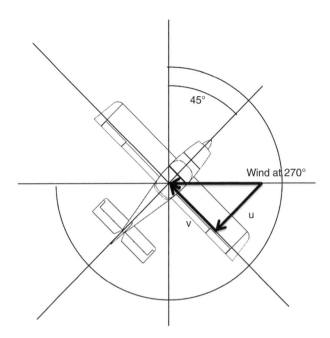

Figure 6.9 Determining the components of wind acting on the UAV. *Source*: Henchey, Matthew J.; Batta, Rajan; Karwan, Mark; Crassidis, Agamemnon, "A Flight Time Approximation Model for Unmanned Aerial Vehicles: Estimating the Effects of Path Variations and Wind," *MORS Journal*, Vol. 19, No. 1, 2014, John Wiley and Sons, LTD.

6.7 Computational Expense of the Final Model

The final flight time approximation model is able to produce flight times for thousands of $i, j,$ k combinations in a real-time scenario. Furthermore, it is able to include practical surveillance tasks that a UAV may be required to perform in a mission setting. Finally, it contains a method accounting for the effects of wind on the UAV. In this section, we present a model runtime analysis along with a comparison of "actual" and approximated flight times.

6.7.1 Model Runtime Analysis

For the model runtime analysis we considered a set of 10, 20, and 30 waypoints that would require 720, 6840, and 24 360 i, j, k combinations, respectively. As stated previously, the radius of sight, radius of avoidance, and loitering tasks require additional iterations of the flight time approximation model. This determines the expected, minimum, and maximum flight times. We use a comma-separated file to enter each of the waypoints such as the example in Table 6.2.

The example considers 30 waypoints distributed over an area of approximately 60 miles by 60 miles. For all cases, the fuel burn rate model that calculates the weighted average fuel burn rate for the segmented Dubins curve path uses a wind speed of 2 ft/s and wind heading of 270°. All of the assumptions are followed concerning the heading of the UAV into j and k of the $i, j, k,$ combination, calculating expected flight times for the radius of sight, radius of avoidance, and loitering tasks.

Table 6.3 presents the breakdown of the task types considered for each of the three cases. Table 6.4 is an example of the database that is output from running the model.

Table 6.5 presents the runtime results after running the flight time approximation for each of the three cases. The cases when there were 10 and 20 waypoints clearly were able to meet the requirements set forth in the problem statement. As the number of waypoints increases to 30 waypoints, the flight time approximation begins to take longer to compute all of the combinations. This is a very large improvement over the Dubins aircraft model that required over 500 s computing 720 combinations. Furthermore, the final flight time approximation model outperforms the MATLAB version that had no additional task types or fuel burn rate model implemented. In the 30-waypoint case the flight time approximation model is actually being run more than 365 000 iterations of snapshot tasks if each radius of sight, radius of avoidance, and loitering task is considered to be four separate waypoints.

Table 6.2   Waypoint data used in the flight time approximation model.

#	Longitude	Latitude	Radius	Heading	Resource	Task	Duration
1	−117.03967	35.43222	0	0	PIONEER	Snapshot circle	0
2	−117.46150	35.55427	2000	0	PIONEER	Loitering path	500
...	...	...	...	...	...	...	...
9	−117.6517	35.61547	2000	0	PIONEER	Radius of avoidance	0
10	−117.0562	35.84219	0	0	PIONEER	Snapshot	0

Table 6.3   Summary of the cases used in the runtime analysis.

Tasks	10 waypoints	20 waypoints	30 waypoints
Snapshot tasks	4	8	12
Radius of sight tasks	2	4	6
Radius of avoidance tasks	2	4	6
Loitering tasks	2	4	6

Table 6.4   Waypoint data obtained from the flight time approximation model.

i	j	k	Flight time (s)	Fuel burn rate (gal/h)
1	2	3	749.7920269	3.028399503
1	2	4	523.4798975	3.028823362
1	2	5	907.2068602	3.032716132
...	...	...	...	...
10	9	6	1319.96323	3.031289848
10	9	7	664.637497	3.031411687
10	9	8	1628.940727	3.031256851

Table 6.5 Runtime analysis results for all three cases.

Case	Combinations	Runtime results (seconds)
10-waypoint case	720	0.334
20-waypoint case	6 840	2.036
30-waypoint case	24 360	6.786

Considering the added level of complexity, the final flight time approximation model with additional task types and fuel burn rate model is able to quickly calculate thousands of i, j, k combinations in a time span that meets our requirements. It is likely that in a scenario like the 30-waypoint case, the number of combinations can be limited, as we have previously discussed. Incorporating these flight time and fuel burn rate approximations in the mission planning or dynamic resource management model both will be able to aid decision makers in managing resource assignments more effectively. The following section will demonstrate an example of a small mission and look at the actual flight time approximations calculated.

6.7.2 Actual versus Expected Flight Times

Using an eight-waypoint scenario along with a resource base as the example for this section, a small mission plan was created to demonstrate how the expected flight time approximations compared to the flight time approximation of what the UAV is most likely to do in order to complete the mission as quickly as possible. We assumed there was a runway at the base in which the UAV must take off at a heading of 270° (East) and land at a heading of 270° (East). We do not consider acceleration and deceleration at the runway. For this example, there were two waypoints for each of the four task types: the snapshot task, the radius of sight task, the radius of avoidance task, and the loitering task. The radius of sight and radius of avoidance task both had a radius of 2000 ft (0.38 miles), as did the loitering task, which also had duration of 500 s. We set the UAV to maintain airspeed of 167.86 ft/s (114.45 mph) in the presence of a 2 ft/s (1.36 mph) wind with a heading of 270° (East). Running the flight time approximation model for the entire example, we extracted the flight time and fuel burn rate data for the i, j, k combinations needed for the mission plan. This included the expected, minimum, and maximum flight times also for the radius of sight, radius of avoidance, and loitering tasks. Figure 6.10 demonstrates the mission example with the UAV flying the best path based on the i, j, k flight time approximations. Note that waypoints one and six are radius of avoidance tasks, two and five are snapshot tasks, three and eight are loitering tasks, and four and seven are radius of sight tasks. This example demonstrates the similarity in the radius of avoidance and loitering tasks. Upon closer examination, one can see that the loitering tasks do in fact loop around the entire loitering circle while the radius of avoidance tasks immediately begin to move on to the next waypoint.

Table 6.6 presents the results from this example. The "actual" time of arrival and fuel burn rates are the approximations if the UAV had chosen the best path from j to k for each i, j, k combination in our flight time approximation model as were shown in Figure 6.10. Although it appears that the "actual" time of arrival is not far from the minimum time of arrival calculated from the radius of sight, radius of avoidance, and loitering tasks it is important to remember

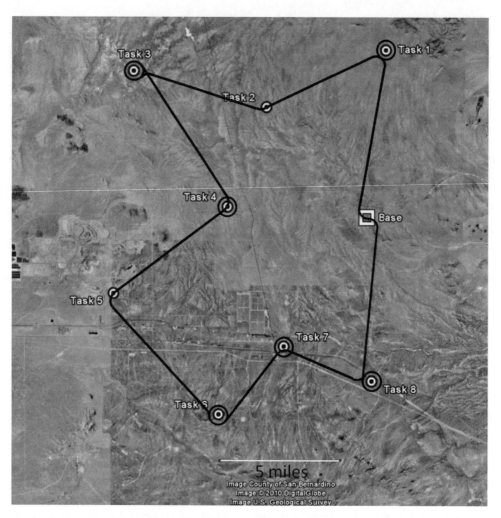

Figure 6.10 An eight-waypoint mission with a resource base. *Source*: Henchey, Matthew J.; Batta, Rajan; Karwan, Mark; Crassidis, Agamemnon, "A Flight Time Approximation Model for Unmanned Aerial Vehicles: Estimating the Effects of Path Variations and Wind," *MORS Journal*, Vol. 19, No. 1, 2014, John Wiley and Sons, LTD.

that the loitering task is the task that influences the differences. This is because we choose the best path for entering the loitering task, leaving the loitering path is completely dependent upon the duration of loitering. If more of the tasks had been loitering tasks it is likely there may be an even greater difference between the "actual" and the minimum. It is also important to note that we add the 500 seconds for the duration of the loitering tasks into the time of arrival as well.

Another important aspect of the results is the difference between the average fuel burn rate calculated from the approximations and the fuel burn rates from the "actual" best paths. In most cases, the two rates are nearly identical. In other cases, the best path may have had a greater head wind or a greater tail wind than the average of the other paths. Overall the average

Table 6.6　Results of the simple mission plan with flight times and fuel burn rates.

i	j	k	Time of arrival at k (seconds)				Fuel burn rate (gal/h)	
			Actual	Minimum	Mean	Maximum	Actual	Average
0	0	1	215.2366	215.2366	217.6190	221.4823	3.0543	3.0545
0	1	2	379.5656	379.5656	403.4974	429.2611	3.0576	3.0287
1	2	3	550.9886	550.9887	577.1532	606.0333	3.0555	3.0554
2	3	4	1258.132	1248.779	1295.487	1397.341	3.0216	3.0270
3	4	5	1447.890	1425.298	1523.677	1675.780	3.0155	3.0327
4	5	6	1654.497	1631.905	1733.427	1890.438	3.0549	3.0557
5	6	7	1796.655	1734.265	1863.572	2053.454	3.0474	3.0306
6	7	8	1918.383	1837.735	1981.639	2185.569	3.0148	3.0143
7	8	0	2646.792	2553.755	2704.036	2913.979	3.0524	3.0352

fuel burn rates are able to give a good approximation of what can be expected based on the location of j and k of the i, j, k combination.

The approximation model provides the ability to choose a more liberal approximation using the minimum flight or a more conservative approximation using the maximum flight time. Using the mean of the flight time approximations provides a compromise between the minimum and maximum for a set of the best possible flight paths. In doing so this also provides the mission planning [1] and dynamic resource management [2] models the ability to use this knowledge determining how likely it is a UAV will either arrive too early or too late to begin its task and will be able to more effectively manage the UAV assignments.

6.8　Conclusions and Future Work

The final approximation model, including surveillance tasks and fuel burn rates, is able to meet the requirements as presented in the problem statement of this chapter. Creating a model to use in mission planning [1] and dynamic resource management [2] models to allow them to effectively and efficiently route and reroute UAV resources was our initial goal. Prior to this work, using Euclidean distances did not consider the flight kinematics of a UAV or the effects of wind on fuel burn rates and flight time. By using the Dubins set, an efficient approximation model, we developed a model that accounts for the idealized flight kinematics of a UAV and provides a large number of flight time approximations within seconds. Although the developed model met the goals, there is still work required to extend this model.

Complexity constraints limit the current mission planning and dynamic resource management models to using the i, j, k triplet. If these models could instead include more waypoints for each combination, it may be possible to optimization of the flight times for larger sets of waypoints. Savla *et al.* [17] and Kenefic [18] have begun to do this using a model similar to the Dubins aircraft model. Savla *et al.* used an algorithm called the alternating algorithm to force the UAV to fly at a straight-line heading between the two waypoints between every other pair of waypoints. Although the main goal was determining bounds on the traveling salesperson problem (TSP), they demonstrate the ability to determine flight paths that minimize the overall flight time of a mission. Kenefic [18] takes this work even further and uses particle swarm optimization (PSO) to optimize the headings at each of the waypoints to minimize the overall flight time. The PSO uses the alternating algorithm as an initial starting point and then optimizes the headings at each of the waypoints to smooth out the entire flight path so that there are as few large turns or complete loops as possible. Although these approaches use computationally expensive simulation models to determine the flight path of the UAV, using our flight time approximation model in the optimization routine would provide a faster solution. During iteration, it would not be necessary to compute the entire path, instead determining the necessary flight paths at the end after finalizing the optimal headings. This not only saves computation time but also allows the UAV to save time and fuel given that time windows were still satisfied.

Myers *et al.* [12] also discuss the possibility of using an i', i, j, k combination. In their model, which determines shortest paths between waypoints in the presence of obstacles, they claim that this larger combination set allows for more accurate flight time approximations. This is a result of dropping some of the necessary assumptions, most importantly the assumption of the heading at waypoint i. Although this idea would be beneficial to explore, further research into resource management models is needed, as using i', i, j, k combinations would make current models intractable in a real-time situation.

Although we used the Dubins set from [5] as the basis for the approximation model, additional approaches for estimating flight times may provide better results. For our approximation model, using the minimum, maximum, and mean flight times obtained from the surveillance tasks together may provide a better approximation. For example, a regression equation that uses the three approximated flight times to match the "actual" flight time in the mission plan example could be developed. Furthermore, it may be important with continuing development of mission-planning models to consider confidence intervals for the flight time approximations. The use of confidence intervals for the flight time approximations allows optimization models to solve the VRPTW with a greater sense of the likelihood that a UAV will be able to arrive within the time windows. Üster and Love [19] show how to develop these confidence intervals for estimated distances. Although their work focuses mainly on determining confidence intervals for estimated distances in road networks it may be possible to extend this work for use in flight time approximations. This was not possible in developing the flight time approximation model in this work as access to actual UAVs was not possible.

Finally, conducting further research using the fuel burn rates can aid in minimizing the overall cost of a mission. Optimizing fuel consumption while a UAV is in the presence of wind is possible over the flight path of the UAV. The current planning models use fuel consumption only as a constraint. Finding optimal airspeeds between two waypoints rather than assuming constant airspeed is possible through optimization routines. Although certain airspeeds showed the UAV entering a waypoint too soon given the required time window, slower airspeeds would enable the UAV to complete the task, allowing us to determine additional and possibly better mission plans. Harada and Bollino [20] show in the case of a periodic circling flight in constant wind, the circle can be broken into maximum thrust and minimum thrust arcs, or "boost arc" and "coast arc" respectively. This demonstrates an improved fuel burn rate, the UAV beginning the coast arc at the point in which the wind becomes a tail wind. This also requires the use of additional previous points in order to know the exact heading before the UAV flies from j to k in the i, j, k combination, making the i', i, j, k combination a valuable future research area.

Acknowledgments

The authors would like to thank the two anonymous reviewers from the *Military Operations Research Journal*. The two reviewers provided us with excellent advise and guidance in order to strengthen and improve our paper. The Office of Naval Research Optimization Planning and Tactical Intelligent Management of Aerial Sensors (OPTIMAS) grant N00173-08-C-4009 and Air Force Research Laboratories Platform and Data Fusion Technologies for Cooperative ISR, through IAVO, grant FA9550-09-C-0066, PO#30,40 provided funding for this research.

References

1. Mufalli, F., Batta, R., and Nagi, R. (2012). Simultaneous sensor selection and routing of unmanned aerial vehicles for complex mission plans. *Computers and Operations Research*, **39**(11):2787–2799.
2. Murray, C. and Karwan, M. H. (2010). An extensible modeling framework for dynamic reassignment and rerouting in cooperative airborne operations. *Naval Research Logistics*, **57**(3):634–652.
3. Weinstein, A. L. and Schumacher, C. J. (2007). UAV Scheduling via the Vehicle Routing Problem with Time Windows. Technical Report AFRL-VA-WPTP-2007-306, Air Force Research Laboratory.
4. Grymin, D. J. and Crassidis, A. L. (2009). Simplified Model Development and Trajectory Determination for a UAV using the Dubins Set. In *AIAA Guidance, Navigation, and Control Conference and Exhibit*, Montreal, Canada.
5. Shkel, A. M. and Lumelsky, V. (2001). Classification of the Dubins set. *Robotics and Autonomous Systems*, **34**:179–202.
6. Dubins, L. E. (1957). On curves of minimal length with a constraint on average curvature, and with prescribed initial and terminal positions and tangents. *American Journal of Mathematics*, **79**(3):497–516.
7. Jeyaraman, S., Tsourdos, A., Zbikowski, R., White, B., Bruyere, L., Rabbath, C.-A., and Gagnon, E. (2004). Formalised Hybrid Control Scheme for a UAV Group Using Dubins Set and Model Checking. In 43rd IEEE Conference on Decision and Control, Atlantis, Paradise Island, Bahamas.
8. Enright, J. J. and Frazzoli, E. (2006). Cooperative UAV Routing with Limited Sensor Range. In AIAA Guidance, Navigation, and Control Conference and Exhibit, Keystone, Colorado.
9. Sujit, P. B., Sinha, A., and Ghose, D. (2006). Multiple UAV Task Allocation Using Negotiation. In Proceedings of Fifth International Joint Conference on Autonomous Agents and Multiagent sytems, Tokyo, Japan.
10. Schumacher, C. J., Chandler, P., Pachter, M., and Pachter, L. (2007). Optimization of air vehicles operations using mixed-integer linear programming. *Journal of the Operational Research Society*, **58**:516–527.
11. Schouwenaars, T., Valenti, M., Feron, E., and How, J. (2005). Implementation and Flight Test Results of MILP-based UAV Guidance. In Aerospace Conference, 2005 IEEE, pages 1–13, Big Sky, MT.
12. Myers, D., Batta, R., and Karwan, M. (2010). Calculating Flight Time for UAVs in the Presence of Obstacles and the Incorporation of Flight Dynamics. In INFORMS Annual Meeting, Austin, Texas.
13. Ceccarelli, N., Enright, J. J., Frazzoli, E., Rasmussen, S. J., and Schumacher, C. J. (2007). Micro UAV Path Planning for Reconnaissance in Wind. In Proceedings of the 2007 American Control Conference, pages 5310–5315, New York.
14. Rasmussen, S. J., Mitchel, J. W., Chandler, P. R., Schumacher, C. J., and Smith, A. L. (2005). Introduction to the MultiUAV2 Simulation and its Application to Cooperative Control Research. In Proceedings of the 2005 American Control Conference, Portland, OR.
15. McNeely, R. L., Iyer, R. V., and Chandler, P. (2007). Tour planning for an unmanned air vehicle under wind conditions. *Journal of Guidance, Control, and Dynamics*, **30**:629–633.
16. Crassidis, A. L. and Crassidis, J. L. (2010). Personal communication with Agamemnon and John Crassidis for advice on the effects of wind on a UAV's fuel burn rate.
17. Savla, K., Frazzoli, E., and Bullo, F. (2005). On the point-to-point and traveling salesperson problems for Dubins vehicle. In American Control Conference, volume **2**, pages 786–791, Portland, Oregon.
18. Kenefic, R. J. (2008). Finding good Dubins tours for UAVs using particle swarm optimization. *Journal of Aerospace Computing, Information, and Communication*, **5**:47–56.
19. Üster, H. and Love, R. F. (2003). Formulation of confidence intervals for estimated actual distances. *European Journal of Operations Research*, **151**:586–601.
20. Harada, M. and Bollino, K. (2008). Minimum Fuel Circling Flight for Unmanned Aerial Vehicles in a Constant Wind. In AIAA Guidance, Navigation, and Control Conference and Exhibit, Honolulu, Hawaii.

7

Impacts of Unmanned Ground Vehicles on Combined Arms Team Performance

Fred D. J. Bowden, Andrew W. Coutts, Richard M. Dexter, Luke Finlay, Ben Pietsch, and Denis R. Shine
Land Capability Analysis Branch, Joint and Operations Analysis Division, DSTO-E, Edinburgh, SA, Australia

7.1 Introduction

Land Capability Analysis Branch (LCA) of Australia's Defence Science and Technology Organisation (DSTO) uses analytical campaigns [1, 2] as a means to explore key Australian Army modernisation issues [3]. These campaigns span problems such as: major systems procurement [4, 5]; changing structures; and developing future concepts [6] as well as how these areas are sensibly brought together [7, 8].

The key consideration in such campaigns is being able maintain rigorous analysis and relevance to the real world. That is, how are the conflicting paradigms of internal and external validity balanced to provide confidence in the final answer(s)?

As part of one such campaign into the Australian Army's future requirements, LCA conducted an exploratory analysis on the impact of unmanned ground vehicles (UGVs) in combination with different types of manned fighting vehicles within different war-fighting contexts. The intent was to consider whether there was any evidence that a theoretical UGV capability could contribute to operational performance of a Combined Arms Team (CAT) and identify possible areas warranting further investigation. To do this, a multi-method analytical campaign was constructed to determine the potential operational impacts of equipping a CAT with options based on mixes of these manned vehicles and UGVs. The methods included within the campaign were analysis of real events, human-in-the-loop (HIL) simulation and closed-loop (CL) simulation.

Operations Research for Unmanned Systems, First Edition. Edited by
Jeffrey R. Cares and John Q. Dickmann, Jr.
© 2016 John Wiley & Sons, Ltd. Published 2016 by John Wiley & Sons, Ltd.

Careful selection of the types of manned vehicles and UGVs equipping the CAT options tested in the campaign allowed not only the impact of UGVs to be evaluated, but also the possible trade-offs between the quality of manned combat vehicles and level of UGV support. In total six different CATs were considered: a medium weight manned combat vehicle with 'no', light and heavy UGV options, and light and heavy vehicle options with the light UGV option which were compared to a baseline reflecting vehicles similar to those in the Australian Army.

This chapter describes the analytical approach used to explore the operational impact of UGVs and their ability to supplement manned vehicles in complex war-fighting environments. The chapter provides insights gained from the study but focuses on how the approach used can be improved for future studies.

7.2   Study Problem

The Australian Army's Future Operating Concept states that future land force will be 'optimised for close combat in the complex terrain' [9]. With the ever increasing ability of ground unmanned systems it is important that the Australian Army understands how such systems might improve its ability to conduct close combat. However, in the medium term this is unlikely to simply be a replacement of manned with unmanned systems. What is more likely, as is current the case with unmanned aerial vehicles (UAVs), is a mix of both manned and unmanned systems. Thus, it is important to think about how unmanned systems can complement manned systems with different capabilities. So, it is not simply a matter of looking at the capabilities of different UGVs to determine what roles they might fulfil. Rather there is a need to begin to understand the operational impact such systems will have within a CAT conducting a close combat mission.

Within this study there were two independent variables considered. These were the complexity of the terrain and the impact of the manned and unmanned vehicle options. There were two different terrains considered and six combinations of vehicle options. The other variables, including the Opposition Force (OPFOR), other parts of the Experimental Force (EXFOR) and the mission, were kept constant.

7.2.1   Terrain

The first meta-trend of the Australian Army's Future Land Warfare Report [10] describes the ongoing urbanisation of the environment in which land forces need to operate. Given this, both terrain types considered within the study were urban based with changes in the density of the buildings being encountered by the EXFOR as the principal difference.

The terrain used for the first part of the study was based on a town that was the site of a major Australian battle during the Vietnam War in 1969. The urban centre comprised a series of small, single-storey houses with long, wide streets providing relatively long lines of sight and was identified as being a representative of a low density urban environment (Figure 7.1). This was used in the initial study as there was a direct link to a real event involving Australian forces and provided excellent external validity as results could be compared to historical studies, see [5, 11].

The second terrain used was based on a Middle East Representative Town, similar to what was encountered in combat operations during the Iraq War in 2004. This terrain is characterised by dense, multi-storey dwellings with very restricted lines of sight (Figure 7.2).

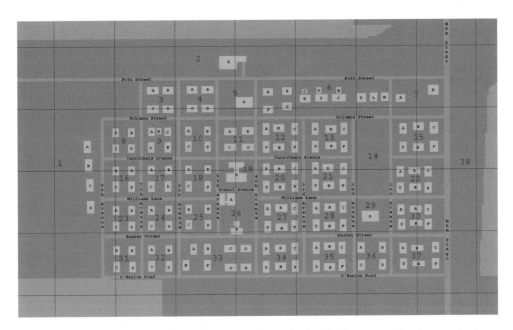

Figure 7.1   Low density urban environment. (*For color detail, please see color plate section.*)

Figure 7.2   High density urban environment. (*For color detail, please see color plate section.*)

Both terrain types are seen as being likely to occur within future areas of operation for the Australian Army.

7.2.2   Vehicle Options

The aim of this study was to explore not only the impact of unmanned vehicles but also how they might complement different types of manned vehicles. So the options considered included variations in both manned and unmanned vehicles. There were four manned vehicle options considered:

- Similar vehicles to those being used by the Australian Army (the baseline);
- A Heavy vehicle option including heavy tanks and infantry fighting vehicles (IFVs) with similar protection to that of a main battle tank;
- A Medium vehicle option with heavy tanks and medium IFV; and
- A Light vehicle option with medium tanks[1] and IFV.

There were two UGV options considered:

- A small UGV similar to the iRobot PackBot UGV, with a limited capability optical sensor and no armour.
- A large UGV similar to the Armed Robotic Vehicle[2] armed with a 30mm cannon and protected against small arms fire.

These two capability dimensions provide for a two-dimensional set of nine scenarios, plus the baseline. However, time constraints during the activities allowed only five of the nine options to be studied; these are highlighted in Table 7.1. Although all five options were studied in both terrain sets the baseline was only considered in the low density terrain reducing the total options considered to 11.

7.2.3   Forces

To reduce the number of independent variables the forces of both sides were kept constant, with the exception of EXFOR manned and unmanned vehicle variations.

Table 7.1   Experimental options, with the options explored highlighted.

		UGV options		
		A. No UGV	B. Light UGV	C. Heavy UGV
Vehicle options	1. Light option	A1	B1	C1
	2. Medium option	A2	B2	C2
	3. Heavy option	A3	B3	C3
	4. Baseline	BASE		

---

[1] The medium tank was a medium IFV possessing a 120mm main armament.
[2] At the time of this study, the Armed Robotic Vehicle was being considered by the US Army as part of the Future Combat System (FCS) program. It has since been cancelled.

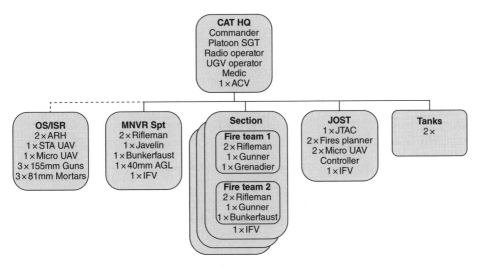

Figure 7.3   Experimental force.

7.2.3.1   Experimental Force

The EXFOR CAT was built around a mechanised infantry platoon[3] which consisted of three infantry sections, each of which was divided into two fire teams of four. In addition, it contained an attached tank section (two tanks), manoeuvre support section and Joint Offensive Support Team (JOST). In support, but not directly under the control of the commander, were offensive support (OS) and the Tier 2 UAV used for intelligence surveillance and reconnaissance (ISR). The UGV options considered were placed directly under the control of the sections except for the heavy variant which was under the command of the commander. The force is shown in Figure 7.3.

7.2.3.2   Opposition Force

The OPFOR represents a complex enemy and comprises two main elements: the conventional element and the unconventional, insurgent, element. These two OPFOR elements contain no uniform command element and as such operate mostly independently with little cross communication (Figure 7.4).

The conventional elements were represented as a military style conventional force and, as such, fought to prevent EXFOR achieving its objectives. However, they also aimed to preserve themselves should the situation become untenable. Insurgents represented a religious-led force and had the same objective as the conventional force but without concern about casualties.

7.2.3.3   Civilian Elements

Given the terrain was urban there was a need to also represent the civilian elements that could be encountered in the operational environment. This consisted of three types of civilians.

---

[3] These series of studies focused on a platoon-sized CAT which was referred to as a micro-combat team.

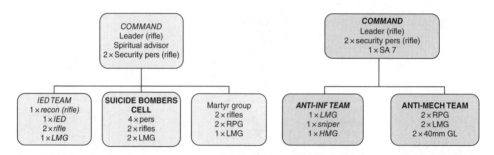

Figure 7.4 Opposition force.

- *Static Civilians*: 30 civilians who stayed in houses, representing people hiding from the fighting.
- *Mobile Civilians*: 20 civilians in 8 groups who moved around the village, representing people on urgent errands. The civilians were controlled by a set of rules designed to ensure that a similar number of civilians were moving within the environment throughout the duration of each wargame.
- *OPFOR/Sympathetic Civilians*: 8 civilians sympathetic to OPFOR, providing them with another situational awareness feed for locating EXFOR.

7.2.4   Mission

The mission used was a clearance operation. External support could also be called for in terms of OS and UAV ISR assets. The mission was to locate, seize and secure identified locations within the area of operation.

Three distinct phases were identified in the missions:

1. *ISR Phase*: A preliminary phase using ISR assets to search for OPFOR and engage identified targets.
2. *Break-In*: The initial penetration of IFVs and tanks into the urban area. This phase was considered complete after the first house of the clearance was secured.
3. *Clearance*: This phase is complete after the last house has been cleared.

As expected the detailed schemes of manoeuvre within each of these phases varied depending on the option being considered. In regard to the employment of the UGV assets it was identified that they were used to perform the following tasks:

- Reconnaissance during ISR phase;
- Securing movement routes;
- Forward surveillance;
- Forward observer for indirect fire;
- Overwatch;
- Fire support (Heavy UGV); and
- Flank guard (Heavy UGV).

7.3 Study Methods

In building an analytical campaign that balanced both internal and external validity [3] two key tools were chosen; HIL and CL simulation. These were complemented by real event analysis, in particular looking at the use of historical events to baseline the tools [5, 11], and after action reviews (AARs), used to explore how best to use the UGV capabilities being introduced. The use of historical analysis and AARs ensured the overall campaign retained its link to the real world (external validity) despite its heavy reliance on CL simulations to establish cause and effect (internal validity).

The study was run in two parts. The first considered the low density terrain, the second considered the high density terrain. Both parts used the same structure, which was developed and refined over a series of similar previous studies such as in [5, 12]. For each part a two-week event was held where LCA analysts were paired with appropriate Army subject matter experts (SMEs) to conduct a series of HIL simulations within a competitive, simulation-supported environment. These activities would start with a series of development runs to allow the participants to familiarise themselves with both the environment and the capabilities within the option.

The HIL simulation allowed the players to implement their mission plans in real time. This competitive environment accounted for the nature of warfare; as the Blue players implemented mission plans for each capability set, the Red players were actively attempting to defeat them, through mitigating strengths and isolating weaknesses of any capability set.

The outcome was the capturing of the 'best' Blue and Red plans for each capability set. The plans were then converted into CL simulations, allowing them to be executed within a Monte Carlo simulation process as defined in [13]. These simulation runs were built and executed in the weeks after the HIL simulation event. The overall process is shown in Figure 7.5.

7.3.1 Closed-Loop Simulation

The main analytical tool used within this study was a CL combat simulation. These simulations apply a range of models to simulate the various aspects of combat operations allowing opposing combat units to move and interact with each other in a 'realistic' manner without human players. The greatest strength of a CL simulation is its capacity to generate multiple replications (100+) of the research scenario (vignette) and hence large quantities of data for subsequent statistical analysis. This offers a high level of internal validity allowing the analyst to isolate and identify the cause and effect of specified outcomes. However, they have reduced external validity as the human decision-making aspects of the vignette are supplied through a series of limited models. This is why it is important to include other methods, such as HIL simulation and real-event analysis, with direct links to the CL simulation as part of the analytical campaign.

The tool used for this study allows for the ready transfer of what was done in the HIL simulation into the CL simulation. It also has the following advantages:

- The scale of the tool is on an engagement at lower than a company level.
- Offers excellent urban simulation capabilities consistent with the requirements of the environment, including detailed representations of buildings and destroyed buildings.
- Provides detailed results, enabling analysis of many aspects of the HIL simulation.
- The LCA branch contains the skilled staff necessary to develop and analyse the detailed scenarios required for this work.

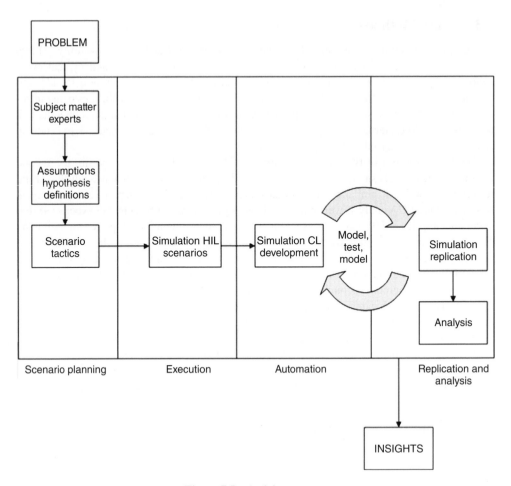

Figure 7.5   Activity process.

7.3.2   *Study Measures*

The study required a range of Measures of Effectiveness (MoE) in order to compare the operational performance of the capability options in a complex environment (high human and high physical complexity). However, as the Australian Army's Future Operating Concept stresses, there is a need to consider the broader impacts of actions than just focusing on the outcome of combat [9, 14, 15]. Thus, the measures developed for this study go beyond traditional combat outcomes to include the impact of the different options on the environment more generally. This extends the work of previous Army studies, such as [5, 6, 11, 12, 16–20] where the focus was primarily on the combat impact of given options. Therefore, the following categories of measures were incorporated into this study (note they are provided here in order of importance):

• *Mission Success*: Mission success was defined as completing the allotted clearance task without losing a critical asset, such as a tank, which could be exploited as part of an enemy's information operations campaign.[4] An additional mission success condition was that EXFOR did not become combat ineffective. Where combat ineffective was defined as losing a complete infantry section or two sections losing more than half of their initial strength.

- *EXFOR Casualties*: EXFOR casualties were defined by the casualties caused to EXFOR soldiers and excluded vehicle casualties. Vehicle casualties were part of the Mission Success and Other Manned EXFOR Vehicle metrics. This activity assumed that force preservation will remain a key performance measure for a future Australian combat force. Future contributions to coalition operations, such as the scenario used for this campaign, would likely increase the importance of minimising Australian casualties. The preservation of Australian forces remains a key performance measure for the future Australian combat force.

- *Civilian Casualties*: Civilian casualties were defined by the casualties caused to any of the three civilian groups represented within the scenario. As the scenario for this activity occurs within an urban environment, civilian casualties must be avoided. Successful counterinsurgency [21] requires building trust within local populations and this can be undermined through civilian casualty incidents caused by counterinsurgent forces.

- *Damage to Sensitive Buildings*: A key part of the player briefings was an emphasis on minimising damage to buildings in and around the objective, especially for designated sensitive sites – a school, a hospital, a religious building and the home of a religious leader.

- *OPFOR Casualties*: OPFOR casualties were defined as any casualties taken by either the regular or insurgent OPFOR elements. These were considered to be one group for this metric. In combating an insurgency, long-term success is unlikely to be achieved through killing insurgents [21]. While mission success for tactical operations may still require the ability to destroy an enemy, this cannot occur at the expense of high civilian and EXFOR casualties or significant infrastructure damage.

- *Non-mission Critical Manned EXFOR Vehicles*: Minimising the loss of IFVs was considered of equal importance to maximising OPFOR casualties. Unmanned vehicles were deliberately not included in the performance measures as the emphasis was on EXFOR employing the unmanned vehicles to improve their mission performance without increased risk of casualties.

- *Damage to Non-sensitive Buildings*: A key part of the player briefings was an emphasis on minimising damage to buildings in and around the objective, including buildings not designated as sensitive.

Based on the analyst's interpretation of the requirements for complex warfighting identified in the Future Land Operating Concept [9], weightings were developed for these sets of metrics. While stakeholders were informally consulted in relation to final weightings, no formal consultation occurred to establish these rankings. The highest weighted measures were mission success, EXFOR casualties, civilian casualties and damage to sensitive buildings. The lower weighted measures were OPFOR casualties, other manned EXFOR vehicles and damage to non-sensitive buildings. The highest measures were assigned a weighting of double that of the other measures in the multi-criteria performance calculations.

---

[4] Informational complexity is a characteristic of complex environments and refers to the prevalence of news media capable of monitoring, recording and reporting on operations from an enemy's perspective.

7.3.3   System Comparison Approach

There are many multi-criteria decision-making methods that are available to allow the comparison of options with multiple metrics [22]. Many of these allow for the aggregation of measures such as those presented in Section 7.3.2. However, for this study, what was found was that much of the data was not normally distributed,[5] making it difficult to apply many of these approaches. In the end, two different methods were combined to generate a statistical ranking of the options.

First a non-parametric statistical test was carried out, using the tool BReakpoint Analysis with Nonparametric Data Option (BRANDO) [23]. BRANDO uses the Kruskal–Wallis test [24] to statistically compare k data sets without assuming that the data is normal, non-skewed and does not contain outliers. For each test a null hypothesis stating that all k data sets are statistically the same and an alternate hypothesis stating that at least one had values that are systematically larger [25] were used. If there was insufficient evidence to reject the null hypothesis (for this study 95% confidence level was used, that is a p value of greater than 0.05) then the null hypothesis that there was no statistical difference between the samples was accepted. If there was a difference, then the breakpoint was calculated and recursively solved for the two subgroups independently. In this way it was possible to compare two sets of data and test whether, within a specified confidence level, one set of data had significantly higher, lower or similar values than the other. Thus, the options were ranked against each of the seven measures.

To aggregate these individual rankings the Multi-criteria Analysis and Ranking Consensus Unified System (MARCUS) [26] algorithm was applied. MARCUS takes the seven rankings, one for each measure, and combines these using the weightings described in Section 7.3.2 to generate an overall ranking of the options being considered.

7.4   Study Results

This section outlines the results of the study. It does this by first presenting results based on the EXFOR and OPFOR casualties providing a more traditional assessment of the options. The results using the full measures are then presented and compared to the casualty-only assessment. Finally the results of the two terrains are compared.

7.4.1   Basic Casualty Results

Here the outcomes of the study are presented based on the use of only the EXFOR and OPFOR casualty values. In aggregating these values equal weighting was given to both.

7.4.1.1   Low Density Urban Terrain Casualty Only Results

Tables 7.2 and 7.3 summarise the average casualty figures resulting from the CL simulations for each of the options. It is important to note in comparing the average values in these tables that any differences do not necessarily indicate they are statistically significant. Calculating confidence internals is complicated by the non-normal distributions of some of the results.

---

[5] Many of the distributions display skewness and contain outliers on a normal quantile plot.

Table 7.2 Summary of EXFOR replication casualties.

EXFOR casualties		UGV options		
		A. No UGV	B. Light UGV	C. Heavy UGV
Vehicle options	1. Light option		11.42	
	2. Medium option	12.78	11.51	5.78
	3. Heavy option		10.23	
	4. Baseline	13.205		

Table 7.3 Summary of OPFOR CL simulation casualties.

OPFOR casualties		UGV options		
		A. No UGV	B. Light UGV	C. Heavy UGV
Vehicle options	1. Light option		32.93	
	2. Medium option	31.14	34.47	33.78
	3. Heavy option		31.16	
	4. Baseline	29.625		

Table 7.4 Ranking of casualties by CL simulation.

Measure	Ranking
EXFOR casualties	C2 < B3 < B2 = B1 < A2 = BASE
OPFOR casualties	B2 > C2 > B1 > A2 > B3 > BASE
Overall casualties	C2 > B2 > B1 = B3 > A2 > BASE
	Best ⟶ Worst

Using the non-parametric approach described in Section 7.3.3, Table 7.4 provides the option rankings.

From these tables it can be seen that the following options achieved the best result for each of the casualty types based on average casualties alone:[6]

- *EXFOR*: the lowest average EXFOR casualties occurred with C2.
- *OPFOR*: the highest average OPFOR casualties occurred with B2.
- *Overall*: when combining EXFOR and OPFOR casualties, the best overall result was given by C2.

If only the UGV Augmentation Options (A2, B2 and C2) are compared it was found that:

- *EXFOR*: the lowest average EXFOR casualties occurred with C2.
- *OPFOR*: the highest average OPFOR casualties occurred with B2.
- *Overall*: when combining EXFOR and OPFOR casualties, the best overall result was given by C2.

---

[6] This means it is not possible to say that any differences are statistically significant.

If only the Vehicle Protection Options (B1, B2 and B3) are compared it was found that:

- *EXFOR*: the lowest average EXFOR casualties occurred with B3.
- *OPFOR*: the highest average OPFOR casualties occurred with B2.
- *Overall*: when combining EXFOR and OPFOR casualties, the best overall result was given by B2.

7.4.1.2   Dense Urban Terrain Casualty-Only Results

Tables 7.5 and 7.6 summarise the average casualty figures resulting from the CL simulation for the study options. Once more it is important to note in comparing the average values in these tables that any differences do not necessarily indicate they are statistically significant. Using the non-parametric approach described in Section 7.3.3, Table 7.7 provides the option rankings.

From this table, it can be seen that for the replications, the following options achieved the best result for each of the casualty types based on average casualties alone:[7]

- *EXFOR*: the lowest average EXFOR casualties occurred with C2.
- *OPFOR*: the highest average OPFOR casualties occurred with B3.

Table 7.5   Summary of EXFOR replication casualties.

EXFOR casualties		UGV options		
		A. No UGV	B. Light UGV	C. Heavy UGV
Vehicle options	1. Light option		14.32	
	2. Medium option	12.77	9.39	8.86
	3. Heavy option		11.15	

Table 7.6   Summary of OPFOR replication casualties.

OPFOR casualties		UGV options		
		A. No UGV	B. Light UGV	C. Heavy UGV
Vehicle options	1. Light option		28.49	
	2. Medium option	29.27	25.12	27.36
	3. Heavy option		29.54	

Table 7.7   Ranking of casualties by replications.

Measure	Ranking
EXFOR casualties	C2 < B2 < B3 < A2 < B1
OPFOR casualties	B3 > A2 > B1 > C2 > B2
Overall casualties	B3 = C2 > A2 > B1 = B2
	Best ⟶ Worst

---

[7] This means we cannot say that any differences are statistically significant.

- *Overall*: when combining EXFOR and OPFOR casualties it was not possible to statistically separate B3 and C2.

If only the UGV Augmentation Options (A2, B2 and C2) are compared it was found that:

- *EXFOR*: the lowest average EXFOR casualties occurred with C2.
- *OPFOR*: the highest average OPFOR casualties occurred with A2.
- *Overall*: When combining EXFOR and OPFOR casualties, the best overall result was given by C2.

If only the Vehicle Protection Options (B1, B2 and B3) are compared it was found that:

- *EXFOR*: the lowest average EXFOR casualties occurred with B2
- *OPFOR*: the highest average OPFOR casualties occurred with B3.
- *Overall*: When combining EXFOR and OPFOR casualties, the best overall result was given by B3.

7.4.2 Complete Measures Results

In this section the outcomes of the study using all seven measures are presented. In aggregating these values to get the overall rankings of options the weightings used are as given in Study Measures.

7.4.2.1 Low Density Urban Terrain Results

In a low density urban environment the overall rankings found for all of the metrics considered are shown in Table 7.8. Note that in the low density terrain study, the CL simulation was not able to provide the building damage measures or the mission success 2 measure.

Table 7.8 Option ranking by single measures for low density urban terrain.

Measure	Weight	Ranking
EXFOR casualties	4	C2<B3<B2=B1<A2=BASE
OPFOR casualties	2	B2>C2>B1>A2>B3>BASE
Civilian casualties	4	B1<B3=BASE<C2<B2<A2
No mission-critical manned EXFOR vehicles	2	B3=B2<B1=A2=C2<BASE
Damage to sensitive buildings	4	NA
Damage to non-sensitive buildings	2	NA
Mission success 1[a]	4	A2<B1=B3=BASE<B2<C2
Mission success 2[b]	4	NA
		Best ──────────────→ Worst

[a] Mission failure occurs only if a critical vehicle is lost.
[b] Mission failure occurs if overall losses make CAT combat ineffective.

By using this ranking data and the MARCUS method it was found that the overall ranking of the options was as shown in Table 7.9.

Using the results from the MARCUS data it can be concluded that:

- The best-performing UGV augmentation option was C2 and B2–the medium manned vehicle and the heavy or light UGV options, respectively.
- The best-performing vehicle protection option was B1 and B3–the light or heavy manned vehicle option, respectively and the light UGV option.
- The best-performing options overall were B3–the heavy vehicle protection and light UGV augmentation option–and B1–the light vehicle protection and light UGV augmentation option.
- The worst-performing options overall were A2–medium vehicle protection and no UGV augmentation–and the baseline option–baseline vehicles and no UGVs.

7.4.2.2 Dense Urban Terrain Results

In a high density urban environment overall rankings of the options for each measure are given in Table 7.10.

By taking this ranking data and applying the performance ranking method (MARCUS) it was found that the overall ranking of the options was as shown in Table 7.11.

From this result it can be concluded that:

- The best-performing manned vehicle options were B2–the medium vehicle protection vehicle and the light UGV augmentation option.

Table 7.9 Options ranked by complete measures for low density urban terrain.

				Ranking				
Low density urban	B1	=	B3	>	B2	=	C2	>	A2=BASE
			Best			→Worst			

Table 7.10 Ranking by single measures for dense urban terrain.

Criterion	Weight	Ranking
EXFOR casualties	4	C2<B2<B3<A2<B1
OPFOR casualties	2	A2=B3>B1>C2>B2
Civilian casualties	4	B1=B2>A2>C2>B3
No mission-critical manned EXFOR vehicles	2	B2=C2<B1=B3<A2
Damage to sensitive buildings	4	B1=C2<A2<B2<B3
Damage to non-sensitive buildings	2	C2<B2<B1<A2<B3
Mission success 1[a]	4	B2>B1=C2>A2=B3
Mission success 2[b]	4	B2=C2>A2=B1=B3
		Best ——————→ Worst

[a] Mission failure occurs only if a critical vehicle is lost.
[b] Mission failure occurs if overall losses make CAT combat ineffective.

Table 7.11 Options ranked by complete measures for dense urban terrain.

	Ranking								
High density urban	B2	>	B1	=	C2	>	A2	=	B3
	Best ───────────────────────────────▶ Worst								

Table 7.12 Performance ranking using complete measures.

	Ranking								
Low density urban	B1	=	B3	>	B2	=	C2	>	A2
High density urban	B2	>	B1	=	C2	>	A2	=	B3
	Best ───────────────────────────────▶ Worst								

- The best performing UGV augmentation options were B2 – the light UGV augmentation option with medium vehicle protection vehicles – and C2 – the heavy UGV augmentation option with medium vehicle protection vehicles.
- The best-performing options overall were B2 – the medium vehicle protection vehicle and the light UGV augmentation option
- The worst performing options overall were A2 – the medium vehicle protection vehicle and the no UGV augmentation option – and B3 – heavy vehicle protection vehicle and the light UGV augmentation option.

We also observed similar trends between low to high density environments with the impact of vehicle option characteristics and performance – that is both environments tended to favour high lethality options with some level of UGV augmentation – with the exception that higher survivability may have compensated for reduced lethality in the low density urban terrain study. However, this is based solely on qualitative comparisons of the data and should be treated with a lower confidence.

7.4.2.3 Comparison of Low and High Density Urban Results

The implicit null hypothesis for this campaign was that the level of urban complexity would not affect the relative performance, and hence the ranking of the vehicle options. To do this the two experimental performance outcomes, calculated using the MARCUS multi-criteria ranking method (Table 7.12), were compared.

Clearly, the results are different based on the rank order of the options. However, this provides a very shallow view of the impact of more complex urban terrain on combat performance. In order to understand more about the impact of complex terrain on different types of vehicles, it may be useful to examine the relationship of different vehicle characteristics and the environment. These generic experimental options were originally established to represent different levels of capability in three dimensions: lethality, survivability and the

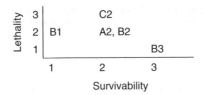

Figure 7.6   Plot of ordinal rankings of key manned vehicle characteristics.

level of UGV augmentation. If we explicitly identify these characteristics for each experimental option, we may be able to more effectively establish trends in the data. The relative level of UGV augmentation is independent of the manned vehicles in use and is already established in the experimental design (UGV option) as none, light and heavy. Similarly, survivability has also been established qualitatively in the experimental design (vehicle protection option) and was based on the mix of hull types used in each option. Hence B3 was considered a high survivability (heavy protection option) with a value of 3, B1 was a low survivability option with a value of 1 and the remainder were deemed to have medium survivability with a value of 2. What remains to support our analysis is to characterise the lethality of the options.

Figure 7.6 displays a plot of lethality against survivability.[8] The qualitative values for lethality are based on the following assumptions:

- The option (C2) that included the 120mm cannon (tank), the 35mm cannon (IFV) and the 30mm cannon (heavy UGV option) was considered the most lethal and assigned a value of 3.
- The options that included the 120mm cannon and the 35mm cannon but not the 30mm cannon (A2, B1, B2) were considered medium lethality and assigned a value of 2.
- The option (B3) that included the 120mm cannon but did not include either the 35mm cannon or the 30mm cannon was considered to have a low lethality and assigned a value of 1.

By comparing the rankings in Table 7.12 we can see that, aside from option C2 being ranked in the middle and A2 being ranked last in both cases, change has occurred between the rankings for the low density terrain and the high density terrain. The key differences are:

- The low lethality, high survivability and light UGV option (B3) dropped from first in the low density terrain to third (last) place in the high density terrain.
- The medium lethality, medium survivability and light UGV augmentation option (B2) rose from second place (middle) in the low density terrain to first place in the high density terrain.
- The medium lethality, low survivability and light UGV option (B1), dropped from first place in the low density terrain to second place (middle) in the high density terrain.

If we consider the implications of these rankings, we could infer that the environmental characteristics of highly complex terrain appeared to favour higher lethality options – B2 followed closely by B1 and C2 – that include at least some level of UGV support. The ranking of C2 in second place, which included the armed UGV that increased both its lethality and

---

[8] The scale used here is ordinal as the numbers refer to relative SME/analyst rankings rather than actual calculated values of lethality and survivability.

level of UGV augmentation, appears to counter this trend. However, sensitivity calculations indicate that the ranking separation between B2 (first place) and B1 and C2 (equal second place) is relatively small, when compared to the separation of the second place options and the two last place options, and therefore the distinction between first and second place is less certain. There is no clear indication that performance is strongly linked to the level of UGV augmentation as evidenced by C2, with the highest level of augmentation being ranked second and last place being shared by A2, with no UGV augmentation and B3 with light UGV augmentation. The level of survivability of each option also did not appear to directly explain the performance given that the lowest survivability option (B1) was ranked second, while the highest survivability option (B3) was ranked last. We cannot, however, infer from these results that a high level of survivability was negative to performance; instead we might infer that increased survivability did not compensate for reduced lethality.

By comparison, the trends in the low density terrain are even less clear. While the three high lethality options (B1, B2 and C2) were ranked in the first two places, the low lethality option, B3, was ranked equal first. The impact of increased survivability is also unclear, with the highest survivability option (B3) being ranked equal first place with the lowest survivability option (B1). Possibly, increased survivability compensated more for the reduced lethality in the lower complexity of the low density terrain than the high density terrain. There does appear to be a trend in the relationship between performance and the level of UGV augmentation with all of the options ranked in first and second place including some level of UGV augmentation, and the high lethality, medium survivability and no UGV augmentation (A2) being ranked last.

Summarising the two experiments, the high density terrain outcomes tended to favour high lethality options with some level of UGV augmentation, while the low density terrain outcomes demonstrated a weaker trend towards high lethality and UGV augmentation. These observations do not, on their own, provide sufficient evidence to reject the experimental null hypothesis that the level of complexity does not affect the relative performance of the vehicle options. While clearly the rankings have changed, the overall trends of the performance of vehicle characteristics remain similar.

This discussion assumes that the vehicle characteristics were the primary discriminator between the various options, however, in reality a range of factors could have affected the performance. These include player learning and subsequent modifications to their approach to game play and differences in the military experience level of the different team of players used for the low density terrain and the high density terrain experiments.

7.4.3   Casualty versus Full Measures Comparison

As discussed earlier one of the directions of the Australian Army is towards a broader consideration of the battlefield environment to take into account not only EXFOR and OPFOR casualties. This study provides an illustration of the potential impact of such an approach. The results from both the casualty-only and the complete metrics rankings are given in Table 7.13. As can clearly be seen there are significant changes in these rankings. Consider, for example, the ranking of option B3, the heavy option, in the dense urban terrain. Based on casualty data alone this option was ranked as equal best. However, when considering the full suite of metrics it was ranked equal last. Further changes in rankings can be seen in the remainder of the data showing the importance of the metrics used.

Table 7.13 Performance rankings comparing casualty and complete measures.

		Ranking								
Casualty	Low density urban	C2	>	B2	>	B1	=	B3	>	A2
Only	Dense urban	B3	=	C2	>	A2	>	B1	=	B2
Complete	Low density urban	B1	=	B3	>	B2	=	C2	>	A2
	Dense urban	B2	>	B1	=	C2	>	A2	=	B3

Best ────────────────────────────→ Worst

7.5 Discussion

The challenges involved in studying the operational impact of a novel capability such as UGVs revolve around the need to manage internal and external validity of the analytical design. Trade-offs will necessarily occur with the scope of the study, the resources available and the desired fidelity of the results and the internal and external validity measures adopted in the design. This exploratory study was constrained in terms of: scope – it was exploratory in nature; and resources – there was limited access to SMEs. Additionally, the problem space was broader than more traditional capability analysis problems given that there were no equivalent operational examples on the employment of the UGVs in direct support of a CAT. Consequently more emphasis was required in considering the impact of the capability on friendly and enemy tactics. As a result, the analytical traded off validity for analytical breadth on these issues.

The HIL games are time consuming but provide a high level of engagement with SMEs and enable them through cycles of planning and executing these plans in an HIL to provide insights into how UGVs might be employed. While this necessarily has a confounding effect on the results from the narrow point of view of option comparisons, it strengthens the value of the activity to understand what the key variables may be and inform follow-on experimental design. Without this time for SMEs to 'play' with a novel capability in a realistic setting, with near real-time consequences against an intelligent and adaptive human enemy, it is unlikely that the likely collective employment of UGVs would have been appreciated and this would have impacted negatively on future experimental design.

However, there are limitations in employing HIL for more focused options analysis. The extensive time and resources required for HIL could be spent once the problem space, and indeed the employment, of UGVs supporting CATs is better understood. Since this study was conducted, LCA has improved the way we conduct options analysis campaigns. These advancements are focused on increasing the breadth of techniques we use to get a better balance between internal and external validity. They also have included the idea of philosophical validity [3] which aims to ensure all relevant perspectives are captured.

One of the techniques that LCA has been making greater use of within our more recent analytical campaigns is analytical seminar wargaming [6, 27, 28]. By conducting this type of activity earlier in the campaign it allows for a higher level of exploration of the problem space, helping to better define those roles which would be best suited for UGVs. Unfortunately this was not possible for this campaign but would have provided a better focus for the HIL simulation and would have allowed for more options to be explored. This approach also allows for additional external validity to be established by providing the opportunity to explore 'real-world' outcomes at a higher level. Analytical seminar wargaming is also a useful tool to help bring together a number of focused studies to examine the overall impact of individual changes and this helps to ensure that any impacts determined in methods with lower external validity, such as CL simulation, are also likely to be reflected into the real world. Most times it is sufficient to conduct this type of activity as part of an AAR activity with the participants, stakeholders and other interested parties. We have found this to be a very useful way to ensuring the results seen in methods with high internal validity, such as CL simulation, can be correctly interpreted and provided with external validity.

Another method employed by LCA to improve the analytical validity of these studies is confirmatory 'triangulation' [29, 30] – in which multiple methods or data sources are employed to identify a level of confidence in a study's findings [31, 32]. This could range from contrasting of CL results with historical outcomes and other qualitative data or the deliberate employment of parallel methodologies in the experimental design (e.g. wargaming, facilitated seminars and CL simulation). Confidence can then be assigned to findings based on the level of convergence of results from the multiple data sources.

References

1. The Technical Cooperation Program, *Guide for Understanding and Implementing Defence Experimentation (GUIDEx)*. 2006.
2. Kass, R.A., *The Logic of Warfighting Experiments*. 2006: CCRP Press, Washington, DC.
3. Bowden, F.D.J. and P.B. Williams, *A framework for determining the validation of analytical campaigns in defence experimentation*, in *MODSIM2013, 20th International Congress on Modelling and Simulation*, J. Piantadosi, R.S. Anderssen and J. Boland, Editors. 2013, Modelling and Simulation Society of Australia and New Zealand: Adelaide, Australia. p. 1131–1137.
4. Bowley, D.K., T.D. Castles and A. Ryan, *Constructing a Suite of Analytical Tools: A Case Study of Military Experimentation*, ASOR Bulletin, vol. 22, 2003. p. 2–10.
5. Shine, D.R. and A.W. Coutts, *Establishing a Historical Baseline for Close Combat Studies – The Battle of Binh Ba*, in *Land Warfare Conference*. 2006, Adelaide, Australia.
6. Bowden, F.D.J., Finlay, L., Lohmeyer, D. and Stanford, C., *Multi-Method Approach to Future Army Sub-Concept Analysis*, in *18th World IMACS Congress and MODSIM09 International Congress on Modelling and Simulation*. 2009: Cairns, Australia.
7. Bowden, F.D.J., Coutts, A., Williams, P. and Baldwin, M., *The Role of AEF in Army's Objective Force Development Process*, in *Land Warfare Conference*. 2007: Brisbane, Australia.
8. Whitney, S.J., Hemming, D., Haebich, A., and Bowden, F.D.J., *Cost-effective capacity testing in the Australian Army*, in *22nd National Conference of the Australian Society for Operations Research (ASOR 2013)*, P. Gaertner et al., Editors. 2013, The Australian Society for Operations Research: Adelaide, Australia. p. 225–231.
9. Directorate of Army Research and Analysis, *Australia's Future Land Operating Concept*. 2009: Head Modernisation and Strategic Planning - Army.
10. Directorate of Army Research and Analysis, *The Future Land Warfare Report 2013*. 2013.
11. Hall, R. and A. Ross, *The Effectiveness of Combined Arms Teams in Urban Terrain: The Battles of Binh Ba, Vietnam 1969 and the Battles of Fallujah, Iraq 2004*. 2007: ADFA, Canberra, Australia.
12. Bowley, D., T.D. Castles and A. Ryan, *Attrition and Suppression: Defining the Nature of Close Combat*. 2004, DSTO, Canberra, Australia.
13. Law, A.M. and W.D. Kelton, *Simulation, Modeling and Analysis*. 3 ed. McGraw-Hill International Series. 2000, McGraw-Hill: Singapore.
14. Bilusich, D., F.D.J. Bowden and S. Gaidow, *Influence Diagram Supporting the Implementation of Adaptive Campaigning*, in *The 28th International Conference of The Systems Dynamics Society*. 2010. Seoul, Korea.
15. Bilusich, D., et al., *Visualising Adaptive Campaigning – Influence Diagrams in Support of Concept Development*. Australian Army Journal, vol. IX, p. 41, 2012. Autumn.

16. Chandran, A., G. Ibal and Z. Tu, *Developing ARH troop tactics using operations research constructive simulation*, in *MODSIM 2007 International Congress on Modelling and Simulation*, L. Oxley and D. Kulasiri, Editors. 2007, Modelling and Simulation Society of Australia and New Zealand, Canberra, Australia.

17. Chau, W. and D. Grieger, *Operational synthesis for small combat teams: exploring the scenario parameter space using agent based models*, in *MODSIM2013, 20th International Congress on Modelling and Simulation*, J. Piantadosi, R.S. Anderssen and J. Boland, Editors. 2013, Modelling and Simulation Society of Australia and New Zealand: Adelaide, Australia.

18. James, S., Coutts, A., Stanford, C., Bowden, F. and Bowley, D., *Measuring Kill Chain Performance in Complex Environments, Proceedings of the SPIE 6578*, in *Defense Transformation and Net-Centric Systems*. 2007, Washington, DC.

19. Millikan, J., M. Wong and D. Grieger, *Suppression of dismounted soldiers: towards improving dynamic threat assessment in closed loop combat simulations*, in *MODSIM2013, 20th International Congress on Modelling and Simulation*, J. Piantadosi, R.S. Anderssen and J. Boland, Editors. 2013, Modelling and Simulation Society of Australia and New Zealand: Adelaide, Australia.

20. White, G.A., R.A. Perston and F.D.J. Bowden, *Force flexibility modelling in bactowars*, in *MODSIM 2007 International Congress on Modelling and Simulation*, L. Oxley and D. Kulasiri, Editors. 2007, Modelling and Simulation Society of Australia and New Zealand, Canberra, Australia.

21. Nagl, J.A., *Learning to East Soup with a Knife: Counterinsurgency Lessons from Malaya and Vietnam.* 2002, University of Chicago Press, Chicago, IL, p. 280.

22. Yoon, K.P. and C.-L. Hwang, *Multiple Attribute Decision Making: An Introduction.* 1995, Sage, New York.

23. Emond, E.J. and A.E. Turnball, *BRANDO- BReakpoint Analysis with Nonparamteric Data Option.* 2006, D CORA, Ottawa, ON.

24. Kruskal, W.H. and W.A. Wallis, *Use of Ranks in One-Criterion Variance Analysis.* Journal of the American Statistical Association, 1952. 47(260): p. 583–621.

25. Moore, D.S. and G.P. McCabe, *Introduction to the Practice of Statistics.* 2006, W.H. Freeman, New York.

26. Emond, E.J., *Developments in the Analysis of Rankings in Operational Research.* 2006, C. DRDC, Ottawa, ON.

27. Coutts, A. and S. James, *Seminar War Games as an Objective Evaluation Tool*, in *18th National ASOR Conference 2005.* 2005: Perth, Australia.

28. Holden, L., D. Lohmeyer and C. Manning, *Seminar Wargaming Enabled by jSWAT*, in *82nd MORS Symposium.* 2014, Alexandria, VA.

29. Webb, E.J., Campbell, D.T., Schwartz, R.D. and Sechrest, L., *Unobtrusive Measures: Nonreactive Research in the Social Sciences.* 1966: Rand McNally.

30. Kelle, U., *Computer Assisted Analysis of Qualitative Data, in A Companion to Qualitative Research.* 2004, SAGE Publications: London, UK.

31. Coutts, A., *A Structured Approach to Extract Strategic Objective Categories from Textual Sources*, in *22nd National Conference of the Australian Society for Operations Research.* 2013, ASOR: Adelaide, Australia.

32. Cao, T., A. Coutts and G. Judd. *A Multi-Criteria Decision Analysis Framework for Operational Assessment of the Battle Group Command, Control and Communication System*, in *Defence Operation Research Symposium.* 2012, Adelaide, Australia.

8

Processing, Exploitation and Dissemination: When is Aided/ Automated Target Recognition "Good Enough" for Operational Use?

Patrick Chisan Hew
Defence Science and Technology Organisation, HMAS Stirling, Garden Island, WA, Australia

8.1 Introduction

In 1984, a distinguished aerospace writer opined that (then) future unmanned vehicles would not "saturate the data-link with high-quality images of passing trees." Rather, the vehicle would have a target recognition capability, and transmit only when it found something significant [1]. The potential remains unrealized some 30 years on. In recent operations, human operators have continually monitored the "raw" information collected by sensors, prototypically as full-motion video. Conveying information in this manner requires the vehicle to communicate at high data rates [2]. Said communication may render the vehicle susceptible to counter-acquisition and attack, a matter of considerable concern in anti-access/area-denial scenarios.

Designers of unmanned vehicle systems thus have a conundrum when contemplating options for processing, exploiting, and disseminating the information collected by the vehicle. The vehicle might perform target recognition by some onboard means, increasing its survivability but (likely) at lower performance than if performed by a human. Conversely, communicating

© Commonwealth of Australia. Reproduced with permission of Defence Science and Technology Organisation.

Operations Research for Unmanned Systems, First Edition. Edited by
Jeffrey R. Cares and John Q. Dickmann, Jr.
© 2016 John Wiley & Sons, Ltd. Published 2016 by John Wiley & Sons, Ltd.

the information back to a human will increase performance but compromise survivability. How should designers choose?

This chapter addresses the question of whether aided/automatic target recognition ought to be used as part of an unmanned vehicle system, in the trade space of performance versus survivability. Our model and solution are simple, but they may be valid as a first approximation in many situations, and otherwise show the way to more sophisticated treatments.

Our discussion is organized as follows: We start with a background overview of the operational context, important technical issues, and previous research. We then formulate the question in quantitative terms and solve the mathematics. We finish with a discussion of implications for operations research in the design of unmanned vehicle systems.

8.2 Background

8.2.1 Operational Context and Technical Issues

We are motivated by the perennial problem of acquiring difficult, high-value targets in contested environments. The problem is central to military operations, be they on land, on or under the water, in the air or space, or even in the spectrum or cyber dimensions. For our purposes, the problem's defining characteristics are that the targets will cede few (if any) opportunities to the unmanned vehicle's sensors, and are sufficiently important to risk the loss of the vehicle (albeit the risk should be managed).

Target recognition refers to the recognition of targets based on data from sensors. It can be applied to detect objects of interest, identify and classify them, track their movements, and assess the effectiveness of actions taken against them. That is, we say that a target has been *acquired* if the information about the target is of sufficient quality and quantity to cue effective actions. Recognition is the essential precursor to such actions, be they offensive or defensive, or by the vehicle or networked parties.

Systems engineers would describe our investigation as being an exercise in *activity allocation*. In the lexicon of the US Department of Defense Architecture Framework (or the equivalent UK Ministry of Defence Architecture Framework), an *activity* is something to be done, without specifying how. A *function* is something that is done via some specified means. Thus activities are *implemented* as functions that are *performed* by one or more humans or machines, thereby *allocating* the activities to those humans or machines.

Automatic target recognition allocates the target recognition activity to machines only. A definitive review of automatic target recognition technology is far beyond our scope. Here, we merely accept that systems exist at varying levels of technological maturity and performance, depending on targets and environments (see [3] for an overview of contemporary issues and opportunities). Of the appropriate examples, perhaps the most mature is the radar warning receiver, a system that warns of the presence of radar. While its performance is less than that of a suitably skilled and equipped human, it is physically and financially unviable to provision every platform in this way. A radar warning receiver may thus be the only way to provide warning of a radar-guided attack, against the alternative of having no warning at all. The success of these systems also highlights their challenges. Radar warning receivers can achieve a high performance, as measured by low missed target and false alarm rates, because the targets are relatively easy to recognize at tactically useful distances. The radar emissions have structure and regularities that the system can discern easily from noise, so long as the system has an up-to-date library of the radars' characteristics [4]. Keeping those libraries supplied is an enormous effort that, in many ways, is the mark and domain of first-rate military forces.

We contrast with *extended human sensing*, where target recognition is allocated entirely to humans. This requires that the vehicle sensors' information be conveyed to the humans as if they were at the sensors' location. In this sense, the information is "raw" or "unprocessed." Images and video are exemplars – the humans see what the camera sees, with their own eyes. The same can be said for audio into human ears.

Aided target recognition refers to machine processing of the sensor information, as an aid to human recognition. More formally, the target recognition activity is implemented as a sequence of functions, with a human on the critical path. Examples include energy detection in passive sonar and electronic support measures, and facial recognition in imagery. Both serve to cue a human operator to a signal or object of potential interest.

We assume that aided/automatic target recognition is performed on machines carried by the unmanned vehicle. This is consistent with our interest in reducing the transmissions made by the vehicle.

8.2.2 Previous Investigations

Our interest is in whether aided/automatic target recognition ought to be used given its performance. The pertinent literature is sparse, and distinct from the massive efforts to develop the systems, establish their performance and optimize their employment. The largest concentration of research is in cognitive engineering, and the question of whether activities ought to be allocated to machine or human remains central to human factors/ergonomics (see [5] for a recent survey). (Note that human factors researchers invariably refer to "allocation of functions," as their literature predates the modern formalisms of systems engineering and the jargon has stuck. For precision, we continue with the systems engineering terminology.)

In this vein, we scope our investigation via the Parasuraman–Sheridan–Wickens model of automation in information processing systems [6]. The said model proposed four activities: information acquisition, information analysis, decision and action selection, and action implementation. Any or all of the activities could be automated to some degree, under a 10-point scale ranging from the machine doing nothing to the machine doing everything. Accepting the caveat that the model is "almost certainly a gross simplification," we describe target recognition as an instance of information analysis. Extended human sensing corresponds to automation at level 1, aided target recognition is at levels 2–4, and automated target recognition is at levels 5 or higher.

In the accepted terminology, automation is said to be *imperfect* or *unreliable* if it misses a stimulus that a suitably skilled human would catch, or responds to a stimulus that the human would ignore. Three themes have dominated the literature on activity allocation to date. The first rejects the proposition that automation should be applied wherever it is technologically possible (discredited as the "substitution myth" [7]). The second seeks structural or qualitative factors that weigh for or against automation (as exemplified by the Fitts list [8]). The third improves the means for operators to choose automation (naming "adaptive automation" [9] as one example).

We instead follow a fourth, lesser explored path, on the rational allocation of activities [10]. We use *rational* in the sense understood by operations research, of measuring or prognosing a candidate allocation in a scenario, performing a calculation and thereby deducing whether the allocation is warranted or not. In this way, our immediate predecessors are [11, 12], who looked at the average payment that we should expect from automation, given its probabilities of responding to stimuli and the payments (or costs) garnered from responses. Analysis of this kind is well suited to an industrial setting, where dollar values for payments are evident. It is less suited to military settings for the same reason.

We similarly delineate our question, "Should an activity be automated?" from the question commonly posed in human factors research, "Will a human choose to automate?". In particular, we acknowledge but leave aside the hypothesis of trust, as in reliable performance predicts trust in automation, which in turns predicts whether automation will be used [13, 14]. We are equally intrigued by, but again leave aside, experiments finding that when reliability drops below 0.70, unreliable automation is worse than no automation at all ([15, 16] for a meta-analysis, [17] as one of many experiments). Said experiments are a study in human behavioral choices, as the humans balance their workload across multiple demands. An open question is on whether the 0.70 threshold is rational, and how the capabilities of machine and human combine to set a threshold. Moreover, the human must have sufficient skill to recognize that it is the automation that is unreliable. Designers will find useful guidance in [18] on gaining acceptance for automation by addressing its apparent reliability, as distinct from actual.

We comment briefly on our investigation's connection with situation awareness, due to that concept's prominence in cognitive engineering. Our position is that we can study the activity of target recognition as a necessary input to situation awareness. Target recognition is the inference that an object exists in the battlespace, and of its properties now and into the future. It generates a representation of reality, expressed in some symbolic representation. In her seminal treatment of situation awareness, Endsley takes the symbolic representation as her input to situation awareness. Given the representation, how well has it been taken up into a person's thinking? The fidelity of that representation to the real world is framed as an external factor [19].

We are otherwise interested in analyses of the operational utility of having a target recognition capability that performs at some given level of proficiency (and comparing against the utility of employing a human). There is a voluminous literature on how aided/automated target recognition can cue operators and thus improve their performance, and again this is largely seated in cognitive engineering (see, for example [20]). A niche approach characterizes utility in terms of reduction in information entropy from having recognized a target [21] (and see [22] for more on information theory approaches). That said, it would be desirable to link the performance of target recognition to effects in the battlespace, rather than stopping at the operator. Likewise, if we wish to study performance versus cost [23], then we should allow that costs may be acceptable if the effect is commensurate.

The usual problem with connecting system performance to battlespace outcomes is that humans can make decisions that are sometimes astonishing and therefore difficult to model. This consideration is less important when studying unmanned vehicle operations – decision making can be assumed to be optimal (albeit mechanistic) under bounded assumptions. Predecessors have thus studied the impact of automatic target recognition in unmanned vehicle operations, and especially in search and destroy missions [24, 25]. Probabilities of missed targets and false alarms were the key factors (we further recommend [25] for its review of literature on searching for targets). In this context, scenarios have been proposed as a bridge from technological performance to warfighting utility [26].

That aided/automatic target recognition can be used to reduce susceptibility to counter-acquisition does not yet appear to have been addressed in detail in the literature. The risk of being counter-acquired as a result of communicating is understood as an issue for operations [27]. Indeed, unmanned vehicle data links can and have been counter-detected, intercepted and exploited, as demonstrated with video feeds from aircraft [28] and anticipated with sono-buoys [29]. Systems analysis merely quantifies the capacity for communication that is feasible under a given probability of counter-detection (see [30], for example). Our interest is in whether higher capacity communication is actually needed, in the risk–reward of performance in target recognition versus susceptibility to counter-detection. That is, to improve a vehicle's survivability, we would seek to reduce the signature arising from communications (as one contributor). However, if the vehicle's information must be conveyed back to a human in a "raw" state, then at some point it will be impossible to reduce the communications signature further. We will either have to accept that level of signature, or process the information so that lower capacity communications can suffice.

8.3 Analysis

The main result of our analysis is the following guidance to designers of unmanned vehicle systems.

Criterion for choosing aided/automated target over extended human sensing

Unmanned Vehicle's Mission: To acquire a difficult target in an environment where communicating makes the vehicle susceptible to counter-acquisition.

Specific Concept of Operations: The vehicle alternates between two states: searching for targets until the vehicle is counter-acquired, and then hiding until it can return.

Parameters: In the following, subscript $\cdot_* = \cdot_A$ refers to aided/automated target recognition, and $\cdot_* = \cdot_H$ refers to extended human sensing:

z	Mean time between target ceding glimpses
p_*	Probability of recognizing the target when given a glimpse
b_*, s_*^2	Mean and variance of the time to counter-acquire the vehicle
a, r^2	Mean and variance of the time to resume search after being counter-acquired

Criterion: Choose automation if it reduces the expected probability of missing the target when compared to a human. This is equivalent to choosing to automate when $A \geq 0$ where

$$A = D_H - D_A$$

and

$$D_* = \begin{cases} \zeta_* c_* \left(\zeta_* \left(1 - c_*\right) w_* - 1 \right) & \text{if } \zeta_* \leq \widehat{\zeta_*} \\ \widehat{\zeta_*} c_* \left(\widehat{\zeta_*} \left(1 - c_*\right) w_* - 1 \right) & \text{otherwise} \end{cases}$$

$$\zeta_* = p_* \zeta \text{ where } \zeta = \frac{1}{z}$$

$$\widehat{\zeta_*} = \frac{1}{2\left(1 - c_*\right) w_*}$$

$$c_* = \frac{b_*}{a + b_*}$$

$$w_* = \frac{a^2 s_*^2 + b_*^2 r^2}{2ab_* \left(a + b_*\right)}$$

8.3.1 Modeling the Mission

As notation, we use subscripts $\cdot_A$ and $\cdot_H$ to denote aided/automated target recognition versus extended human sensing. Our analysis could be used to compare any two proposed sets of capability, but we use $\cdot_A$ and $\cdot_H$ for definiteness. We use subscript $\cdot_*$ when discussing a matter that applies equally to both possibilities (so $\cdot_* = \cdot_A$ for automation and $\cdot_* = \cdot_H$ for human). Following usual conventions, we use Roman capital letters for random variables and lower letters for particular values. If we need to denote a rate then we will use a lower case Greek letter, and its corresponding Roman letter will be the corresponding time between events.

The unmanned vehicle's mission is to acquire a difficult target, where "difficult" is quantified here and below. We quantify the general problem as four factors: mean time z between the target ceding glimpses, the probability p_* of recognizing a target when given a glimpse, and the mean b_*, and variance s_*^2 of the times to counter-acquire the vehicle. We require z, b_* and s_*^2 to be finite (and positive). In detail, the target cedes glimpses as a homogeneous Poisson process with mean inter-arrival duration z. The vehicle likewise cedes glimpses in some stochastic process, and the durations between glimpses have a distribution with mean b_* and variance s_*^2 (we do not need to know the actual distribution). We frame our analysis so that the cited values are a good approximation to reality over some epoch, as it is unwise to perform analyses across widely varying circumstances [12].

We set $\zeta = \frac{1}{z}$ as the rate at which the target cedes glimpses, and then $\zeta_* = p_*\zeta$. That is, we can talk in two mathematically equivalent forms: the target cedes glimpses at rate ζ, and each glimpse is at probabilities p_A by automation versus probability p_H by human *or* the target cedes glimpses at rate ζ_A to the automation versus rate ζ_H to the human. Desirably, ζ would be as large as possible, but with "difficult" targets we would expect ζ to be quite small. On foreseeable technologies, we expect $p_A < p_H$ (extended human sensing outperforms aided/automated target recognition in recognizing glimpses when ceded) but unfortunately $b_H < b_A$ (extended human sensing makes the vehicle more susceptible to counter-acquisition).

Glimpse rates are commonly used in operations research to abstract the performance of sensors acquiring targets from the details of technology [31, 32]. We must, however, acknowledge that a substantial amount of work is needed to calculate or measure the glimpse rate that can be attained against a target under a given set of circumstances. Our purpose is to establish that said efforts are worthwhile for unmanned vehicle sensors and counter-sensors, by seeing how the information about sensors would apply to analyses of operations. We further emphasize that the assumption of a homogeneous Poisson process is highly simplistic, and is only valid if the operational circumstances are stable over the epoch to which analysis is being applied.

We recognize the issue of false alarms, and take them as tacitly incorporated into the parameters p_*, b_* and s_*^2. In studying recognition systems, we typically need to consider the probability of missing a valid target ($= 1 - p_*$ in our notation) and the rate at which the system reports a target when none is present (a false alarm). The two quantities are connected by *sensitivity* – increased sensitivity should reduce the probability of missed targets but will likely increase the rate of false alarms. We do not explicitly include false alarm rates in our analysis here. Instead, we assume that the vehicle's reports are filtered by the vehicle's (human) controllers, and that this adds to the communications overhead. Thus a low rate of false alarms will allow the vehicle to be operated at lower susceptibility to counter-acquisition (parameters b_* and s_*^2), or we reduce the system's sensitivity and accept a lower probability of recognizing the target (parameter p_*).

8.3.2   Modeling the Specific Concept of Operations

Figure 8.1 illustrates our concept of operations. The vehicle alternates between two states:

1. *Searching* for the target until it is counter-acquired (or assessed as such).
2. *Hiding* from the adversary's sensors until it can resume its search.

We measure effectiveness as the *probability of acquiring the target during a time interval* $[0, t]$. The concept is a plausible response to anti-access/area-denial capabilities when technological options for reducing the vehicle's susceptibility have been exhausted. It is loosely based on the "Wild Weasel" suppression of enemy air defense missions flown by US Air Force aircraft against surface-to-air missile batteries in North Vietnam.

We thus introduce two more factors, namely the mean a and variance r^2 of the times until the vehicle can resume its search. We require a and r^2 to be finite (and positive). Again in detail, the opportunities for the vehicle to resume its search arrive stochastically, and the durations between glimpses have a distribution with mean a and variance r^2 (we do not need to know the actual distribution). The factors apply equally to both aided/automated target recognition and extended human sensing.

We model the vehicle as instantaneously flipping between the states of searching or hiding. The states may, of course, be decomposed into further sub-states (for example, evading a counter-acquisition, waiting out some duration, re-inserting, etc.), but our analysis does not require this level of detail.

We may restrict our attention to scenarios where both a and b_* are much smaller than t, as the other scenarios do not need to be analyzed in depth. Namely, if b_* is comparable to or

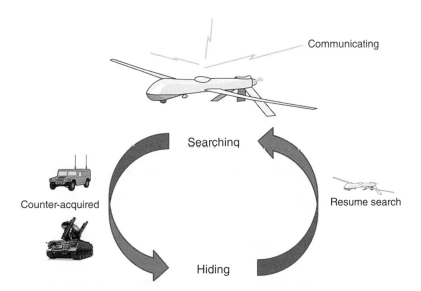

Figure 8.1   Operations to acquire a difficult target in an anti-access/area-denial scenario. When searching, the unmanned vehicle must communicate and is thus susceptible to counter-acquisition. The vehicle may intermittently have to hide and then resume its search.

larger than t, then the chance of being counter-acquired is low and thus the vehicle can search with impunity. Alternately, if a is comparable to or larger than t then there is little prospect of resuming a search after being counter-acquired.

In positing our concept of operations, we have left undefined the trigger that causes the vehicle to hide. Notionally the unmanned system is informed of threats to the vehicle, and the vehicle has a capacity to evade. We should not underestimate the practical challenges in achieving such a capability, especially against threats that are passive and/or late in unmasking. Our interest is in understanding whether the concept of operations can be operationally useful (and especially with aided/automated target recognition), to build the case for equipping unmanned vehicles in the manner required.

8.3.3   Probability of Acquiring the Target under the Concept of Operations

We calculate the probability Q_* of *missing (not acquiring) the target during a time interval* $[0, t]$. The probability of acquiring the target is then $1 - Q_*$. It transpires that our assumptions permit us to calculate Q_* using mathematical expressions that can be evaluated by commonly available software (a so-called "closed form solution"). Otherwise we would apply stochastic simulation.

First, we have the vehicle as intermittently searching for the target during *sojourns* $U_1, U_2, \ldots$. If we assume that the probabilities of acquiring the target are uncorrelated across sojourns, then Q depends only on the *cumulative sojourn* $U_* = U_1 + U_2 + \ldots$ and in fact

$$Q_* = e^{-\zeta_* U_*} \tag{8.1}$$

(See Appendix 8.A for proof. The key is that "missing the target" equates to capturing zero glimpses during a sojourn.)

Discerning readers will recognize that we have modeled the unmanned vehicle's operations as an alternating renewal process, and that U_* is the process's *uptime* during $[0, t]$. We thus invoke a classic result from the literature [33]: as $t \to \infty$, the distribution for U_* converges to a normal distribution with mean μ_{U_*} and variance $\sigma^2_{U_*}$ where

$$\mu_{U_*} = c_* t$$
$$\sigma^2_{U_*} = 2c_*(1 - c_*)w_* t \tag{8.2}$$

where $c_* = \dfrac{b_*}{a + b_*}$ and $w_* = \dfrac{a^2 s_*^2 + b_*^2 r^2}{2ab_*(a + b_*)}$. Thus the distribution for $-\zeta_* U_*$ also approaches a normal distribution, this one with mean μ_* and variance σ^2_* given by

$$\mu_* = -\zeta_* c_* t$$
$$\sigma^2_* = 2\zeta_*^2 c_* (1 - c_*) w_* t \tag{8.3}$$

Consequently as $t \to \infty$, the distribution of Q_* approaches a log-normal distribution with location parameter μ_* and scale parameter σ_*.

The above equations lose validity as ζ_* becomes large. As $\zeta_* \to \infty$, we would presume that $Q_* \to 0$ surely — if the target is ceding a huge number of glimpses, then we should have a miniscule probability of missing it. Our equations do not behave that way; we have Q_* tending to "minus infinity plus/minus infinity" (to take a grotesque interpretation). The problem is with $\zeta_* U_*$, which corresponds physically to the number of glimpses ceded by the target during U_*. As $\zeta_* \to \infty$, we should see $\zeta_* U_* \to \infty$ surely, but we have "minus infinity plus/minus infinity" instead.

8.3.4 Rational Selection between Aided/Automated Target Recognition and Extended Human Sensing

Let $\mathbb{E}[\cdot]$ denote "expected value of $\cdot$". We will choose to automate if $\mathbb{E}[Q_A] \leq \mathbb{E}[Q_H]$, wherein aided/automated target recognition reduces the expected probability of missing the target. For clarity, $\mathbb{E}[Q_*]$ is the probability that we would expect if we calculated the mean of Q_* over a large number of deployments by the vehicle.

We showed that as $t \to \infty$, the distribution of Q_* approaches a log-normal distribution with location parameter μ_* and scale parameter σ_*. This distribution has a mean value of $e^{\mu_* + \sigma_*^2/2}$. Consequently as $t \to \infty$ we have

$$\left| \mathbb{E}[Q_*] - e^{\mu_* + \sigma_*^2/2} \right| \to 0 \tag{8.4}$$

and we note that

$$\begin{aligned} e^{\mu_* + \sigma_*^2/2} &= \exp\left(-\zeta_* c_* t + \zeta_*^2 c_* (1 - c_*) w_* t \right) \\ &= \exp\left(\zeta_* c_* t \left(\zeta_* (1 - c_*) w_* - 1 \right) \right) \end{aligned} \tag{8.5}$$

Now declare the *effectiveness exponent*

$$D_* = \begin{cases} \zeta_* c_* \left(\zeta_* (1 - c_*) w_* - 1 \right) & \text{if } \zeta_* \leq \widehat{\zeta}_* \\ \widehat{\zeta}_* c_* \left(\widehat{\zeta}_* (1 - c_*) w_* - 1 \right) & \text{otherwise} \end{cases} \tag{8.6}$$

where

$$\widehat{\zeta}_* = \frac{1}{2(1 - c_*) w_*} \tag{8.7}$$

and then declare

$$A = D_H - D_A \tag{8.8}$$

as the *advantage of aided/automated target recognition over extended human sensing*, with aided/automated target being rational if and only if $A \geq 0$. The declarations can be justified as follows:

1. $\mathbb{E}[Q_*] = \exp(D_* t)$ for all $\zeta_* \leq \widehat{\zeta}_*$ and the approximation becomes perfect as $t \to \infty$.
2. $\exp(D_A t) \leq \exp(D_H t)$ if and only if $D_A \leq D_H$, as $\exp(\cdot)$ is a strictly increasing function.
3. For $\zeta_* > \widehat{\zeta}_*$, D_* behaves as it "ought" to. Notice that $\exp(D_* t) = 1$ when $\zeta_* = 0$ and then $\exp(D_* t)$ increases as ζ_* departs from zero, as we would expect. The expected probability of missing the target should continue to decrease as ζ_* increases. Unfortunately the function $f(\zeta_*) = \zeta_* c_* (\zeta_* (1 - c_*) w_* - 1)$ reverses from decreasing to increasing at $\zeta_* = \widehat{\zeta}_*$ (see Appendix 8.A). This behavior is a corollary of the problems described in the previous section, and so we need to compensate.

The calculations are readily performed in a spreadsheet, and a sample implementation has been included at this book's website.

8.3.5 Finding the Threshold at which Automation is Rational

We seek the capabilities of automation such that $A = 0$. We expect a "trade space," where the automation performs less well than a human but regains operational effectiveness from increased survivability. Mathematically, if $p_A < p_H$ (automation is poorer at recognition), we should have $b_A > b_H$ to compensate (greater mean time to be counter-acquired).

The analysis may be simplified by making two further assumptions:

1. We might assume that the times to counter-acquire the vehicle are exponentially distributed, and thus $s_* = b_*$. This equates to the vehicle ceding glimpses to the adversary as a homogeneous Poisson process, in the same way that we modeled the target ceding glimpses.
2. It may be further reasonable to assume that the times to resume search are exponentially distributed, and thus $r = a$. The assumption is reasonable in that this grants the adversary no information about when the next insertion will come, from the memorylessness properties of the exponential distribution.

Consequently, we can fix z, a, p_H and b_H and then look at how p_A varies with b_A (or vice versa).

8.3.6 Example

We illustrate using the values listed at Table 8.1, as broadly indicative of an unmanned aerial vehicle searching for a mobile target on land when there is a counter-air threat. Figure 8.2 shows the results in this case. It confirms the intuition that we posited at the start of the chapter: that automation can perform less well than a human in terms of recognizing targets, but can regain operational effectiveness from increased survivability. We see a "trade space"

Table 8.1 Parameter values, for example scenario, broadly indicative of an unmanned aerial vehicle searching for a mobile target on land when there is a counter-air threat.

Target	
Mean time between target ceding glimpses z	50 s
Unmanned vehicle operations	
Mean time to resume search a	500 s
Variance in times to resume search r^2	$r = a$ (exponentially distributed)
Human capabilities	
Probability recognize target when given a glimpse p_H	90%
Mean time to counter-acquire b_H	30 s
Variance in times to counter-acquire s_H^2	$s_H = b_H$ (exponentially distributed)
Automation capabilities	
Mean time to counter-acquire b_A	30 s and upwards
Variance in times to counter-acquire s_A^2	$s_A = b_A$ (exponentially distributed)
Probability recognize target when given a glimpse p_A	Solve for $A = 0$

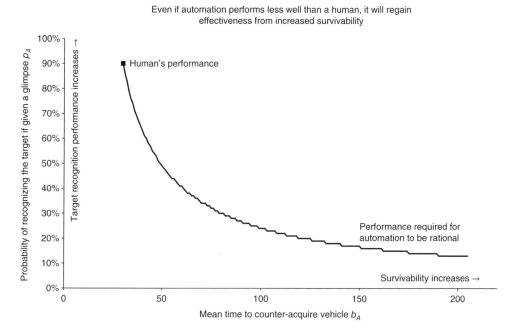

Figure 8.2   Performance required for automation to be rational in the example scenario. Even if automation performs less well than a human, it will regain effectiveness from increased survivability.

between performance (vertical axis) and survivability (horizontal axis). The performance required for automation to be rational decreases as survivability increases.

We consider two "what if" events. Figure 8.3 examines the effect of reducing the mean time to resume search (setting $a = 100s$). Doing so raises the performance threshold for automation to be rational. The advantage of automation is in reducing the need to hide. As the cost of hiding reduces, this advantage is eroded.

Figure 8.4 examines the effect of increasing the time to between the target ceding glimpses (setting $z = 500s$). Doing so raises the performance threshold for automation to be rational. If the target cedes fewer glimpses, then automation has to do better with those glimpses to compensate.

8.4   Conclusion

Our analysis should be regarded as a proof-of-feasibility of allocating activities to automation or human, in the design of unmanned systems. We have considered one activity (*target recognition*), through one mission (*acquiring a difficult target in a contested environment*), under one concept of operations (*alternately searching then hiding*) and applying one criterion (*reducing the expected probability of missing the target*). Obviously we should take many such perspectives when designing a real-world system.

That said, the question of whether a given activity ought to be automated is fundamental in the design of unmanned vehicle systems, and longstanding in cognitive engineering. We have shown that the question is amenable to operations research, and indeed to classical methods

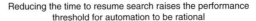

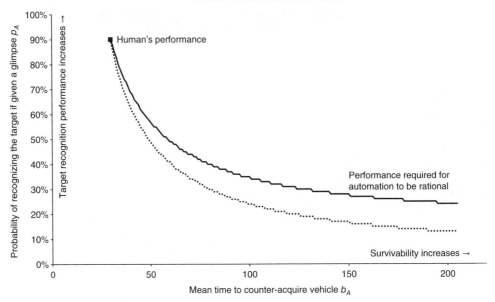

Figure 8.3 Reducing the time to resume search raises the performance threshold for automation to be rational (solid line sets $a = 100$s, dotted line sets $a = 500$s). The advantage of automation is in reducing the need to hide. As the cost of hiding reduces, this advantage is eroded.

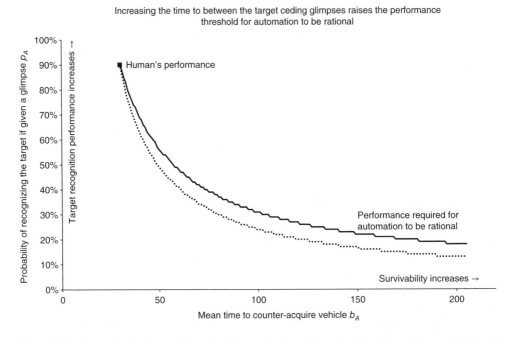

Figure 8.4 Increasing the time to between the target ceding glimpses raises the performance threshold for automation to be rational (solid line sets $z = 500$s, dotted line sets $z = 50$s). If the target cedes fewer glimpses, then automation has to do better with those glimpses to compensate.

from that field. We further have an intriguing coupling between technology and operations as follows: we took a technology that had potential to unmanned vehicle systems (namely aided/ automated target recognition at some level of performance) and *proposed* a concept of operations that harnessed that potential for effect.

Such analyses place operations researchers in a unique and influential position, as *rational proponents* of emergent technologies and *conceptual developers* of future operations. The importance of this role can be seen when compared to alternatives, of which two are of note. The first is advocacy of emergent technologies purely on their "newness," and without (quantitative) analysis. There is certainly a case for paying attention to such technologies for their potential to be disruptive, but this is not the same as making a case for adoption. The second alternative we contrast with is in leaving the conceptualization of future operations to warfighters. Instead, we treat the unmanned vehicle's operations as being open to design, and developed in conjunction with the vehicle's hardware and software.

Acknowledgments

The author thanks Thanh Ly and Nick Redding for their feedback comments. Microsoft clip art used under the Microsoft Services Agreement (31 July 2014, paragraph 8.1). "predator" and "comm-hwmwv" images produced by Joseph B Kopena (tjkopena.com), used under a Creative Commons Attribution 3.0 Unported License.

This article is UNCLASSIFIED and approved for public release. Any opinions in this document are those of the author alone, and do not necessarily represent those of the Australian Department of Defence.

Appendix 8.A

Calculating Q

Lemma

Suppose the vehicle is intermittently searching for the target during *sojourns* $U_1, U_2, \ldots$. If we assume that the probabilities of acquiring the target are uncorrelated across sojourns, then Q depends only on the *cumulative sojourn* $U = U_1 + U_2 + \ldots$ and in fact

$$Q - e^{-\varsigma U}$$

Proof

We note Q equals the probability of having zero glimpses during all of the sojourns, or

$$Q = Pr \begin{pmatrix} \text{Zero glimpses during} U_1 \text{ and} \\ \text{Zero glimpses during} U_2 \text{ and} \\ \vdots \end{pmatrix}$$

By assuming the probabilities of acquiring the target are uncorrelated across sojourns, we can decompose Q into a product.

$$Q = \prod_k Pr\big(\text{Zero glimpses during } U_k\big)$$

The target cedes glimpses at rate ζ so using the equation for the Poisson model

$$Q = \prod_k e^{-\zeta U_k}$$
$$= e^{-\zeta(U_1 + U_2 + \ldots + U_k + \ldots)}$$
$$= e^{-\zeta U}$$

Ensuring $\mathbb{E}[Q_*]$ Decreases as ζ_* Increases

We have $\mathbb{E}[Q_*] = \exp(\zeta_* c_* t(\zeta_*(1 - c_*)w_* - 1))$ and require that $\mathbb{E}[Q_*]$ decreases as ζ_* increases. This occurs when $f(\zeta_*) = \zeta_* c_* t\big(\zeta_*(1 - c_*)w_* - 1\big)$ is decreasing with ζ_*. We recognize $f(\zeta_*)$ as a quadratic on ζ_*, concave up with minimum at $\widehat{\zeta_*} = \dfrac{1}{2(1 - c_*)w_*}$. We thus require $\zeta_* \le \widehat{\zeta_*}$.

References

1. Sweetman, B., *Aircraft 2000: The Future of Aerospace Technology.* 1984, Golden Press: Sydney.
2. Buxham, P., Tackling the bandwidth issue, in *Tactical ISR Technology.* 2013, KMI Media Group.
3. Ratches, J.A., Review of Current Aided/Automatic Target Acquisition Technology for Military Target Acquisition Tasks. *Optical Engineering,* 2011. 50(7): p. 072001–072001-8.
4. *Electronic Intelligence: The Analysis of Radar Signals.* Second Edition ed, Artech House Radar Library, D.K. Barton, Editor. 1993, Artech House: Boston.
5. Challenger, R., C.W. Clegg and C. Shepherd, Function Allocation in Complex Systems: Reframing an Old Problem. *Ergonomics,* 2013. 56(7): p. 1051–1069.
6. Parasuraman, R., T.B. Sheridan and C.D. Wickens, A Model for Types and Levels of Human Interaction with Automation. *IEEE Transactions on Systems, Man and Cybernetics, Part A: Systems and Humans,* 2000. 30(3): p. 286–297.
7. Sarter, N.B., D.D. Woods and C.E. Billings, Automation Surprises, in *Handbook of Human Factors & Ergonomics,* G. Salvendy, Editor. 1997, Wiley: New York.
8. Winter, J.C.F. and D. Dodou, Why the Fitts List has Persisted Throughout the History of Function Allocation. *Cognition, Technology & Work,* 2011. 16(1): p. 1–11.
9. Scerbo, M.W., Adaptive Automation, in *Neuroergonomics: The Brain at Work: The Brain at Work,* R. Parasuraman and M. Rizzo, Editors. 2007, Oxford University Press: New York. p. 239–252.
10. Sheridan, T.B. Allocating Functions Rationally between Humans and Machines. *Ergonomics in Design: The Quarterly of Human Factors Applications,* 1998. 6(3): p. 20–25.
11. Sheridan, T.B. and R. Parasuraman, Human Versus Automation in Responding to Failures: An Expected-Value Analysis. *Human Factors: The Journal of the Human Factors and Ergonomics Society,* 2000. 42(3): p. 403–407.
12. Inagaki, T., Situation-Adaptive Autonomy: Dynamic Trading of Authority between Human and Automation. *Proceedings of the Human Factors and Ergonomics Society Annual Meeting,* 2000. 44(13): p. 13–16.
13. Lee, J.D. and K.A. See, Trust in Automation: Designing for Appropriate Reliance. *Human Factors: The Journal of the Human Factors and Ergonomics Society,* 2004. 46(1): p. 50–80.
14. Merritt, S.M. and D.R. Ilgen, Not All Trust Is Created Equal: Dispositional and History-Based Trust in Human-Automation Interactions. *Human Factors: The Journal of the Human Factors and Ergonomics Society,* 2008. 50(2): p. 194–210.
15. Wickens, C.D. and S.R. Dixon, The Benefits of Imperfect Diagnostic Automation: A Synthesis of the Literature. *Theoretical Issues in Ergonomics Science,* 2007. 8(3): p. 201–212.

16. Wickens, C.D. and S.R. Dixon, *Is There a Magic Number 7 (to the Minus 1)? The Benefits of Imperfect Diagnostic Automation: A Synthesis of the Literature*. 2005, Institute of Aviation, Aviation Human Factors Division, University of Illinois: Urbana-Champaign.

17. Wickens, C.D., S.R. Dixon and N. Johnson, Imperfect Diagnostic Automation: An Experimental Examination of Priorities and Threshold Setting. *Proceedings of the Human Factors and Ergonomics Society Annual Meeting*, 2006. 50(3): p. 210–214.

18. Kessel, R.T., *Apparent Reliability: Conditions for Reliance on Supervised Automation*. 2005, Defence R&D Canada–Atlantic.

19. Endsley, M.R., Toward a Theory of Situation Awareness in Dynamic Systems. *Human Factors: The Journal of the Human Factors and Ergonomics Society*, 1995. 37(1): p. 32–64.

20. Chancey, E.T. and J.P. Bliss, Reliability of a Cued Combat Identification Aid on Soldier Performance and Trust. *Proceedings of the Human Factors and Ergonomics Society Annual Meeting*, 2012. 56(1): p. 1466–1470.

21. Hintz, K.J., A Measure of the Information Gain Attributable to Cueing. *IEEE Transactions on Systems, Man and Cybernetics*, 1991. 21(2): p. 434–442.

22. Washburn, A.R., Bits, Bangs, or Bucks? The Coming Information Crisis. *PHALANX*, 2001. 34(3): p. 6–7 (Part I) & 4 : p. 10–11(Part II).

23. He, J., H.-Z. Zhao and Q. Fu, Approach to Effectiveness Evaluation of Automatic Target Recognition System. *Systems Engineering and Electronics*, 2009. 31(12): p. 2898–2903.

24. Kress, M., A. Baggesen and E. Gofer, Probability Modeling of Autonomous Unmanned Combat Aerial Vehicles (UCAVs). *Military Operations Research*, 2006. 11(4): p. 5–24.

25 Kish, B.A., *Establishment of a System Operating Characteristic for Autonomous Wide Area Search Vehicles, in School of Engineering and Management*. 2005, Air Force Institute of Technology: Ohio.

26. Bassham, B., K.W. Bauer and J.O. Miller, Automatic Target Recognition System Evaluation Using Decision Analysis Techniques. *Military Operations Research*, 2006. 11(1): p. 49–66.

27. Ghashghai, E., *Communications Networks to Support Integrated Intelligence, Surveillance, Reconnaissance, and Strike Operations*. 2004, RAND: Santa Monica, CA.

28. Gorman, S., Y.J. Dreazen and A. Cole, Insurgents Hack U.S. Drones, in *The Wall Street Journal*. 2009.

29. Small Business Innovation Request, Low Cost Information Assured Passive and Active Embedded Processing, Department of Defence (US). 2014.

30. Bash, B.A., D. Goeckel and D. Towsley, Limits of Reliable Communication with Low Probability of Detection on AWGN Channels. *IEEE Journal on Selected Areas in Communications*, 2013. 31(9): p. 1921–1930.

31. Koopman, B.O., *Search and Screening. General Principles with Historical Applications*. 1980, Pergamon Press: New York.

32. Wagner, D.H., W.C. Mylander and T.J. Sanders, eds. *Naval Operations Analysis*. Third Edition ed. 1999, Naval Institute Press: Anapolis, MD.

33. Takács, L., On a sojourn time problem in the theory of stochastic processes. *Transactions of the American Mathematical Society*, 1959. 93: p. 531–540.

9

Analyzing a Design Continuum for Automated Military Convoy Operations

David M. Mahalak
Applied Logistics Integration Consulting, Dallas, PA, USA

9.1 Introduction

Most people are familiar with artificial intelligence and automation within science fiction, but technological advances may result in the fulfillment of things that once seemed impossible, such as Amazon using drones to deliver purchases to your house within 30 min, or self-driving vehicles that provide operators with the flexibility to focus their attention on other tasks.

These advances in semi-automation and automation technologies are poised to drastically change military operations. Unmanned Aerial Systems (UAS) and Unmanned Ground Systems (UGS) are an increasingly attractive option for the US military, especially in an era of reduced manpower and fiscal constraints. Automation has the potential to reduce accidents, function in more austere conditions than a human driver can endure, and reduce fuel consumption and maintenance requirements. Furthermore, automated vehicles can mitigate or eliminate crew or driver rest requirements, allowing for continuous operations or reducing the vehicle fleet required to move a given amount of cargo. In addition to those benefits, tasks that can be completed by a "robot" can remove soldiers from dangerous situations, reducing the risk of injury or death. While some tasks will not translate between the commercial and defense sectors, much of the technological innovation can be transposed. This transparency reduces the costs to the Department of Defense (DoD) for research and development, and allows for future commercial development to be incorporated into defense platforms.

Over the past decade, the DoD has conducted major robotics experiments and demonstrations. Some of the milestone events include the Defense Advanced Research Projects Agency (DARPA) challenge, the Convoy Active Safety Technologies (CASTs) Warfighter I–III Leader-Follower

Operations Research for Unmanned Systems, First Edition. Edited by
Jeffrey R. Cares and John Q. Dickmann, Jr.
© 2016 John Wiley & Sons, Ltd. Published 2016 by John Wiley & Sons, Ltd.

experiments, the Autonomous Mobility Appliqué System (AMAS) Joint Capability Technology Demonstration (JCTD), and the AMAS Capabilities Advancement Demonstration (CAD). The capabilities of the automated vehicles increased over the years as technology matured and became more affordable. Recently, the AMAS JCTD, and the AMAS CAD have displayed a full suite of automated system configurations and successfully demonstrated the feasibility of executing military local and line haul convoy operations with the use of unmanned vehicles.

Discussions involving automation often result in debates about specific terms, and ways to best define a system's "smartness." Various solutions have been proposed, and agencies each have their own definitions. This can make communication within military acquisitions and between industry participants difficult. The DoD acquisition system is designed to acquire systems with required capabilities, or software to fill a particular need. Semi-automation and automation are unique and different problem sets, because they are in effect a hardware *and* software solution to a problem. Requirements must not only address a system's physical capabilities, but also the *actions of the system* under certain conditions.

This chapter focuses on the analysis of automation in military convoy operations under a variety of conditions, but consisting of only three vehicles for a 60-min mission. This model does not address combinations of automation categories, such as Leader-Follower combined with Driver Assist (DA), "teaming" situations where individual automated vehicles become "smarter" when networked with other vehicles, nor learning curves where a system may require less input over time as the system "learns." The purpose of this research is to establish an automation continuum for military convoy operations and to discuss how analyzing the trade-space between human input and robotic decision-making ability can inform important decisions about system design configurations. The framework described in this chapter is a way to consider the various implementations of semi-automation and automation and will assist engineers, military acquisition personnel, and military logistics professionals to discuss both the minimal requirements and the maximum boundaries of automation that should be fielded as technology advances.

9.2   Definition Development

This framework models and discusses "Human Input Proportion" (H) and "Robotic Decision-Making Ability" (R). H is defined as the ratio of elapsed time of human input for mission critical parameters to the total operating time of the mission. This can never be zero, as any system requires a human being to turn it on for the first time, no matter how "smart" it is, and to provide the information critical to accomplishing the mission. *Input* is the transmission of instructions, while *interaction* is simply the transmission of data such as location or video feeds. A system can simultaneously receive input, interact with the operator, and execute its assigned task. R is defined as the ability of the system to perform a given set of tasks independent of human input. This task list can change based on the system, but was primarily focused on its potential application in vehicles.

9.2.1   Human Input Proportion (H)

Human input proportion (H) is the percentage of time throughout a mission that an operator must provide instructions to the robotic vehicle. This is a ratio of the total time of human input divided by the total time of mission accomplishment, and will produce a value between (0, 1].

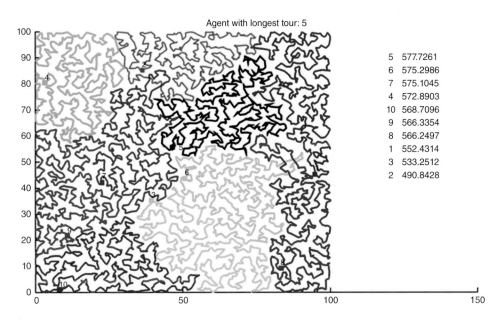

Figure 3.6 A *Min-Max* solution for 10 UAVs and 5000 targets. Runtime was about 12 000 s on a computer with i7, 3.4 GHz CPU with 16 GB RAM.

Operations Research for Unmanned Systems, First Edition. Edited by
Jeffrey R. Cares and John Q. Dickmann, Jr.
© 2016 John Wiley & Sons, Ltd. Published 2016 by John Wiley & Sons, Ltd.

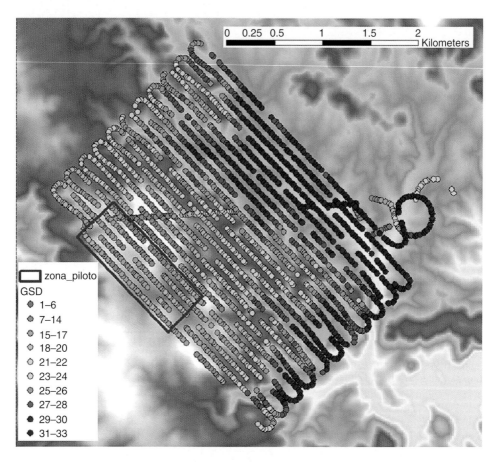

Figure 5.5 Results of the pixel size (cm) analysis control. Location of the pilot area marked in blue.

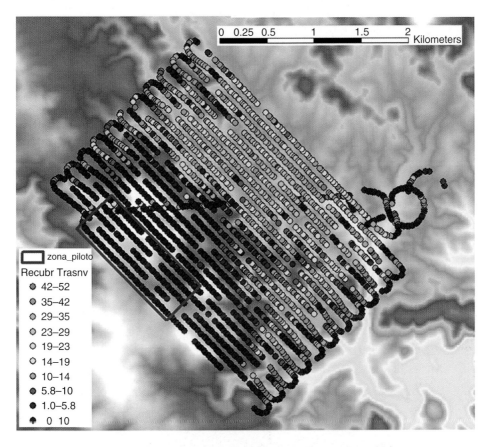

Figure 5.6 Results of the overlap between flight paths analysis.

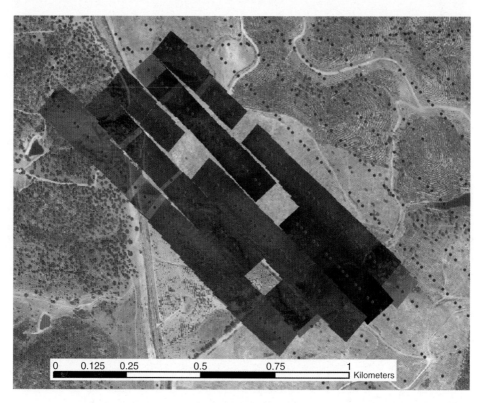

Figure 5.7 Pilot area mosaic (153 images from five flight paths, on PNOA orthophotograph).

Figure 5.8 Positional differences and geometric limitations between one of the geo-referenced images (red elements) versus PNOA orthophotograph (green).

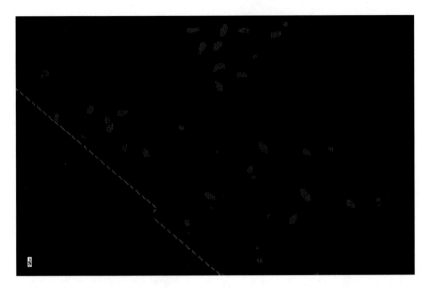

Figure 5.9 Imaging results from the pilot mosaic. Elements (animals) detected through digital processing and classification.

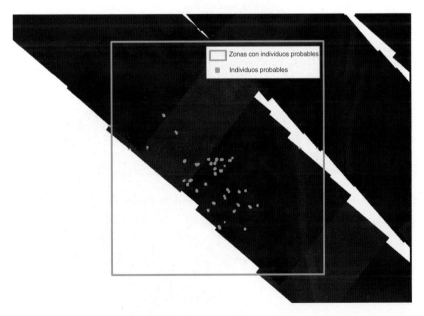

Figure 5.10 Results on the pilot area. Potential individuals detected on the thermal images mosaic and the PNOA orthophotograph.

No probable

Probable

Figure 5.11 Results on the pilot mosaic of locating probable (in green) and non-probable (in red) individuals in the study area (refined vectorization).

Zonas con individuos probables

Zonas con individuos no probables

Figure 5.12 Location of areas with probable presence of animals after the classification, processing, and interpretation of the images.

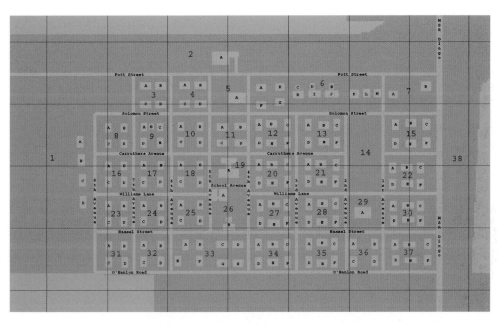

Figure 7.1 Low density urban environment.

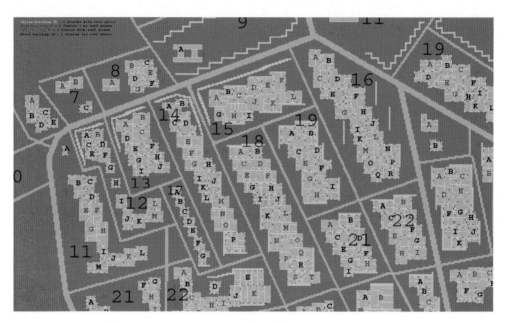

Figure 7.2 High density urban environment.

H was scored based on Subject Matter Expert (SME) input, and then the data were used in a Monte Carlo simulation to validate the methodology and results.

$$H = \frac{\text{Elapsed Time of Human Input for Mission Critical Parameters}}{\text{Elapsed Time for the Mission}} \tag{9.1}$$

9.2.2   Interaction Frequency

Interaction frequency is the frequency of communication of any type between the operator and the system. It can be initiated by either the operator or the system.

9.2.3   Complexity of Instructions/Tasks

Complexity of instructions/tasks is a function of the skill and experience level (human and/or automated system) required to accomplish a task under certain conditions. This is not dependent on the machine's intelligence. Do not assume that a system is "dumb" because it requires complex instructions.

9.2.4   Robotic Decision-Making Ability (R)

Robotic decision-making ability (R) is a scoring range [0, 1) based on the number of variables that the system can perceive and is allowed *and* able to control. This was scored using the following 15 factors: ignition, acceleration, deceleration, direction/turning, planning a route, traversing a route, dynamically re-routing, terrain negotiation, formation integrity/intervals, obstacle recognition, obstacle avoidance, obstacle passing, identifying signage, properly reacting to signage, and obeying traffic patterns. Similar to H, R was scored based on SME input and incorporated into a Monte Carlo simulation to validate the methodology and results.

$$R = \frac{\sum \textit{Average of task accomplishment probability for convoy}}{\text{Number of tasks}} \tag{9.2}$$

In Eq. (9.2), a probability for each vehicle in the convoy to accomplish a task independent of human input is calculated for a given task with respect to system configuration. The numerator of Eq. (9.2) then averages the probability of task accomplishment for each vehicle, which produces a holistic probability of task accomplishment for the convoy. Lastly, summing the holistic convoy task accomplishment probabilities and dividing by the number of tasks under consideration produces the overall R probability, or value, for the given system configuration.

9.3   Automation Continuum

Automation can be partitioned into sub-categories. This research focused on the logistical platforms that conduct military convoy operations. A convoy is defined to consist of three or more Tactical Wheeled Vehicles (TWVs) moving supplies and/or personnel from one location to another as a group. Calculations were based on a convoy of three TWVs executing a 60-min mission. The various levels of automation for conducting military convoy operations are presented next and discussed throughout this chapter in terms of increasing R values, that is, robotic decision-making ability, along the automation continuum. Appendix 9.A provides a detailed description of the manpower requirement for the given mission, and the benefits, cons, and additional costs associated with the automation technology are displayed in Figures 9.A.1–9.A.13.

9.3.1   Status Quo (SQ)

Status Quo (SQ): vehicles are operated entirely by human operators. This is how military convoy operations are currently conducted.

9.3.2   Remote Control (RC)

Remote Control (RC): a system that allows an operator to drive a vehicle or platform from a location other than inside the vehicle. This system configuration can be wired or wireless, but the operator does not require sensory data from the system. It is assumed that a single operator can control the operation of one logistical platform.

9.3.3   Tele-Operation (TO)

Tele-Operation (TO): a system which allows an operator to drive a vehicle or platform from a location other than inside the vehicle. This system configuration can be wired or wireless, but the operator does require sensory data from the system. It is assumed that a single operator can control the operation of one logistical platform. The main difference between TO and RC is that TO is used during non-line-of-sight missions, whereas RC requires the operator to have line-of-sight.

9.3.4   Driver Warning (DW)

Driver Warning (DW): a system within the vehicle which provides visual, tactile, auditory, or other forms of notification to platform drivers of potential dangers, obstacles, and accidents. This system configuration is similar to the SQ, but the system provides increased situational awareness to the operators.

9.3.5   Driver Assist (DA)

Driver Assist (DA): a system within the vehicle that provides assistance to the drivers by assuming temporary control of one or more vehicle functions, for example, acceleration control or braking. This system configuration is similar to DW, but the system has the ability to control critical driving functions when a dangerous situation is anticipated.

9.3.6 Leader-Follower (LF)

Leader-Follower (LF): a system involving two or more vehicles that allows a "follower" vehicle to mimic the behavior of the "leader" vehicle in the order of movement.

9.3.6.1 Tethered Leader-Follower (LF1)

Tethered (LF1): vehicles within the convoy are physically connected. The lead vehicle has a human operator physically present inside its cab, but follower vehicles do not.

9.3.6.2 Un-tethered Leader-Follower (LF2)

Un-tethered (LF2): vehicles within the convoy are connected via non-physical means (radio, etc.). The lead vehicle has an operator physically present inside its cab, but follower vehicles do not.

9.3.6.3 Un-tethered/Unmanned/Pre-driven Leader-Follower (LF3)

Un-tethered/Unmanned/Pre-driven (LF3): vehicles within the convoy are connected via non-physical means. No operators are present in any of the vehicles. The lead vehicle operates in a waypoint mode and is controlled through the Operator Control Unit (OCU), while the remaining vehicles are followers and mimic the behavior of the lead vehicle. It is assumed that a single operator can control the movement of four leader platforms through the OCU; however, further analysis is needed to validate this assumption. The OCU is used to communicate with the convoy vehicles and provide critical input needed to successfully complete the mission.

9.3.6.4 Un-tethered/Unmanned/Uploaded Leader-Follower (LF4)

Un-tethered/Unmanned/Uploaded (LF4): vehicles within the convoy are not tethered, and digital map data is uploaded from a database. No operators are present in any of the vehicles. The lead vehicle operates in a waypoint mode and is controlled through the OCU. The main difference between LF3 and LF4 is that LF4 does not require the route to be pre-driven. This will save time, and increase the unit's operational tempo and effectiveness.

9.3.7 Waypoint (WA)

Waypoint Mode (WA): a mode of operation where the system follows a path generated by a series of waypoints. The difference between leader-follower and waypoint navigation is that in waypoint each vehicle is an independent entity and can move within formation, while the follower vehicles in leader-follower are dependent and cannot function if the lead vehicle is disabled.

9.3.7.1 Pre-recorded "Breadcrumb" Waypoint (WA1)

Pre-recorded "Breadcrumb" Waypoint (WA1): the system must be initially "taught" by a pre-recorded route driven by an operator. No operators are present in the vehicles after the initial teaching. The OCU is used to communicate with the convoy vehicles. Note that in this scenario each vehicle must be programmed with the "breadcrumbs," whereas in LF3 and LF4 only the lead vehicle required the "breadcrumb" programming. This results in an increased burden on a single operator to control the movement of three logistics platforms. Thus, it is critical to analyze how many vehicles a single operator can reasonably control via the OCU, and how long a single operator can operate an OCU without experiencing fatigue.

9.3.7.2 Uploaded "Breadcrumb" Waypoint (WA2)

Uploaded "Breadcrumb" Waypoint (WA2): digital map data is provided to the system and waypoints are plotted by an operator. No operators are present in the vehicles during mission execution. The difference between WA1 and WA2 is that WA2 does not require a human to pre-drive the route. This will save a significant amount of time, thus increasing the operational tempo and effectiveness of the unit.

9.3.8 Full Automation (FA)

Full Automation (FA): a mode of operation in which the vehicles are now becoming "self-aware/intelligent." In the waypoint mode, a human is required to provide all of the input parameters, for example, maps, "breadcrumbs," travel speeds, interval distances, and so on, but in the full automation mode the vehicle/system is capable of providing most of the parameters, which decreases the amount of human input required. A human will use the OCU to communicate and control the fully automated vehicles before, during, and after the mission.

9.3.8.1 Uploaded "Breadcrumbs" with Route Suggestion Full Automation (FA1)

Uploaded "Breadcrumbs" with Route Suggestion Full Automation (FA1): digital map data is provided to the system, with specified origin and destination grid coordinates. The system will provide suggested routes, and the human will select which route to travel.

9.3.8.2 Self-Determining Full Automation (FA2)

Self-Determining Full Automation (FA2): the system determines its own route given that the grid coordinates of the origin and destination are provided, performs all safety-critical driving functions, monitors roadway conditions, interprets sensory information to identify obstacles and relevant signage, and dynamically re-routes to traverse alternate routes through uncharted, all-weather environments for the entire mission. No operators are present in the vehicles during the mission. This is fully automated because the vehicles are now "self-aware/intelligent." In the FA1 mode, the vehicles still require human input, for example, which route to travel from the suggested options. However, in FA2 the vehicles are able to travel to their destination independently. Keep in mind that a human will always have the ability to control the vehicles through the OCU, but the option exists to allow the vehicles to travel from origin to destination independently.

9.4 Mathematically Modeling Human Input Proportion (H) versus System Configuration

9.4.1 Modeling H versus System Configuration Methodology

Analysis began by calculating the human input proportion required for each system configuration with respect to the given scenario, which states that three tactical wheeled vehicles are conducting a 60-min mission. However, automation is a revolutionary technology that will change the way future wars are fought, and the one disadvantage of modeling cutting-edge technology is that there is a limited amount of empirical data available. Therefore, SME data were compiled in the following manner.

First SMEs were presented the automation continuum from Section 9.3 and corresponding figures from the appendix. Then, with respect to each system configuration, the SMEs were asked to provide a triangular distribution estimate for the amount of time (in minutes) they would expect human input to be required throughout a single 60-min mission for each vehicle. The three parameters that describe a triangular distribution are a, the minimum value; b, the maximum value; and c, which equals the mode. After collecting the SME data for each system configuration, a triangular distribution was developed to estimate the amount of human input that would be required for each of the three vehicles throughout a single mission.

Next, the SME data was incorporated into a Monte Carlo simulation with 150 trials. Tables 9.1–9.3 depict the first 10 trials of the Monte Carlo simulation for the system configurations {DA, LF1, LF2}, {LF3, LF4, WA1}, and {WA2, FA1, FA2}, respectively. Notice that neither SQ, RC, TO, nor DW are depicted in Figures 9.1–9.3. These system configurations were evaluated in the Monte Carlo simulation, but since each vehicle in all of the configurations requires a human operator to control it at all times, it follows that H equals 1, or 100%, and need not be listed here.

Table 9.1 The first 10 trials of the Monte Carlo simulation for the DA, LF1, and LF2 configurations.

Trial number	U(0,1) vehicle 1	U(0,1) vehicle 2	U(0,1) vehicle 3	DA			LF1			LF2		
1	0.08	0.43	0.09	57.86	58.31	57.87	60.00	5.07	4.31	60.00	4.07	3.31
2	0.27	0.15	0.40	58.09	57.94	58.27	60.00	4.43	5.01	60.00	3.43	4.01
3	0.18	0.93	0.38	57.98	59.42	58.24	60.00	7.00	4.95	60.00	6.00	3.95
4	0.72	0.53	0.57	58.82	58.47	58.53	60.00	5.36	5.45	60.00	4.36	4.45
5	0.15	0.29	1.00	57.94	58.12	59.96	60.00	4.74	7.93	60.00	3.74	6.93
6	0.05	0.68	0.18	57.82	58.74	57.97	60.00	5.82	4.49	60.00	4.82	3.49
7	0.89	0.65	0.24	59.27	58.67	58.05	60.00	5.70	4.62	60.00	4.70	3.62
8	0.44	0.75	0.69	58.32	58.88	58.75	60.00	6.06	5.84	60.00	5.06	4.84
9	0.39	0.94	0.82	58.25	59.44	59.04	60.00	7.03	6.34	60.00	6.03	5.34
10	0.73	0.79	0.82	58.84	58.97	59.05	60.00	6.22	6.35	60.00	5.22	5.35

Table 9.2 The first 10 trials of the Monte Carlo simulation for the LF3, LF4, and WA1 configurations.

Trial number	U(0,1) vehicle 1	U(0,1) vehicle 2	U(0,1) vehicle 3	LF3				LF4				WA1		
1	0.08	0.43	0.09	60.00	8.58	4.07	3.31	14.94	4.07	3.31	60.00	8.58	9.93	8.61
2	0.27	0.15	0.40	60.00	9.26	3.43	4.01	15.47	3.43	4.01	60.00	9.26	8.82	9.82
3	0.18	0.93	0.38	60.00	8.93	6.00	3.95	15.21	6.00	3.95	60.00	8.93	13.27	9.71
4	0.72	0.53	0.57	60.00	11.46	4.36	4.45	17.21	4.36	4.45	60.00	11.46	10.42	10.58
5	0.15	0.29	1.00	60.00	8.81	3.74	6.93	15.12	3.74	6.93	60.00	8.81	9.35	14.88
6	0.05	0.68	0.18	60.00	8.45	4.82	3.49	14.83	4.82	3.49	60.00	8.45	11.22	8.92
7	0.89	0.65	0.24	60.00	12.80	4.70	3.62	18.26	4.70	3.62	60.00	12.80	11.02	9.14
8	0.44	0.75	0.69	60.00	9.96	5.06	4.84	16.02	5.06	4.84	60.00	9.96	11.63	11.26
9	0.39	0.94	0.82	60.00	9.76	6.03	5.34	15.87	6.03	5.34	60.00	9.76	13.32	12.13
10	0.73	0.79	0.82	60.00	11.52	5.22	5.35	17.26	5.22	5.35	60.00	11.52	11.92	12.14

Table 9.3 The first 10 trials of the Monte Carlo simulation for the WA2, FA1, and FA2 configurations.

Trial number	U(0,1) vehicle 1	U(0,1) vehicle 2	U(0,1) vehicle 3	WA2			FA1			FA2		
1	0.08	0.43	0.09	14.94	16.00	14.96	3.29	3.86	3.31	1.68	2.38	1.70
2	0.27	0.15	0.40	15.47	15.12	15.92	3.58	3.39	3.82	2.03	1.81	2.33
3	0.18	0.93	0.38	15.21	18.63	15.83	3.44	5.27	3.77	1.87	4.10	2.27
4	0.72	0.53	0.57	17.21	16.39	16.52	4.51	4.07	4.14	3.17	2.63	2.72
5	0.15	0.29	1.00	15.12	15.55	19.90	3.39	3.62	5.95	1.80	2.08	4.94
6	0.05	0.68	0.18	14.83	17.02	15.20	3.24	4.41	3.44	1.62	3.05	1.86
7	0.89	0.65	0.24	18.26	16.86	15.38	5.07	4.32	3.53	3.86	2.95	1.97
8	0.44	0.75	0.69	16.02	17.34	17.05	3.88	4.58	4.42	2.40	3.26	3.07
9	0.39	0.94	0.82	15.87	18.67	17.74	3.79	5.29	4.79	2.30	4.13	3.52
10	0.73	0.79	0.82	17.26	17.57	17.74	4.53	4.70	4.79	3.20	3.41	3.52

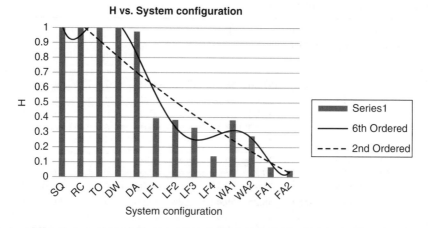

Figure 9.1 Mathematical model of human input frequency (H) vs. system configuration. Series 1 represents the H value for each system configuration, and the solid and dashed trend lines represent sixth ordered and second ordered polynomial regression equations, respectively.

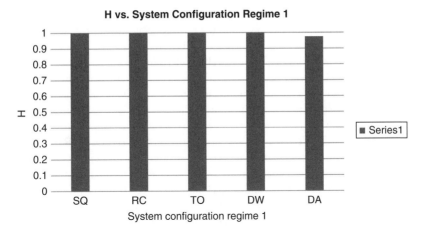

Figure 9.2 Mathematical model of human input frequency (H) vs. system configuration for Regime 1. Series 1 represents the cumulative H value for each system configuration.

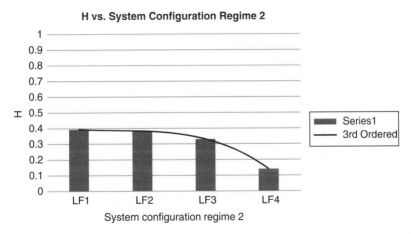

Figure 9.3 Mathematical model of human input frequency (H) vs. system configuration for Regime 2. Series 1 represents the cumulative H value for each system configuration and the solid trend line represents a third ordered polynomial regression equation.

In Figure 9.2, all of the vehicular, H_i values were estimated using a triangular distribution that will be annotated as follows, TRIA(a, c, b). The DA system configuration was modeled using a TRIA(55, 59, 60), based on the SME input. This implies that the human operator will have control over the vehicle for 55–60 min of the 60-min mission with a mode of 59 min. This seems logical since the human operator is present within the cab of the vehicle at all times and controlling the vehicle's driving functions. The system will only take control of the vehicle's driving functions in the event of a potential accident, collision, and so on. For LF1, notice that the lead vehicle is occupied with a human operator, while vehicles two and three use a tether to mimic the behavior of the lead vehicle. Therefore, the H_2 and H_3 values for vehicles two and three were modeled based on a TRIA(3, 5, 8) distribution, which incorporates the time needed to connect the tethers between the vehicles. Lastly, LF2 once again requires a human operator to be present in the lead vehicle at all times, but the second and third vehicles are connected via non-physical means, for example, radio frequency. The H_2 and H_3 values for the second and third vehicles were modeled based on a TRIA(2, 4, 7) distribution, which incorporates the time needed to program the follower vehicles to mimic the behavior of the lead vehicle.

Similarly, in Table 9.2, the H_i values for LF3 were calculated as follows. First, you will notice that there are four vehicles used in the calculation for LF3. This is because we need to include the time required to pre-drive the route, which is accounted for in the first entry of LF3. Once the route data has been captured, it was estimated that the leader vehicle, H_1, could be programmed with the route data, and mission parameters based on a TRIA(6, 10, 15) distribution. As in LF2, the second and third vehicles are mimicking the behavior of the leader vehicle via non-physical means. Therefore, to have consistency throughout the various system configurations, the second and third vehicles' H_2 and H_3 values for LF3 were modeled using a TRIA(2, 4, 7) distribution. For LF4, the lead vehicle is provided a digital map and does not require the route to be pre-driven. Thus, the leader vehicle H_1 value for LF4 was modeled based on a TRIA(13, 16, 20) distribution, which incorporates extra programming time needed to "teach" the vehicle the route. Once again, the second and third vehicles' H_2 and H_3 values were modeled using a TRIA(2, 4, 7) distribution. Finally, WA1 requires a human operator to pre-drive the route, which is captured in the first entry. However, the difference between LF and WA is that each vehicle in WA is an independent entity that is capable of completing the mission, which implies that each vehicle is a leader vehicle and must be programmed accordingly. Thus, the H_i values for the vehicles in the convoy are modeled based on TRIA(6, 10, 15) distribution, which is consistent with the parameters for the leader vehicle in LF3.

In Table 9.3, the H_i values for WA2 are modeled based on a TRIA(13, 16, 20) distribution, which is consistent with the parameters for the leader vehicle in LF4. Once we enter the FA system configurations the vehicles are becoming self-aware and capable of executing many of the mission's essential functions. For FA1, the system only requires the operator to input the grid coordinates for the origin and destination. The system will then provide a list of suggested routes for the operator to select from. Once a route has been selected the vehicle will traverse the route independent of human input. The H_i values for each vehicle in the FA1 system configuration were modeled using a TRIA(2, 4, 6) distribution. Lastly, FA2 is similar to FA1 except that FA2 does not require the human operator to select a route to traverse. Instead, once the system is provided its origin and destination grid coordinates, the vehicle will begin executing its mission. The H_i values for FA2 were modeled based on a TRIA(1, 2, 5) distribution.

Once the SME data were compiled, a Monte Carlo simulation was created to run multiple trials, or missions. For the execution of the Monte Carlo simulation three uniformly distributed random numbers, U_1, U_2, U_3, were created with parameters $a=$ minimum value $=0$ and $b=$ maximum value $=1$. Note that random numbers U_1, U_2, U_3 correspond with vehicles 1, 2, 3, respectively. Given the $U_i \sim U(0, 1)$. Then an approximate value H_i for each vehicle is determined using the inverse triangular distribution function depicted in Eqs. (9.3) and (9.4).

$$H_i = a + \sqrt{Ui(b-a)(c-a)} \quad \text{for} \quad 0 < U_i < \frac{(c-a)}{(b-a)} \tag{9.3}$$

$$H_i = b - \sqrt{(1-Ui)(b-a)(b-c)} \quad \text{for} \quad \frac{(c-a)}{(b-a)} < U_i < 1 \tag{9.4}$$

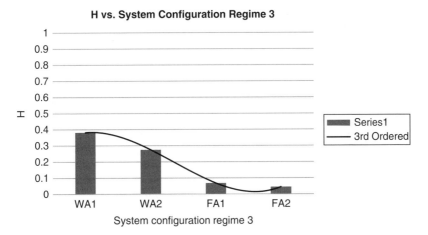

Figure 9.4 Mathematical model of human input frequency (H) vs. system configuration for Regime 3. Series 1 represents the cumulative H value for each system configuration and the solid trend line represents a third ordered polynomial regression equation.

Therefore, instead of executing a single mission, data are replicated for a total of 150 missions, which increases the model validity and provides a better approximation of the H_i values for each TWV in the convoy with respect to the given system configuration.

After compiling the data from 150 trials, the Central Limit Theorem is invoked with a sample size of $n = 5$ to transform the triangularly distributed data into normally distributed data. In doing so, the 150 trials were reduced to a total of 30 trials, from which the data for each H_i can now be modeled by a normal distribution. Then the average $\bar{H}_i$ were calculated for vehicles 1, 2, and 3, by taking the average of the 30 normally distributed H_i values. Finally, Eq. (9.5) was used to compute the overall H value for each system configuration by averaging the three $\bar{H}_i$, values for vehicles 1, 2, and 3 and dividing by the total mission duration, which in this case was 60 min.

$$H\left(\text{system configuration}\right) = \frac{\text{Average of vehicular H values}}{\text{Total Mission Duration}} \tag{9.5}$$

Note that when calculating the average of $\bar{H}_i$ for LF3 and WA1, a total of four vehicles are included in the calculation, to account for the time required to pre-drive the route. Figure 9.4 displays the computed H value for each system configuration of the automation continuum, and Figure 9.1 shows a graphical representation of the data in Table 9.4 with regression analysis.

9.4.2 Analyzing the Results of Modeling H versus System Configuration

Using regression analysis a sixth ordered polynomial curve was fit (Eq. (9.6)), which is shown by the solid trend line in Figure 9.1, to model H versus system configuration. Let y represent the human input proportion required over the duration of the mission and x represent the various system configuration levels.

$$y = 0.00006x^6 - 0.0026x^5 + 0.0435x^4 - 0.3387x^3 + 1.2527x^2 - 2.0239x + 2.0807 \tag{9.6}$$

Table 9.4 Human input proportion (H) values from the Monte Carlo simulation with respect to configuration.

System configuration	Human input proportion (H)
SQ	1
RC	1
TO	1
DW	1
DA	0.975197552
LF1	0.393718536
LF2	0.382607425
LF3	0.330677436
LF4	0.140635348
WA1	0.381694492
WA2	0.274640029
FA1	0.068627109
FA2	0.044909546

Table 9.5 Human input proportion (H) versus system configuration, computed from Eq. (9.6).

	x	$y = 0.00006x^6 - 0.0026x^5 + 0.0435x^4 - 0.3387x^3 + 1.2527x^2 - 2.0239x + 2.0807$
SQ	1	1.01183
RC	2	0.95102
TO	3	1.07447
DW	4	1.07198
DA	5	0.94295
LF1	6	0.83558
LF2	7	0.92927
LF3	8	1.36022
LF4	9	2.19023
WA1	10	3.4187
WA2	11	5.03783
FA1	12	7.13102
FA2	13	10.01447

   This single equation accurately models the relationship between H and the system configurations with a coefficient of determination, R^2, equal to 0.9528. Although the above equation fits the data well, its applicability is limited because x, the independent variable, is subjective in nature. Suppose that SQ=1, RC=2, TO=3, DW=4, DA=5, LF1=6, LF2=7, LF3=8, LF4=9, WA1=10, WA2=11, FA1=12, and FA2=13. It would follow that if a design engineer wanted to analyze a military convoy operation using LF3, then x=8 should be substituted into Eq. (9.6), and a value between 0 and 1 should be returned that would denote the amount of human input proportion required throughout the mission. However, the results from the analysis, which are depicted in Table 9.5, are undesirable, since they violate our (0, 1] constraint for H.

Therefore, in order to create a single equation that could be useful in design engineering it is necessary to solve for x, thus making the system configurations the dependent variable. Based on manpower constraints an engineer can estimate the availability of resources throughout the mission and the new equation with x as the dependent variable will provide a recommended system configuration.

To better conceptualize this new methodology (and to alleviate the need to perform unwieldy sixth root calculations from arbitrary data) assume that regression analysis was used to develop a second ordered polynomial equation to model the relationship between H and system configurations, which is shown by the dashed trend line in Figure 9.1 and calculated in Eq. (9.7).

$$y = 0.0022x^2 - 0.1239x + 1.2648 \qquad (9.7)$$

Note that the representations of x and y still remain the same, but the coefficient of determination, R^2, for Eq. (9.7) equals 0.8521. There is an increase in the sum of the squared residuals, which means that Eq. (9.7) does not fit the data as well as Eq. (9.6), however, the applicability of Eq. (9.7) is more useful. Suppose now that Eq. (9.7) was solved for x; the equations calculated in Eqs. (9.8) and (9.9) result.

$$x = 21.3201\left(\sqrt{y + 0.479656} + 1.32078\right) \qquad (9.8)$$

$$x = -21.3201\left(\sqrt{y + 0.479656} - 1.32078\right) \qquad (9.9)$$

The results of varying y at intervals of 0.1 between 0 and 1 are shown in Table 9.6. Notice that the values for Eq. (9.8) do not provide much value for the engineer. On the other hand, the results from Eq. (9.9) align well with the system configuration scale where SQ=1, RC=2, TO=3, and so on.

For example, if a systems engineer decided there was enough "manpower budget" available such that H=0.40 (human input proportion required for the mission was 40%), then the result from the above analysis is x=8.16, which correlates with LF3. Thus, the mission should be

Table 9.6 Equations (9.8) and (9.9) to computation for values for x, as y varies between 0 and 1.

y	$x = 21.3201\left(\sqrt{y + 0.479656} + 1.32078\right)$	$x = 21.3201\left(\sqrt{y + 0.479656} - 1.32078\right)$
0	42.92486636	13.393457
0.1	44.3912503	11.92707305
0.2	45.73571901	10.58260435
0.3	46.98441158	9.333911776
0.4	48.15527881	8.163044548
0.5	49.26127927	7.05704409
0.6	50.31213062	6.006192733
0.7	51.31534227	5.002981089
0.8	52.27685987	4.04146349
0.9	53.2014865	3.11683686
1	54.09316828	2.225155071

designed for the LF3 system configuration mode with the allocated resources. Note that because Eq. (9.7) does not fit the data perfectly, there is some variability in the possible results for x, but the method itself is valid (since the original approximation would be more accurate).

9.4.3 Partitioning the Automation Continuum for H versus System Configuration into Regimes and Analyzing the Results

We can consider the previous approach a first approximation to the trade-space question. In general we might say the data suggest that roughly speaking, human input is reduced by a square root relationship as automation is increased. Innovation is not always a smooth function; there are typically discontinuous jumps in capability as a technology improves. It is possible to identify "breaks" in Figure 9.1 where increased automation drives significant discontinuities in H. Suppose that we partition the automation continuum into the following regimes: SQ, RC, TO, DW, and DA form the first regime; LF1, LF2, LF3, and LF4 form the second regime; and WA1, WA2, FA1, and FA2 form the third regime.

The results of analyzing the first regime are depicted in Figure 9.2. Notice that H equals or is very close to a value of one for each of the system configurations in Regime 1. This implies that although automation technology is incorporated into the platforms, a decrease in H is not being realized. This tell us that in this regime in the human input-automation trade-space, technology is not decreasing human input requirements so much as it is providing other benefits, such as increased safety to the soldiers operating the platforms. Thus, if the DoD were to invest in Regime 1 technologies, it is apparent that human operators will still be required, almost on a full-time basis, to control the movement of the TWVs and successfully complete the mission.

The results of analyzing the second regime are depicted in Figure 9.3. Using regression analysis a third ordered polynomial equation was developed in Eq. (9.10), which is shown by the solid trend line in Figure 9.A.11. Let y represent the human input proportion required over the duration of the mission, notice that $0.14 < y < 0.39$, and x represents the various system configuration levels of Regime 2, where $x = 1 = LF1$, $x = 2 = LF2$, $x = 3 = LF3$, and $x = 4 = LF4$.

$$y = -0.0162x^3 + 0.0769x^2 - 0.1283x + 0.4613 \qquad (9.10)$$

Equation (9.10) models Regime 2 perfectly with an R^2 equal to 1. Contrary to Regime 1, Regime 2 shows a monotonically decreasing value for H over the automation continuum. Furthermore, a significant decrease in human input proportion should be expected because the technology enables soldiers to be removed from the vehicles. For LF1 and LF2, the lead vehicle in the convoy requires a soldier to operate the platform. However, the second and third vehicles only need human input under certain circumstances, for example, if the tether between the vehicles disconnects or communication is blocked due to jamming. In these cases, a human operator may need to re-establish connection of the tether/signal, or physically assume control of the vehicle. A further decrease in H is realized in LF3 and LF4 because these system configurations allow for fully unmanned convoy operations. It may have been expected that LF3 would have a larger decrease in H when compared to LF2, but it is imperative to remain cognizant of the system configuration definitions. LF3 required a soldier to pre-drive the route, which means that the time associated with this task was incorporated into the cumulative H calculation. Notice that a substantial drop-off in H occurs between LF3 and LF4 due to LF4 having the flexibility to upload a digital map into the system and plot digital "breadcrumbs" for the lead vehicle to follow. This saves the unit a significant amount of time, which in turn increases their operational tempo and effectiveness.

Finally, the results of analyzing the third regime are depicted in Figure 9.4. Using regression analysis a third ordered polynomial equation was developed in Eq. (9.11), which is shown by the solid trend line in Figure 9.4. Let y represent the human input proportion required over the duration of the mission, notice that $0.05 < y < 0.38$, and x represents the various system configuration levels of Regime 3, where $x = 1 = WA1$, $x = 2 = WA2$, $x = 3 = FA1$, and $x = 4 = FA2$.

$$y = 0.0469x^3 - 0.3307x^2 + 0.557x + 0.1085 \qquad (9.11)$$

Equation (9.11) models Regime 3 perfectly with an R^2 equal to 1. For the third regime there were predictable results and others that may require further explanation. Certainly, it is reasonable to project that as the automation continuum approaches FA1 and FA2 that minute human input will be required. However, notice that the H values for WA1 and WA2 significantly increased when compared to LF4. The question is why did this happen? Once again, the answer can be found in the definitions. Essentially, LF3 and LF4 correlate with WA1 and WA2, respectively. For LF3 and WA1, the route is pre-driven, and the vehicles are unmanned. Similarly, for LF4 and WA2, a digital map is uploaded, "breadcrumbs" are plotted, and the vehicles are unmanned. The main difference between them is that waypoint was defined such that each vehicle is an independent entity, which would enable the vehicles to complete the mission regardless of the status of the other vehicles in the convoy. Conversely, the follower vehicles in the leader-follower system configuration are dependent on the viability of the lead vehicle. Therefore, it is assumed that programming a follower vehicle requires less time than programming a leader vehicle that is using waypoint navigation. Hence, LF3 and LF4 only require this increased programming time for one vehicle in the convoy, whereas WA1 and WA2 require the same increased programming time for all of the vehicles in the convoy. This explains the increase in H for WA1 and WA2.

This section has explored the relationship between human input proportion and system configuration based on the relative rank of each system in the automation continuum. The next step in examining the human input-automation trade-space is to mathematically model system configurations against their respective robotic decision-making ability.

9.5 Mathematically Modeling Robotic Decision-Making Ability (R) versus System Configuration

9.5.1 *Modeling R versus System Configuration Methodology*

In order to calculate R versus system configuration, a similar methodology was used with respect to the one that was developed in Section 9.4.1 to model H versus system configuration. First, SMEs were asked to once again provide a triangular distribution estimate for the probability, with respect to the system configuration, that a given vehicle would be able to accomplish the following tasks independent of human input: ignition, acceleration, deceleration, direction/turning, planning a route, traversing a route, dynamically re-routing, terrain negotiation, formation integrity/intervals, obstacle recognition, obstacle avoidance, obstacle passing, identifying signage, properly reacting to signage, and obeying traffic patterns.

For the SQ, RC, and TO system configurations the automation systems do not control or contribute to the execution of any of the 15 above tasks. Therefore the R values for SQ, RC, and TO are 0. For DW, SMEs used a TRIA(90, 94, 99) distribution to approximate the percentage of times that a vehicle would be able to identify obstacles. As we will see later, although DW has the ability to identify obstacles, it does not have the ability to act upon its observation. This inability to act will impact its R value. For DA, SMEs used a TRIA(88, 95, 99), TRIA(90, 94, 99), and TRIA(80, 84, 99) distribution to estimate the percentage of time that a vehicle would decelerate, identify obstacles, and avoid obstacles, respectively. This process was continued for the remaining system configurations, and as we moved along the automation continuum the vehicles were able to accomplish more tasks independent of human input with a higher probability of success.

Once again, after collecting the SME data a Monte Carlo simulation was executed using three uniformly distributed random numbers, U_1, U_2, U_3, with parameters a=minimum value=0 and b=maximum value=1. These uniform random numbers along with the SME triangular distribution parameters for each task and vehicle with respect to system configuration were appropriately substituted into the inverse triangular distribution, Eqs. (9.3) and (9.4). By substituting the data into Eq. (9.3) or (9.4), the result would be an integer number between 0 and 100 that represents the percentage of times that a vehicle in a given system configuration would be able to complete a given task independent of human input. Therefore, all of the data were divided by 100 to create a probability of successfully completing a task, p, such that $0 \le p < 1$. Note that p was assumed to never be able to equal 1.0 because any system, no matter how reliable, will have some probability of failure.

With the newly acquired p values, an experiment of 10 independent, Bernoulli trials was created to determine the probability that a given vehicle would be able to successfully accomplish a given task at least 8 out of the 10 times. Therefore, let X be a binomial random variable with parameter p (which was approximated from the above method) that represents the number of times a vehicle will accomplish a given task independent of human input. Then the number of trials, n, equals 10 and $P(X \ge 8)$ can be calculated for each combination of task, vehicle number, and system configuration.

The results of the $P(X \ge 8)$ represented the R_i values for {DW, DA, LF1, LF2}, {LF3, LF4, WA1}, and {WA2, FA1, FA2}. Tables 9.7–9.9 display the R_i results, respectively, for each group. Finally, Eq. (9.12) was used to compute the overall R value for each system configuration by averaging the three R_i values for vehicles 1, 2 and 3 and dividing by the total number of tasks, which in this case is 15. Table 9.10 shows a data table for R versus system configuration, and Figure 9.5 shows a graphical representation of the data in Table 9.10 with regression analysis.

$$R\left(\text{system configuration}\right) = \frac{\text{Average of the three } R_i}{\text{Total Number of Mission Tasks}} \qquad (9.12)$$

As shown in Figure 9.5, regression analysis determined that a sixth degree polynomial equation accurately models the distribution of R versus system configuration well. The one deficiency of these results is that it does not incorporate H into the model. It is necessary to integrate the human input proportion into the calculation for R because it is not sufficient to model the probability of a particular system configuration's ability to execute the established tasks on a first attempt without incorporating the amount of time which the system could potentially experience each task.

Table 9.7 The R_i values for the DW, DA, LF1, and LF2 system configurations.

Task/sub-task	DW			DA			LF1			LF2		
Ignition	0	0	0	0	0	0	0	0	0	0	0	0
Acceleration	0	0	0	0	0	0	0	0.988	0.995	0	0.991	0.996
Deceleration	0	0	0	0.988	0.964	0.997	0	0.963	0.997	0	0.979	0.998
Direction/turning	0	0	0	0	0	0	0	0.812	0.969	0	0.966	0.994
Plan a route	0	0	0	0	0	0	0	0	0	0	0	0
Traverse a route	0	0	0	0	0	0	0	0	0	0	0	0
Dynamically re-route	0	0	0	0	0	0	0	0	0	0	0	0
Terrain negotiation	0	0	0	0	0	0	0	0.015	0.427	0	0.071	0.578
Formation integrity	0	0	0	0	0	0	0	0.952	0.878	0	0.952	0.878
Identify obstacles	0.983	0.999	0.975	0.983	0.999	0.975	0	0	0	0	0	0
Avoid obstacles	0	0	0	0.974	0.838	0.851	0	0	0	0	0	0
Bypass obstacles	0	0	0	0	0	0	0	0	0	0	0	0
Identify signage	0	0	0	0	0	0	0	0	0	0	0	0
React to signage	0	0	0	0	0	0	0	0	0	0	0	0

Table 9.8 The R_i values for the FA3, FA4, and WA1 system configurations.

Task/sub-task	FA3				FA4			WA1			
Ignition	0	0.994	0	0	0.994	0	0	0	0.997	0.999	0.999
Acceleration	0	0.972	0.991	0.996	0.972	0.991	0.996	0	0.993	0.994	0.997
Deceleration	0	0.985	0.979	0.998	0.985	0.979	0.998	0	0.996	0.990	0.999
Direction/turning	0	0.951	0.966	0.994	0.951	0.966	0.994	0	0.975	0.972	0.995
Plan a route	0	0	0	0	0	0	0	0	0	0	0
Traverse a route	0	0.973	0	0	0.973	0	0	0	0.978	0.978	0.987
Dynamically re-route	0	0	0	0	0	0	0	0	0.954	0.903	0.918
Terrain negotiation	0	0.091	0.071	0.578	0.091	0.071	0.578	0	0.325	0.290	0.778
Formation integrity	0	0	0.952	0.878	0	0.952	0.878	0	0.990	0.993	0.984
Identify obstacles	0	0.986	0	0	0.986	0	0	0	0.990	0.999	0.986
Avoid obstacles	0	0.983	0	0	0.983	0	0	0	0.984	0.932	0.938
Bypass obstacles	0	0.788	0	0	0.788	0	0	0	0.975	0.997	0.995
Identify signage	0	0.959	0	0	0.959	0	0	0	0.989	0.988	0.991
React to signage	0	0.811	0	0	0.811	0	0	0	0.952	0.995	0.981

9.5.2 Mathematically Modeling R versus System Configuration When Weighted by H

H was incorporated into the model by using the values from Table 9.4 as a weighted coefficient and multiplicatively combined with the values in Table 9.5 using Eq. (9.13).

$$R(\text{system configuration weighted by H}) = \left(1 - H_{\text{sys.con}}\right) \times R_{\text{sys.con}} \qquad (9.13)$$

Table 9.9 The R$_i$ values for the WA2, FA1, and FA2 system configurations.

Task/sub-task	WA2			FA1			FA2		
Ignition	0.997	0.999	0.999	0.997	0.999	0.999	0.998	0.999	0.999
Acceleration	0.993	0.994	0.997	0.997	0.997	0.999	0.998	0.998	0.999
Deceleration	0.996	0.990	0.999	0.998	0.996	0.999	0.998	0.997	0.999
Direction/turning	0.975	0.972	0.995	0.983	0.981	0.996	0.994	0.993	0.998
Plan a route	0	0	0	0.991	0.986	0.994	0.997	0.995	0.998
Traverse a route	0.978	0.978	0.987	0.990	0.990	0.994	0.995	0.995	0.997
Dynamically re-route	0.954	0.903	0.918	0.990	0.980	0.983	0.994	0.989	0.991
Terrain negotiation	0.325	0.290	0.778	0.442	0.408	0.837	0.811	0.793	0.958
Formation integrity	0.990	0.993	0.984	0.988	0.992	0.981	0.993	0.995	0.989
Identify obstacles	0.990	0.999	0.986	0.992	0.999	0.989	0.991	0.999	0.988
Avoid obstacles	0.989	0.932	0.934	0.996	0.978	0.980	0.998	0.990	0.991
Bypass obstacles	0.975	0.997	0.995	0.984	0.997	0.996	0.992	0.998	0.998
Identify signage	0.989	0.988	0.991	0.993	0.993	0.995	0.995	0.995	0.996
React to signage	0.952	0.995	0.981	0.971	0.996	0.988	0.980	0.997	0.992

Table 9.10 The results of the robotic decision-making ability (R) vs. system configuration.

System configuration	Robotic decision-making ability (R)
SQ	0.0000
RC	0.0000
TO	0.0000
DW	0.0655
DA	0.1919
LF1	0.1717
LF2	0.1758
LF3	0.3107
LF4	0.4142
WA1	0.6439
WA2	0.8586
FA1	0.9345
FA2	0.9405

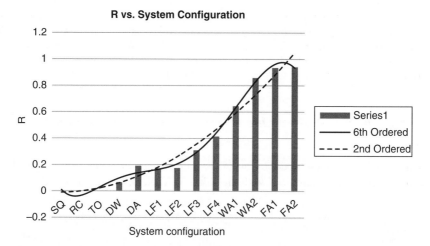

Figure 9.5 Mathematical model of robotic decision-making ability (R) vs. system configuration. Series 1 represents the cumulative R value for each system configuration, and the solid and dashed trend lines represent sixth ordered and second ordered polynomial regression equations, respectively.

From Eq. (9.13) it is clear that if $H_{sys.con}$ equals one, then the robotic decision-making ability of the corresponding system configuration is zero. The results from applying Eq. (9.13) are depicted in Table 9.11 and a graphical representation of the data is displayed in Figure 9.6 with regression analysis.

This model now incorporates the probability of each system configuration being able to execute the identified tasks independent of additional human input and the amount of time that the vehicle is expected to accomplish the identified tasks throughout the mission. Thus, the

Table 9.11 The results of the robotic decision-making ability (R) when weighted by H versus system configuration.

System configuration	R (weighted by H)
SQ	0.0000
RC	0.0000
TO	0.0000
DW	0.0000
DA	0.0047
LF1	0.1033
LF2	0.1093
LF3	0.2048
LF4	0.3506
WA1	0.3968
WA2	0.6207
FA1	0.8685
FA2	0.8978

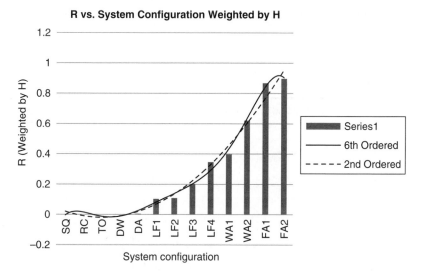

Figure 9.6 Mathematical model of robotic decision-making ability (R) vs. system configuration weighted by H. The bar chart represents the cumulative R value for each system configuration when weighted by H, and the solid and dashed trend lines represent sixth ordered and second ordered polynomial regression equations, respectively.

system must now have the authority to execute a given task, and the ability. Regression analysis was used to develop the sixth ordered polynomial equation in Eq. (9.14), which has a coefficient of determination, R^2, equal to 0.9939.

$$Y = -0.00002x^6 + 0.0007x^5 - 0.0114x^4 + 0.0845x^3 - 0.308x^2 + 0.5005x - 0.2707 \quad (9.14)$$

Let x represent the various system configuration levels of automation and y represent the robotic decision-making ability, that is, "smartness," of the military convoy. Based on the high R^2 value, it appears that Eq. (9.14) accurately models R and the system configuration when weighted by H. However, when we try to implement this equation in an applicable manner, we discover similar problems to when we tried to use Eq. (9.6) to model H versus system configuration in Section 9.4.2. The results of the analysis for Eq. (9.14) are depicted in Table 9.12.

Once again, notice that the results for the dependent variable, y, are illogical because we have defined y such that it ranges between 0 and 1. In order to circumvent this problem, we will again calculate the second ordered polynomial regression equation, and solve for x. Therefore, if based on an assessment of the technological capabilities available, that is, the level of R attainable, then the equation will return the system configuration level that could be used to instantiate the technology.

When R versus system configuration weighted by H is modeled using a second ordered polynomial equation the result is shown in Eq. (9.15).

$$y = 0.0097x^2 - 0.0581x + 0.0679 \quad (9.15)$$

It turns out that the coefficient of determination, R^2, is equal to 0.984 for Eq. (9.15). Unlike when we calculated Eq. (9.6), Eq. (9.15) does not experience a significant drop in how well the equation fits the data. When Eq. (9.15) is solved for x, it follows that Eqs. (9.16) and (9.17) are calculated.

Table 9.12 Using Eq. (9.14), which models R versus system configuration when weighted by H, to compute values of the dependent variable y.

	x	$y = -0.00002x^6 + 0.0007x^5 - 0.0114x^4 + 0.0845x^3 - 0.308x^2 + 0.5005x - 0.2707$
SQ	1	0.00188
RC	2	0.21462
TO	3	1.50332
DW	4	6.37898
DA	5	19.5318
LF1	6	48.62078
LF2	7	105.04892
LF3	8	204.72402
LF4	9	368.80508
WA1	10	624.4343
WA2	11	1005.45468
FA1	12	1553.11322
FA2	13	2316.74972

$$x = 10.1535\left(\sqrt{y + 0.0191} + 0.294958\right) \tag{9.16}$$

$$x = -10.1535\left(\sqrt{y + 0.0191} - 0.294958\right) \tag{9.17}$$

If y, that is, R, is varied at intervals of 0.1 between 0 and 1, the following results are computed in Table 9.13. Notice that the values from Eq. (9.17) do not provide much value for the engineer. On the other hand, the results from Eq. (9.16) align well with the system configuration scale where $SQ = 1$, $RC = 2$, $TO = 3$, and so on.

So, for example, if an engineer assessed a capability at $R = 0.6$, that is, the robotic "smartness" of the convoy is capable of executing 60% of the mission independent of human input, then the result from the above analysis is $x = 10.98$, which correlates with WA2 from Figure 9.6. Thus, the capability should be designed into WA2 system configuration. Note that for $y = R = 0$, Eq. (9.16) calculates $x = 4.398$. This would correlate with DW or DA, but in reality any system configuration below LF1 approximately has an $R = 0$ value.

9.5.3   Partitioning the Automation Continuum for R (Weighted by H) versus System Configuration into Regimes

In a similar manner to Section 9.4.3, where the H versus system configuration was further partitioned into regimes, the same can be done for R (weighted by H) versus system configuration. From Figure 9.6, it is possible to identify "breaks" where the system configurations experience a significant change in R. Suppose that we further stratify the automation continuum into the following regimes: SQ, RC, TO, DW, and DA form the first regime; LF1, LF2, LF3, and LF4 form the second regime; and WA1, WA2, FA1, and FA2 form the third regime.

The results of analyzing the first regime are depicted in Figure 9.7. Notice that R is equal to or very close to zero for Regime 1, which is logical since the system itself does not have the ability to control any vehicular functions, or a very small portion in the case of DA.

Table 9.13   Using Eq. (9.16) (second column) and Eq. (9.17) (third column) to compute values for x, which represents the various system configurations, as y varies between 0 and 1.

y	$x = 10.1535\left(\sqrt{y + 0.0191} + 0.294958\right)$	$x = -10.1535\left(\sqrt{y + 0.0191} - 0.294958\right)$
0	4.398097671	1.591614435
0.1	6.498917021	−0.509204915
0.2	7.747518411	−1.757806305
0.3	8.730460268	−2.740748162
0.4	9.568022104	−3.578309998
0.5	10.31031012	−4.320598011
0.6	10.98392511	−4.994213003
0.7	11.60500011	−5.615288
0.8	12.18419427	−6.194482163
0.9	12.72898637	−6.739274266
1	13.24486334	−7.255151229

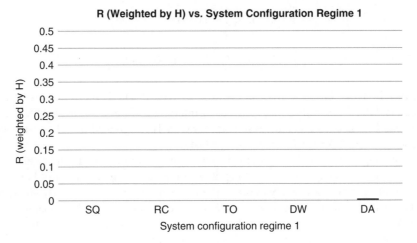

Figure 9.7 Mathematical model of R (weighted by H) vs. system configuration for Regime 1. The bar chart represents the cumulative R value for each system configuration.

The results of analyzing the second regime are depicted in Figure 9.8. Using regression analysis a third ordered polynomial equation was developed in Eq. (9.18), which is shown by the solid trend line in Figure 9.8. Let y represent the human input proportion required over the duration of the mission, notice that $0.10 < y < 0.35$, and x represent the various system configuration levels of Regime 2, where $x = 1 = LF1$, $x = 2 = LF2$, $x = 3 = LF3$, and $x = 4 = LF4$.

$$y = -0.0065x^3 + 0.0839x^2 - 0.1999x + 0.2259 \tag{9.18}$$

Equation (9.18) models Regime 2 perfectly with an R^2 equal to 1. In comparing the H Regime 2 with the R Regime 2, that is, Figures 9.3 and 9.8, respectively, we see that H has a geometric decrease over the Regime 2 continuum while R (weighted by H) has a geometric increase.

Finally, the results of analyzing the third regime are depicted in Figure 9.9. Using regression analysis a third ordered polynomial equation was developed in Eq. (9.19), which is shown by the solid trend line in Figure 9.9. Let y represent the human input proportion required over the duration of the mission, notice that $0.39 < y < 0.89$, and x represent the various system configuration levels of Regime 3, where $x = 1 = WA1$, $x = 2 = WA2$, $x = 3 = FA1$, and $x = 4 = FA2$.

$$y = -0.0404x^3 + 0.02542x^2 - 0.256x + 0.439 \tag{9.19}$$

Equation (9.19) models Regime 3 perfectly with an R^2 equal to 1. In comparing the H Regime 3 with the R Regime 3, that is, Figures 9.4 and 9.9, respectively, we see that H has an almost linear decrease over the Regime 3 continuum while R (weighted by H) has a geometric increase.

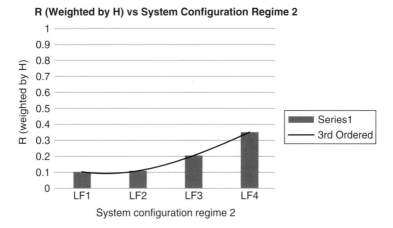

Figure 9.8   Mathematical model of R (weighted by H) vs. system configuration for Regime 2. Series 1 represents the cumulative R value for each configuration and the trend line is a third ordered polynomial regression equation.

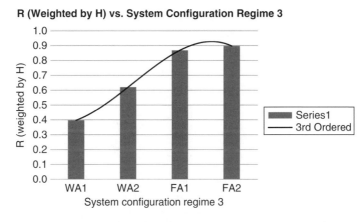

Figure 9.9   Mathematical model of R (weighted by H) vs. system configuration for Regime 3. Series 1 represents the cumulative R value for each system configuration and the solid trend line represents a third ordered polynomial regression equation.

9.5.4   Summarizing the Results of Modeling H versus System Configuration and R versus System Configuration When Weighted by H

Table 9.14 displays the final data calculations for H and R (weighted by H) with respect to system configuration, and Figure 9.10 shows a graphical overlay of the data in Table 9.14. From Figure 9.10, it is clear that H and R meet at two distinct points, LF3 and WA1. Implicitly, these intersection points may indicate the levels of automation that balance H and R when considering military convoy operations. Moving forward, it may be beneficial to invest research, development, testing, and evaluation dollars into comparing LF3 and WA1 in order to quantify costs, benefits, operational effectiveness, and measures of performance.

Table 9.14 The results of calculation H vs. system configuration and R (weighted by H) vs. system configuration.

System configuration	H	R (weighted by H)
SQ	1	0
RC	1	0
TO	1	0
DW	1	0
DA	0.975198	0.004605
LF1	0.393719	0.101899
LF2	0.382607	0.107689
LF3	0.330677	0.203713
LF4	0.140635	0.348738
WA1	0.381694	0.394525
WA2	0.27464	0.617111
FA 1	0.068627	0.866391
FA 2	0.04491	0.89663

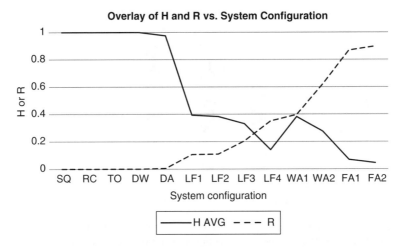

Figure 9.10 Summary of the results from Figure 9.1 and Figure 9.6.

9.6 Mathematically Modeling H and R

After successfully modeling H and R against the various system configurations, the final analysis to pursue was to create a single mathematical equation which accurately models R in terms of H. The model displayed in Figure 9.11 can be used to show the relationship between an automated military convoy's "smartness" and the amount of human input required.

9.6.1 Analyzing the Results of Modeling H versus R

Regression analysis was used to develop the "curve of best fit," which was a sixth ordered polynomial equation. However, a second ordered polynomial regression equation, which is

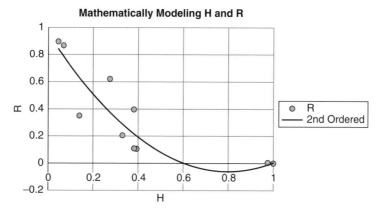

Figure 9.11 Mathematical model of human input frequency (H) and robotic decision-making ability (R).

Table 9.15 Using Eq. (9.20), which models H vs. R, to compute values of the dependent variable y.

x	$y = 1.5962x^2 - 2.5474x + 0.9553$
0	0.9553
0.1	0.716486
0.2	0.509524
0.3	0.334414
0.4	0.191156
0.5	0.07975
0.6	0.000196
0.7	−0.047506
0.8	−0.063356
0.9	−0.047354
1	0.0005

represented by Eq. (9.20), is shown via a solid trend line in Figure 9.11 because it is more logical than the sixth ordered polynomial equation and provides meaningful results for the engineer.

$$y = 1.5962x^2 - 2.5474x + 0.9553 \tag{9.20}$$

Equation (9.20) produces a coefficient of determination, R^2, equal to 0.8434. Let x represent H and y represent R. When x values, ranging from 0 to 1, are substituted into Eq. (9.20), the following results are calculated in Table 9.15. It appears that when x, that is, H, is chosen for values between 0 and 1 then Eq. (9.20) will provide an engineer with the corresponding R value and system configuration. For example, suppose that x=0.1. This would intuitively imply that R would have a high value because there is very little human input required to execute the mission. Based on Eq. (9.20), the corresponding y value for x=0.1 is 0.9553. An engineer can then reference Figure 9.10 to confirm that these numbers are relatively accurate and then determine which system configuration should be used as a design point. From Figure 9.10, when x=0.1 the approximate R value is 0.90 and the corresponding system configuration is FA1.

Note that because Eq. (9.20) has a R^2 value of 0.8434 then the regression equation does not fit the actual data as well as desired. This could lead to errors in approximating R in terms of

H and vice versa, but the overall trend is instructive. With respect to practicality, an engineer may receive more desirable results from referencing the equations that were used to construct Figure 9.10. However, Eq. (9.20) is still a valuable equation because it directly relates R in terms of H, and will generally provide the optimal system configuration neighborhood. More research can be performed to further refine Eq. (9.20). As the R^2 value approaches one, the formula will become more accurate in terms of quantifying H and R, and better at determining the appropriate system configuration for mission execution.

9.7 Conclusion

The potential implications of automation are staggering for national security and defense. Automation may allow for equal or greater combat power with drastically reduced personnel requirements, greater efficiencies in fuel consumption and maintenance, and increased utilization of platforms and assets. This automation continuum and model can serve as the basis for further discussions and analysis as the DoD wrestles with the incorporation of automation into military convoy operations. The increased use of automation will create a variety of challenges ranging from the technological and operational to the legal and moral, but understanding the tradeoffs between human input and robotic control will help engineers manage these challenges.

9.A System Configurations

Status Quo

Status Quo (Only analyzing the impact for the Logistics Platforms)
Manpower Needed: 2 Soldiers per platform → 6 Soldiers needed [Operationally]
                              6 Soldiers in the platforms
Benefits: No additional benefits due to automation
Cons: Soldiers are at risk
Costs: No additional costs due to automation

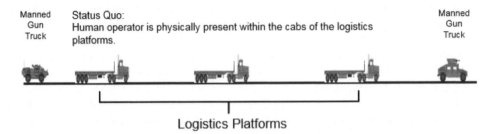

Manned Gun Truck   Status Quo:
Human operator is physically present within the cabs of the logistics platforms.   Manned Gun Truck

Logistics Platforms

Figure 9.A.1   Detailed overview of the status quo system configuration of automation.

Remote Control

Remote Control (Only analyzing the impact for the Logistics Platforms)

Manpower Needed: 1 Soldier per platform → 3 Soldiers needed [Operationally]
                 0 Soldiers in the platforms

Benefits: Reduces the amount of Soldiers needed operationally and removes all the Soldiers from the cabs of the platforms

Cons: Limited to specific mission roles, e.g. used in the FOB for parking, loading, and unloading vehicles; not suitable for local/line haul missions

Costs: Controllers, cords, etc. will be needed

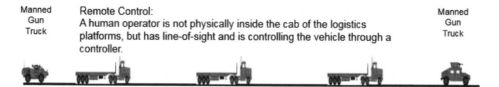

Manned Gun Truck

Remote Control:
A human operator is not physically inside the cab of the logistics platforms, but has line-of-sight and is controlling the vehicle through a controller.

Manned Gun Truck

Figure 9.A.2   Detailed overview of the remote control system configuration of automation.

Tele-Operation

Tele-Operation (Only analyzing the impact for the Logistics Platforms)

Manpower Needed: 1 Soldier per platform → 3 Soldiers needed [Operationally]
                 0 Soldiers in the platforms

Benefits: Reduces the amount of Soldiers needed operationally and removes all of them from the cabs of the platforms

Cons: Limited to specific mission roles, and operators tend to experience fatigue after a couple hours; most likely not suitable for the majority of local/line haul missions

Costs: Will need to have cameras that will provide situational awareness for the operators

Manned Gun Truck

Tele-Operation:
A human operator is not physically present inside the cab of the logistics platforms, but has non-line-of-sight and is controlling the vehicle through a controller. Video feed is needed for the operator to see the surrounding environment.

Manned Gun Truck

Figure 9.A.3   Detailed overview of the tele-operation system configuration of automation.

Driver Warning

Driver Warning (Only analyzing the impact for the Logistics Platforms)

<u>Manpower Needed</u>: 2 Soldiers per platform → 6 Soldiers needed [Operationally]
                          6 Soldiers in the platforms

<u>Benefits</u>: Provides the operators with an increased situational awareness, which may
          prevent potential accidents, roll-overs, etc.

<u>Cons</u>:   Does not remove any of the Soldiers from the platforms

<u>Costs</u>:   Will need additional sensors, signals, etc.

Manned
Gun
Truck

Driver Warning:
A human operator is physically within the cab and controlling the vehicle's
functions. Tactile, auditory, and visual sensors/warnings are used to indicate
potential danger. For instance, a vehicle in your "blind spot" may be indicated by
a blinking red light on your side view mirror. This warns the operator that it is not
safe to change lanes.

Manned
Gun
Truck

Figure 9.A.4   Detailed overview of the driver warning system configuration of automation.

Driver Assist

Driver Assist (Only analyzing the impact for the Logistics Platforms)

<u>Manpower Needed</u>: 2 Soldiers per platform → 6 Soldiers needed [Operationally]
                          6 Soldiers in the platforms

<u>Benefits</u>: Provides the operators with an increased situational awareness, i.e. greater than that
          provided by Driver Warning, which may prevent potential accidents, roll-overs, etc. The
          system has the ability to take control of some of the vehicle's functions, e.g. brakes

<u>Cons</u>:   Does not remove any of the Soldiers from the platforms

<u>Costs</u>:   Will need additional sensors, signals, actuators, etc.

Manned
Gun
Truck

Driver Assist:
A human operator is physically within the cab and controlling the vehicle's
functions. However, the vehicle's semi-automated system has the ability to take
control of certain driving functions in the event that a potentially dangerous
situation may occur. For example, if a pedestrian spontaneously walks in front
of the vehicle and the human operator does not notice or have the required
time to react, the semi-automated system will apply the brakes so injury does
not occur to the pedestrian.

Manned
Gun
Truck

Figure 9.A.5   Detailed overview of the driver assist system configuration of automation.

Leader-Follower 1

Leader-Follower 1 (Only analyzing the impact for the Logistics Platforms)
<u>Manpower Needed</u>: 2 Soldiers per Leader → 2 Soldiers needed [Operationally]
                             2 Soldiers in the platforms

<u>Benefits</u>:   Removes Soldiers from the platforms and may have additional savings, such as fuel consumption due to the system maintaining consistent speeds, braking, and interval distance

<u>Cons</u>:   May not be operationally effective because the Follower vehicles are dependent on the Leader vehicle, as well as the physical connection of the tethered wire. If the Leader is disabled, or the tethered connection is broken, then the convoy will stop

<u>Costs</u>:   Will need additional sensors, signals, actuators, tethered wire, etc.

Manned
Gun
Truck

Leader-Follower 1 (Leader):
A human operator is physically within the cab and controlling the vehicle's functions. A tethered wire will be used for the following vehicle to accurately traverse the desired path.

Leader-Follower 1 (Follower):
No human operator is present inside the cab of the vehicle. This vehicle mimics the behavior of the vehicle in front of it based on parameters, such as angle, slack, etc., of the tethered wire.

Manned
Gun
Truck

     Follower   Follower   Leader

Figure 9.A.6   Detailed overview of the tethered leader-follower system configuration of automation.

Leader-Follower 2

Leader-Follower 2 (Only analyzing the impact for the Logistics Platforms)
<u>Manpower Needed</u>: 2 Soldiers per Leader → 2 Soldiers needed [Operationally]
                             2 Soldiers in the platforms

<u>Benefits</u>:   Removes Soldiers from the platforms and may have additional savings, such as fuel consumption due to the system maintaining consistent speeds, braking, and interval distance

<u>Cons</u>:   The Follower vehicles are dependent on the Leader vehicle, as well as the data transmitted via non-physical means. If the Leader is disabled, or the connection is broken (potentially due to jamming), then the convoy will stop

<u>Costs</u>:   Will need additional sensors, signals, actuators, etc.

Manned
Gun
Truck

Leader-Follower 2 (Leader):
A human operator is physically within the cab and controlling the vehicle's functions. Sensors, GPS, etc. will be used for the following vehicle to accurately traverse the desired path.

Leader-Follower 2 (Follower):
No human operator is present inside the cab of the vehicle. This vehicle mimics the behavior of the vehicle in front of it based on the data that is received.

Manned
Gun
Truck

     Follower   Follower   Leader

Figure 9.A.7   Detailed overview of the un-tethered leader-follower system configuration of automation.

Leader-Follower 3

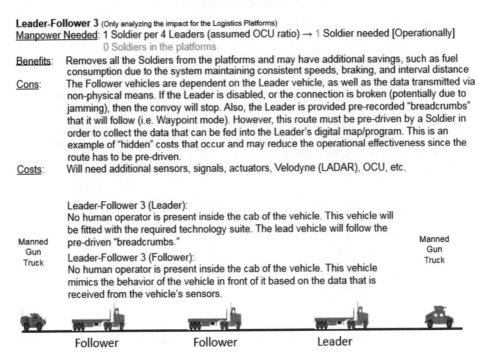

Leader-Follower 3 (Only analyzing the impact for the Logistics Platforms)
Manpower Needed: 1 Soldier per 4 Leaders (assumed OCU ratio) → 1 Soldier needed [Operationally]
        0 Soldiers in the platforms

Benefits: Removes all the Soldiers from the platforms and may have additional savings, such as fuel consumption due to the system maintaining consistent speeds, braking, and interval distance

Cons: The Follower vehicles are dependent on the Leader vehicle, as well as the data transmitted via non-physical means. If the Leader is disabled, or the connection is broken (potentially due to jamming), then the convoy will stop. Also, the Leader is provided pre-recorded "breadcrumbs" that it will follow (i.e. Waypoint mode). However, this route must be pre-driven by a Soldier in order to collect the data that can be fed into the Leader's digital map/program. This is an example of "hidden" costs that occur and may reduce the operational effectiveness since the route has to be pre-driven.

Costs: Will need additional sensors, signals, actuators, Velodyne (LADAR), OCU, etc.

Leader-Follower 3 (Leader):
No human operator is present inside the cab of the vehicle. This vehicle will be fitted with the required technology suite. The lead vehicle will follow the pre-driven "breadcrumbs."

Leader-Follower 3 (Follower):
No human operator is present inside the cab of the vehicle. This vehicle mimics the behavior of the vehicle in front of it based on the data that is received from the vehicle's sensors.

Manned Gun Truck

Manned Gun Truck

Follower   Follower   Leader

Figure 9.A.8 Detailed overview of the un-tethered/unmanned/pre-driven leader-follower system configuration of automation.

Leader-Follower 4

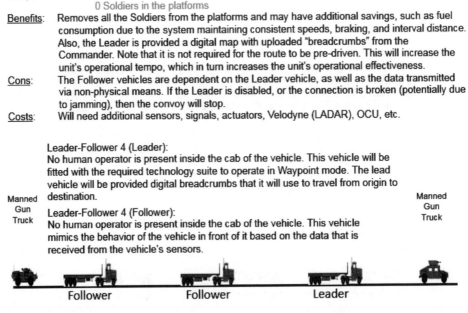

Leader-Follower 4 (Only analyzing the impact for the Logistics Platforms)
Manpower Needed: 1 Soldier per 4 Leaders (assumed OCU ratio) → 1 Soldier needed [Operationally]
        0 Soldiers in the platforms

Benefits: Removes all the Soldiers from the platforms and may have additional savings, such as fuel consumption due to the system maintaining consistent speeds, braking, and interval distance. Also, the Leader is provided a digital map with uploaded "breadcrumbs" from the Commander. Note that it is not required for the route to be pre-driven. This will increase the unit's operational tempo, which in turn increases the unit's operational effectiveness.

Cons: The Follower vehicles are dependent on the Leader vehicle, as well as the data transmitted via non-physical means. If the Leader is disabled, or the connection is broken (potentially due to jamming), then the convoy will stop.

Costs: Will need additional sensors, signals, actuators, Velodyne (LADAR), OCU, etc.

Leader-Follower 4 (Leader):
No human operator is present inside the cab of the vehicle. This vehicle will be fitted with the required technology suite to operate in Waypoint mode. The lead vehicle will be provided digital breadcrumbs that it will use to travel from origin to destination.

Leader-Follower 4 (Follower):
No human operator is present inside the cab of the vehicle. This vehicle mimics the behavior of the vehicle in front of it based on the data that is received from the vehicle's sensors.

Manned Gun Truck

Manned Gun Truck

Follower   Follower   Leader

Figure 9.A.9 Detailed overview of the un-tethered/unmanned/uploaded leader-follower system configuration of automation.

Waypoint 1

Waypoint 1 (Only analyzing the impact for the Logistics Platforms)

<u>Manpower Needed</u>: 1 Soldier per 4 Leaders (assumed OCU ratio) → 1 Soldier needed [Operationally]

      0 Soldiers in the platforms

<u>Benefits</u>:   Removes all the Soldiers from the platforms and may have additional savings, such as fuel consumption due to the system maintaining consistent speeds, braking, and interval distance. Every vehicle is an independent entity that is capable of completing the mission. If one vehicle is disabled, the other vehicles can push forward.

<u>Cons</u>:   Each vehicle is provided pre-recorded "breadcrumbs" that it will follow (i.e. Waypoint mode). However, this route will have to be driven by a Soldier in order to collect the data that can be fed into the Leader's digital map/program. This is an example of "hidden" costs that occur and may reduce the operational effectiveness since the route has to be pre-driven. Also, every vehicle must be fitted with the Velodyne (LADAR), which is a cost driver

<u>Costs</u>:   Will need additional sensors, signals, actuators, Velodyne (LADAR), OCU, etc.

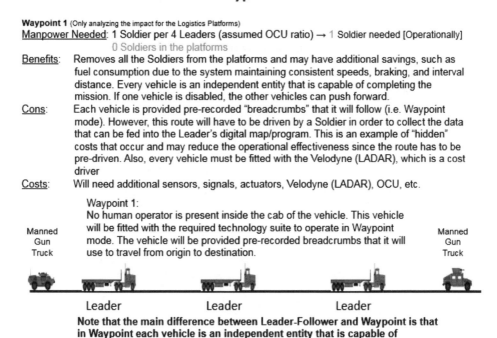

Note that the main difference between Leader-Follower and Waypoint is that in Waypoint each vehicle is an independent entity that is capable of executing the mission.

Figure 9.A.10   Detailed overview of the pre-recorded "breadcrumb" waypoint system configuration of automation.

Waypoint 2

Waypoint 2 (Only analyzing the impact for the Logistics Platforms)

<u>Manpower Needed</u>: 1 Soldier per 4 Leaders (assumed OCU ratio) → 1 Soldier needed [Operationally]

      0 Soldiers in the platforms

<u>Benefits</u>:   Removes all the Soldiers from the platforms and may have additional savings, such as fuel consumption due to the system maintaining consistent speeds, braking, and interval distance. Every vehicle is an independent entity that is capable of completing the mission. If one vehicle is disabled, the other vehicles can push forward. By providing a digital map, the need to pre-record the route will be unnecessary which will reduce costs and increase operational effectiveness

<u>Cons</u>:   Each vehicle will be provided a digital map and require the Velodyne, which are cost drivers

<u>Costs</u>:   Will need additional sensors, signals, actuators, Velodyne (LADAR), OCU, etc.

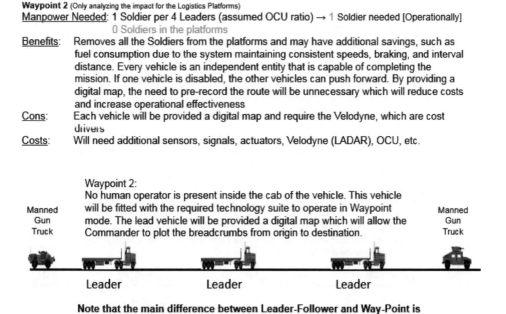

Note that the main difference between Leader-Follower and Way-Point is that in Way-Point each vehicle is an independent entity that is capable of executing the mission.

Figure 9.A.11   Detailed overview of the uploaded "breadcrumb" waypoint system configuration of automation.

Full Automation 1

Full Automation 1 (Only analyzing the impact for the Logistics Platforms)

Manpower Needed: 1 Soldier per 4 Leaders (assumed OCU ratio) → 1 Soldier needed [Operationally]
                  0 Soldiers in the platforms

Benefits: Removes all the Soldiers from the platforms and may have additional savings, such as fuel consumption due to the system maintaining consistent speeds, braking, and interval distance. Every vehicle is an independent entity that is capable of completing the mission. If one vehicle is disabled, the other vehicles can push forward. The Commander will provide origin/destination grid coordinates, and the vehicles will suggest routes to travel, which will reduce the Commander's planning requirements

Cons: High level software/hardware, such as the Velodyne, will be needed to execute the mission, which are cost drivers. Furthermore, more research, development, testing and evaluation will be required

Costs: Will need additional sensors, signals, actuators, Velodyne (LADAR), OCU, etc.

Full Automation 1:
No human operator is present inside the cab of the vehicle. This vehicle will be fitted with the required technology suite to operate in Full Automation mode. The lead vehicle will be provided coordinates of the origin and destination. The system will then generate various routes that could be taken. The Commander will then select the route that will be traveled.

Manned Gun Truck   Manned Gun Truck

    Leader   Leader   Leader

Figure 9.A.12   Detailed overview of the uploaded "breadcrumbs" with route suggestion full automation system configuration of automation.

Full Automation 2

Full Automation 2 (Only analyzing the impact for the Logistics Platforms)

Manpower Needed: 1 Soldier per 4 Leaders (assumed OCU ratio) → 1 Soldier needed [Operationally]
                  0 Soldiers in the platforms

Benefits: Removes all the Soldiers from the platforms and may have additional savings, such as fuel consumption due to the system maintaining consistent speeds, braking, and interval distance. Every vehicle is an independent entity that is capable of completing the mission. If one vehicle is disabled, the other vehicles can push forward. The Commander will provide origin/destination grid coordinates, and the vehicle will select its own route to travel and begin the mission

Cons: High level software/hardware, such as the Velodyne, will be needed to execute the mission, which are cost drivers. Furthermore, more research, development, testing and evaluation will be required

Costs: Will need additional sensors, signals, actuators, Velodyne (LADAR), OCU, etc.

Full Automation 2:
No human operator is present inside the cab of the vehicle. This vehicle will be fitted with the required technology suite to operate in Full Automation mode. The lead vehicle will be provided coordinates of the origin and destination. The system will then select the best route to travel and begin the mission independent of human input.

Manned Gun Truck   Manned Gun Truck

    Leader   Leader   Leader

Figure 9.A.13   Detailed overview of the self-determining full automation system configuration of automation.

10

Experimental Design for Unmanned Aerial Systems Analysis: Bringing Statistical Rigor to UAS Testing

Raymond R. Hill and Brian B. Stone
Department of Operational Sciences, US Air Force Institute of Technology/ENS, Wright-Patterson AFB, OH, USA

10.1 Introduction

This chapter offers a prescription to help achieve statistical rigor in Unmanned Aerial Systems (UASs) test planning and ultimately test execution. Throughout this chapter we use the general term UAS realizing other similar meaning terms include Unmanned or Uninhabited Aerial Vehicles (UAVs), Unmanned Combat Aerial Vehicles (UCAVs), even drones, among potentially many other names.

UASs are seemingly found everywhere. Rarely a day goes by without some news item relating to a UAS, or some impending use of a UAS, even cases of near collisions between a UAS and commercial aircraft. Commercial companies and research consortia are sprouting up to "ride the wave" of interest in UAS applications. Even leading television shows have plots depicting the use of UAS in surveillance and terrorist activities.

Regardless of the application and the UAS platform, testing plays an important role in the development and deployment of any UAS. That testing role, or at least a prescriptive methodology for the planning involved in that testing role, is the focus of this chapter. UAS testing instances range from demonstration tests to operational testing. A demonstration test may be a very focused test of specific UAS capabilities. For instance, a demonstration of a UAS

Operations Research for Unmanned Systems, First Edition. Edited by
Jeffrey R. Cares and John Q. Dickmann, Jr.
© 2016 John Wiley & Sons, Ltd. Published 2016 by John Wiley & Sons, Ltd.

capability to take off, converge to a target, circle and observe a target, and successfully return to base constitutes a viable test for a surveillance UAS. Another test may be to position the UAS in a variety of flight profiles, execute each profile, and collect various response data to help characterize engine performance. In the former case, specific outcomes are required to declare the success of the test. In the latter case, a variety of data are collected and related to flight profile settings with success of the test defined in terms of the quality of the data achieved. The former test is not usually associated as an experimental design situation while the latter is easily identified as benefiting from experimental design.

Although each of the above cases are experimental designs of varied complexity, scope, and objectives, each should be approached with an understanding of experimental design principles and a systems-thinking approach to the statistical design of the experiment – at least from the experimental planning perspective. While our focus will be more on those experiments with objectives to characterize some aspect of UAS performance, our prescriptions apply quite well to demonstration testing since our focus is on the pre-experiment planning.

10.2 Some UAS History

Keane and Carr [1] and Blom [2] provide excellent histories of UAS development and employment, primarily from a defense perspective. In fact, most purported histories of UAS technology have a defense focus. The modern landscape finds ubiquitous use of UAS technology throughout the Department of Defense (DoD) and civilian sector. Modern military UAS applications include:

- intelligence and reconnaissance,
- weapons delivery,
- sensor platforms,
- communications relay nodes.

There is growing interest in the civilian applications as well to include:

- hobbyist,
- urban and park surveillance,
- news and sports telecast,
- security monitoring,
- package delivery.

The growth in commercial and defense use of UAS technology has made this a true growth industry arguably well beyond that ever imagined by the model aircraft hobby industry. While it is not our purpose to relate a full history of UAS use and development, a generalized recounting is beneficial.

Most historians seem to favor early balloon experiences as the start of UAS interest. As early as during the Civil War, balloons were weaponized; historians recount the dropping of incendiary devices over enemy forces. Generally, the record indicates limited success in these endeavors [3].

Powered UAS flights pre-date manned powered flight. As early as 1894 steam powered aircraft were flown, but these were over very short distances that were not nearly far enough for effective deployment [3]. Such inquires were, however, enough to prompt interest in developing "flying bombs" or "aerial torpedoes" as they were called during the early 1900s. This seemed to remain the focus of UAS technology development and employment through the World War I years.

The first US unmanned weapon is credited as the Kettering Bug, developed and tested during World War I in Dayton, Ohio [4]. The device was never fielded and was not a stunning success, but was designed for a reasonable range (up to 75 miles), payload (300 lb), and a defined target point (as calculated by expected engine revolutions to reach the target).

After World War I and into World War II, the primary focus and successes for UAS were in drone operations. Such drones were used by the Navy to release torpedoes, by the Army as anti-aircraft targets, and by the Air Force as aerial targets. Success in these drone operations led to the evolution of these platforms to perform reconnaissance missions [2]. The amount and type of testing involved in these early efforts are unclear. The accounts merely indicate the general approach to testing the concepts.

Weaponeering of a UAS has been a focus since the beginning. Technical problems hindered early success, with examples of problems found in the inaccuracies of the German V1 and V2 rockets, the unreliability and lack of success associated with US attempts to remotely pilot older aircraft during World War II, and various US efforts during the 1950s and 1960s. See Blom [2] for the full history.

The UAS as a viable weapon system appears to have attained real interest with Israeli success against Syria in their 1982 conflict [3]. During that Israeli–Syrian confrontation, a combination of manned and unmanned assets were quite successful in decimating the Syrian Air Force while it was still on the ground.

US interest in UAS capabilities also grew during the 1980s. The strongest evidence to that interest was the publication of a DoD Master Plan for UAS development produced in 1988 [2]. UAS growth in DoD combat campaigns has continued to grow, and systems such as the Air Force Reaper and Global Hawk systems have had dramatic successes documented in leading national news stories. In fact, the growth of UAS missions has promoted controversial debates on the type of personnel operating the systems, the legal ramifications of using the systems, and even the viability of continued use of manned aerial systems. A recounting of these debates is left to other work.

10.3   Statistical Background for Experimental Planning

While our focus is on the planning of a UAS experiment, understanding the statistical planning process for experimental design requires a basic understanding of the terminology used in experimental planning and analysis. Those with a good working knowledge of US DoD Research, Development, Test, and Evaluation (RDT&E) can skip to Section 10.5.

A **factor,** or experimental factor, is an input or controllable parameter for the system. Factors are what researchers believe will influence UAS performance. Factors that we choose to vary or control during a test must relate to the particular objectives of a test. Each factor has prescribed levels defined for the test. These levels are the specific values to which the factor is set and controlled to that value for the experimental run. A **design point** is the specific combination of factor levels set and controlled in a specific experimental run. Combining all the design points planned for the full scope of the experiment yields the experimental design, or the experimental design schedule.

The **response** is the output of the system collected during or at the completion of an experimental run and deemed the output of interest for the system. For a UAS, such a response might be a data stream from the instrumentation aboard the UAS, data from the ground station, or in the case of a demonstration, the determination of a success or failure of the test. The goal of more complex tests is to understand and model how changes to factor level settings influence the system response. An empirical regression model is a powerful approach to quantifying the change in a response variable due to a change in the independent variables.

Noise and **bias** are key concepts in any experimental design and are particularly important in UAS testing. Noise is caused by unknown factors, called lurking variables in some textbooks, that cause variation among responses when obtaining those responses from identical settings. Noise is modeled in our statistical analyses by the random error component. Planning efforts that explicitly define sources of noise in a test can devise strategies to mitigate the effects of that noise in obscuring experimental results.

Bias is the effect of factors influencing the response but not controlled within the experimental design strategy or during experiment execution. For instance, consider a test of computer-based learning skills, in which a suite of n learning scenarios are used and 24 randomly selected participants use each of the learning scenarios in order and their new skill levels are assessed. If the order of the n learning scenarios are the same for each participant, and the latter scenarios are deemed better than the earlier scenarios, one cannot determine if any change in improved skill level among the participants is due to the improved scenarios or to the learning effect of the participants going through all the scenarios sequentially. This learning effect is a bias on the response that masks the effect of the independent variables. Poor test design and execution produce biased results.

Montgomery [5] states that the three fundamental principles of experimental design are **replication**, **randomization**, and **blocking**. A **replicate**, or independent repetition of a design point, allows the experimenter to estimate experimental error. Sources of experimental error include inaccuracies in the measurement system as well as the influence of unknown or uncontrollable factors. Estimating error is crucial to the process of testing whether estimated factor effects have statistical significance. Replicated designs are preferred since there is (i) more data to estimate factor effect levels and (ii) an ability to accurately estimate the pure error component of the experimental noise.

The pure error estimate is used to assess the lack of fit component of noise. Lack of fit refers to the unexplained variation in the response variable unexplained by the existing regression model terms and pure error. Significant lack of fit is a signal to statistical analysts to further refine their empirical model.

Randomization, both of the experimental test runs and the application of experimental units to those runs, is an important aspect of test planning. Randomization allows the effects of unknown and uncontrollable factors to be averaged out over the test runs. It is a very effective method to reduce accidental correlation between a test factor and a nuisance factor. Unfortunately, full randomization is sometimes impractical; more advanced experimental designs (such as nested or split plot designs) accommodate such restrictions on randomization.

Finally, blocking is a technique used to create experimental units that are as homogeneous as possible so that we can more precisely determine influential process factors. Variation in experimental resources such as batches of raw material, differences between personnel interacting with the system under investigation, or changes in the ambient experimental conditions over time induce variation in the measured responses. These variations increase our estimate of the random error in the process. As the estimate of the random error increases, our power to detect active factors decreases. By labeling known sources of variation as **blocking factors** and designing our test matrix so that the factors of interest are uncorrelated with the blocks, we can obtain a more precise estimate of error and more accurately identify active factors.

Other recurring experimental design concerns that are addressed in the Design Of Experiments (DOE) methodology include multicollinearity, model misspecification, prediction variability, and design power. Multicollinearity occurs when the input variables (factors) within the regression model are related or correlated. When factors are highly correlated, the regression model coefficients can be incorrect leading to incorrect conclusions based on the experiment results. Instances of multicollinearity are easily missed to those unaware of its potential existence. Statistical design methods use experimental designs with independent factor levels as much as possible, and use advanced statistical techniques to reduce the effects of multicollinearity when it cannot be avoided. Statistical experts on experimental design teams help recognize and alleviate problems associated with multicollinearity.

Model misspecification occurs when there is a misunderstanding of the underlying system; this leads to incorrect analyses associated with the regression model. Proper planning involves the close cooperation between the system subject matter experts and the statistical experts to ensure the resulting experimental design is constructed according to DOE principles. Good regression models give predictions with small and consistent measures of variability around the predictions. There are designs whose variability properties are not consistent, and this can lead to problems when drawing conclusions based on the test results. Modern computer-based approaches to statistical experimental design often employ optimal designs, which have factor levels chosen to optimize an objective function related to the design quality.

The power of a test deserves special attention, but requires some preliminary explanation regarding the structure of a statistical test. Traditionally, statistical tests are based on a research or test hypothesis consisting of two pieces: a null hypothesis and an alternate hypothesis. The null hypothesis is the expected value of some sample statistic when certain distributional properties are assumed to be true; the null hypothesis is assumed true. In practice, the null hypothesis is what we want to disprove stated in statistical terms. The alternate hypothesis is the conclusion reached when the null hypothesis is proven false by the evidence presented by the experimental data.

Consider the following example as how to construct the hypotheses. A new missile system may require higher reliability than its predecessor system. A null hypothesis would assume no difference in reliability (which we want to disprove) while the alternate hypothesis would conclude greater reliability in the new system (which we want to accept). The data from the experiment are then collected and analyzed to assess the probability that the data came from the assumed null hypothesis. When this probability is too low, falling below some predetermined probability called the α level, the null hypothesis is rejected and the alternate is assumed true. If the probability is not too low, we fail to reject the null hypothesis.

Experimental tests have two common errors, a Type I and Type II error. A Type I error occurs when the data from the experiment leads to the conclusion of rejecting a null hypothesis, when it is actually true. A Type II error, β, occurs when the experiment leads to a conclusion of failing to reject a null hypothesis when in fact that hypothesis is false. In operational terms, a Type I error on a bomb weapon system, for example, means we reject a good bomb. A Type II error on that same system means accepting a dud. In the DoD, Type II errors are quite serious and are generally avoided. The power of a test is the value $(1-\beta)$. A test designed to maximize the power, subject to some value of α, involves reducing the Type II error, by adjusting the size of the test, usually via the number of factor levels considered or replications conducted.

10.4   Planning the UAS Experiment

10.4.1   General Planning Guidelines

In 1977, Hahn noted, "Obtaining valid results from a test program calls for commitment to sound statistical design" [6]. This early paper laid out aspects of experimental design important to the engineer from a statistical perspective. The work also delineated the differing roles of the engineer and the statistician within a team working to arrive at a final experimental plan "tailor made to meet specific objectives and to satisfy practical constraints."

Coleman and Montgomery [7] extended Hahn's work laying out a systematic approach to experiment planning since "the planning activities that precede the actual experiment are critical to successful solution of the experimenter's problem." They emphasize the roles of experimenter versus those of the statistician proposed by Hahn [6], but extend the concept by introducing and discussing the "gap" between the two types of specialists on the planning team.

This knowledge gap between the experimenter and the statistician is due to differing levels of knowledge and experience as brought to bear in the experimental planning task. The experimenter likely has significant system knowledge and little statistical knowledge with the reverse holding for the statistician. The gap is not bad unless its existence is ignored. Bridging the gap via systematic experimental planning brings the strengths of each team member to bear on the planning process and ultimately on the quality of the resulting experimental design. Montgomery [5] lists the steps in experimentation as:

1. Recognition and statement of problem.
2. Choice of factors and levels.
3. Selection of response variable(s).
4. Choice of experimental design.
5. Conduct of the experiment.
6. Data analysis.
7. Conclusions and recommendations.

While arguably the leading text on experimental design, Montgomery's text really does not provide a lot of emphasis on step 1, the planning process for the experiment. This is a shortfall shared by all experimental design texts familiar to the authors. The Coleman and Montgomery [7] paper, some of which can be found as supplemental material in [5], provides a viable planning method complete with full documentation of that planning process.

Guide sheets are the mechanism used in [7] to promote thorough planning by the multifunctional statistical design team. Key takeaways from their discussion include:

- thorough knowledge of pertinent background related to or impacting the current test,
- a test objective that is clearly defined, specific, and measurable,
- responses that are relevant to the objective and measurable,
- a clear understanding of all factors, controlled or uncontrolled, affecting the system or process under study.

The Coleman and Montgomery [7] paper, along with the associated discussion articles in the same issue of the journal, are useful reads for anyone associated with test planning. The 1998 National Research Council (NRC) study on how to improve the statistical aspects of defense testing and acquisition contains similar recommendations [8].

The NRC study [8] focused on how to improve Test and Evaluation (T&E) throughout the DoD. The emphasis was whether statistical methods might aid DoD T&E. The recommendations, if adopted, would in their words give the DoD an environment where "[e]fficient statistical methods are used for decision making based on all available, relevant data." Their detailed chapter on test planning heavily references Coleman and Montgomery [7].

Updates to [7] are found in Johnson *et al.*[9], giving examples from DoD experimentation, and Freeman *et al.*[10], which provides a more general tutorial on the subject of planning experiments. We focus on the Freeman *et al.*[10] tutorial paper.

Freeman *et al.*[10] emphasize the strong linkage between the scientific method and statistical thinking. Quoting Donald Marquardt,

The scientific method is an inherent part of all experimental sciences. Statistics is the discipline responsible for studying the scientific method with the greatest intensity and for providing in-depth expertise to other disciplines.

Another key point made by Freeman *et al.*[10] is that all experiments are sequential, meaning all experiments are built on past experience, either from system knowledge, empirical knowledge, or some combination of the two. This is an important concept in experimental planning – there is always systems knowledge that can be leveraged during experimental planning. Both [7] and [10] provide an emphasis on planning. Items of emphasis added by [10] include:

- Have clearly defined experimental goals.
- Keep scientific questions limited to supporting those defined goals.
- Address how the current experiment supports any overall sequential test strategy.

10.4.2   Planning Guidelines for UAS Testing

The testing of a UAS, whether for demonstration purposes or for some performance characterization purpose, is not unlike the testing of other complex, human-integrated systems. Eventual success of the test event depends upon successful test planning; it is well known that no amount of data analysis can compensate for poor planning. A prescriptive process for the planning of UAS testing builds upon the concepts of experimental design previously discussed. A UAS test planning process involves the following steps:

- Determine the specific research question(s) to be answered by the test.
- Determine the role of the human operator in the study and the test.
- Define and delineate factors of concern for the study.

- Determine response data to collect and correlate to the research question(s).
- Select appropriate design.
- Define test execution strategy.

10.4.2.1 Determine Specific Questions to Answer

The research question to be answered by the test drives all aspects of the test planning and ultimately determines test success or failure. To maintain test focus, there should be a minimum of research questions posed; too many questions can lead to an overly complex test and a dilution of fidelity in the answers to those important research questions. Each research question should be specific to ensure there are clear measures of compliance and clear criteria for success or failure. Ambiguity in the driving research question(s) can, in some cases, allow too wide a range of acceptable results. Ambiguous results are not particularly useful when attempting to draw statistically defensible conclusions for decision-making purposes.

Unfortunately in the UAS domain operational scenarios and performance requirements may be defined before the test planning. The disconnect between system requirements definition and the eventual testing of the UAS system was a driving reason for the myriad failures discussed in the 2003 report by Carr *et al.*[11]. Even operational test events should be structured to measure responses that answer questions that trace back to the system requirements.

The problem gets further complicated when considering a test as one event in a sequence of tests, such as in a developmental or operational test campaign. Planning for the test early in the system development process is necessary to avoid problems later in the process when the system is actually being prepared for test. Warner [12] notes, "the importance of test planning, including the formulation of the evaluation plan and achieving a shared agreement of the plan among stakeholders" as a necessary and difficult component of eventual success in the operational testing domain.

10.4.2.2 Determine Role of the Human Operator

Human operators are crucial in UAS operations as, "the success of a UAV mission relies heavily on human operator, human–human and human–machine communication, and communication of human and machine activities" [13]. The human operator is a significant cause of UAS mishaps [13] and a major source of variability within an experiment. Hodson and Hill [14] indicate that in the context of Live, Virtual, and Constructive (LVC) simulation experiments involving a human operator, "if not the focus of the test, human operators are controlled factors and thus their free will may need to be constrained to obtain the objective data required to estimate system effectiveness and system factor contributions."

What the above means is that when a UAS test objective specifically focuses on the human operator capabilities, the human is part of the test. Thus, human operators are treated as an experiment factor and their input controlled just like any other non-human factor considered in the test. This means including in the test execution plan controls on the operator behavior to ensure they stay within the parameters of the test and they do not introduce bias into the system response.

In other situations, the test is such that the operator is not a focus of the experiment. In these cases, human operators are treated as a held-constant factor and their influence on the system under test should be minimized. This might be a difficult process to define and execute when the test falls within the context of operational assessments, as those are quite often focused on the human operator. The test planning team will need to ensure clear definition of the specific test objectives to specifically define the role of the human operator for that test.

10.4.2.3 Define and Delineate Factors of Concern for the Study

As noted by each of Hahn [6], Coleman and Montgomery [7], and Freeman *et al.*[10], as well as any experimental design textbook, a crucial component of the experimental planning process is identifying those factors of interest believed to influence the system response of interest. Not only must the team identify these factors but they must also determine levels of interest for those factors. These varied factors are called independent variables or control variables.

Two additional types of factors are also considered: held-constant and nuisance factors. These two categories represent factors that can influence the system response but are not a primary focus of the planned experiment. How well an experiment controls each factor determines its category membership.

A held-constant factor can be set to some defined level and maintained at that level for the experiment. By holding the factor constant there should be no variability in the system response caused by that factor. The planning questions are what levels to use and how well those levels are maintained during each experimental run.

A nuisance factor is believed to be an influence on the system but may not be controllable. Its variation during an experiment run will inject variability into the experimental data biasing statistical estimates of error unless precautionary steps are taken. A favored approach is to measure the nuisance factor level during the experiment and use regression techniques to remove any nuisance factor influence from the estimates of experimental error.

A deceptively easy part of experimental planning is determining the levels of the experimental factors. Categorical factors have defined levels but including all possible levels may lead to extremely large factorial experiments when using multiple categorical factors with more than two levels. Test planners should attempt to limit the levels of categorical factors to two or three levels when possible.

Continuous factors may appear to present more of a test-planning challenge, as there are a theoretically infinite number of possible levels for these factors. However, by taking advantage of regression analysis, continuous factors can be adequately tested with two or three levels. A linear model can be fit to data as long as there are two unique values of the independent variable and a quadratic model can be fit to three unique values.

In simple linear regression it can be shown that response data based on input from the highest and lowest feasible value of an independent variable leads to regression models with the lowest coefficient variance. This principle extends to multiple linear regression as long as all extreme factor level combinations are feasible. Subject matter experts should be consulted to determine the highest and lowest levels of the continuous factors that are of interest. If the independent variable's levels are integers, the extreme values should be chosen so that the midpoint of the range is also an integer. The midpoint, or center value, can be used to determine if a linear model is sufficient to model the relationship between the factor and the response or if a quadratic model is more appropriate.

To summarize, it is important for the planning team to identify and categorize factors, explicitly define how well these factors can be set or measured, and consider all such factors in the test execution strategy. For UAS testing, this planning may require specific controls, protocols, and instrumentation for success.

10.4.2.4   Determine and Correlate Response Data

It is generally not hard for a planning team to define a list of responses associated with specified test objectives for a system or process. The challenge is drawing from such a list those responses that are measurable and directly related to the specific objective driving the experimental planning. For instance, a UAS control station display is modified to provide a chat capability among the operators. The response variable is the answer to a questionnaire asking whether the operators like the capability. In this situation the objective of the test is unclear and the response is ambiguous.

A more nuanced question is whether the chat capability improves team performance during operations. While still somewhat vague, the question is more focused than before. The data collection system can monitor the use of chat during crucial phases of the operation within the test to correlate its actual use to the phase in which it was used, and intended to support. The data analysis subsequent to the test can now objectively determine if the operators used the chat and whether their like or dislike of the feature correlates to their use of that feature.

Qualitative responses are quite useful in human-centric systems such as UAS. Full reliance on such responses to produce statistically defensible results can be questionable. In the above scenario, the qualitative questionnaire response could be supplemented with a response variable measuring some aspect of the operation that may be improved by the chat feature, such as the speed of completing a group task.

The statistical planning team should see measurable response variables directly related to the objectives (research questions) of the experiment. Defining such measures allows the planning team to develop specific criteria used to assess achievement of those objectives. These measures can also provide support to the qualitative data collected via questionnaires or operator interviews. The detailed planning guide sheets, and supporting narrative, in [7] and the discussion in [10] provide excellent overviews of response variable definitions for the statistical experiment planning team.

10.4.2.5   Select an Appropriate Design

Johnson *et al.* [9] discuss various experimental designs useful for DoD testing. Textbooks on experimental design specialize in discussing the range of experimental designs available and how to analyze the data collected using that experimental strategy. Designs are often categorized as regular or non-regular designs. Regular designs have either uncorrelated or perfectly correlated effects. Non-regular designs have at least one pair of effects with only partial correlation.

Test strategies based on the principles or replication, randomization, and blocking usually involve the use of 2^k factorials, which are regular designs with k experimental factors set at two possible levels. As the number of factors increases, experimenters use 2^{k-p} fractional factorials, which include only a specially chosen subset of the runs in 2^k factorial designs to reduce the experiment resource requirements. As the notation implies, these designs have run sizes which are a power of 2, which can be limiting in some circumstances.

When resource constraints prevent the use of 2^{k-p} fractional factorials, experimenters use non-regular designs. Categories of non-regular designs in the DOE literature (which are not mutually exclusive) include Plackett–Burman designs, orthogonal arrays, no-confounding designs, and Definitive Screening Designs (DSDs). Non-regular designs can have large correlations between effects or insufficient power to detect active effects and so must be carefully evaluated before use.

Optimal designs are algorithmically generated designs which optimize an objective function related to the design matrix. Software that generates optimal designs may request the number of runs in the experiment, the number and type (categorical or continuous) of factors, the number of levels for each factor, and other details such as the number of desired experimental blocks. The software then produces a design which optimizes an optimality condition determined by either the software of the user. The most common examples are D-optimal and I-optimal designs. The D-optimal design seeks to minimize the determinant of the experimental design matrix. In practical terms this provides improved accuracy when estimating the regression model coefficients. The I-optimal design minimizes the average scaled prediction variance, the definition of which is beyond the scope of this chapter. In practical terms, an I-optimal design improves the accuracy of predictions throughout the design space.

Optimal designs are particularly useful when some combinations of factor levels are infeasible. Other reasons to use optimal designs include the presence of categorical factors with more than two levels or the inability to conduct experiments where the number of runs is a multiple of four. Like non-regular designs, optimal designs generated by software should be evaluated before they are used.

Rather than presenting the details of the above designs, especially since there are a variety of software packages that can generate them, we discuss some of the characteristics of the design selection process the statistician should use as part of the statistical experiment design team. The reader is referred to [5] for a detailed discussion of design matrices and their construction.

Since no test has unlimited resources, the **sample size** of the test should be determined. This may be a function of test cost, test time allowed, or the number of test articles available. This overall sample size will likely need to be spread over many test events.

Experimentation should be thought of as a campaign of sequential test events, where each test event is planned based on information learned from previous experimentation. A general strategy is to first assemble a large list of potentially influential factors and execute a **screening experiment** to narrow down the active factors. Following the analysis of the screening experiment, subsequent experimentation can be conducted with the reduced set of factors to investigate interaction effects and possibly higher order model terms.

The sequential strategy of experimentation has the potential to significantly reduce experimental costs. Regular 2^{k-p} fractional factorials which can estimate unconfounded Main Effects (MEs) and Two-Factor Interactions (2FIs) effects (resolution V designs) have 64 or more runs when the number of factors exceeds 6. For this reason, a simple sequential strategy is often used where the active ME are determined using a screening design, and confounded effects are then de-aliased by augmenting the original test matrix with additional experiments.

The screening experiment is usually a resolution III design having ME correlated with 2FI, or a resolution IV design having ME uncorrelated with 2FI, but 2FI correlated with other 2FI.

Once many of the inactive ME and 2FI are eliminated, the augmentation runs required to deals with the remaining candidate factors, when added to the original design, still result in fewer total runs than executing an initial resolution V design in the original number of factors.

Test results can be used to examine particular factor effects or to estimate a system response surface function. The latter case requires a preliminary estimate of the shape of the response surface. The experimental design selected must be able to provide estimates of the expected terms in the polynomial equation. For instance, a design that only provides estimates of linear effects cannot be used to estimate a quadratic model. Popular designs for estimating quadratic surfaces, such as the central composite designs discussed in Johnson et al.[9] are not as useful when the underlying response surface is of a higher order (a model misspecification error).

Response variables may be discrete or continuous. Naturally, continuous response variables allow more detailed modeling of the interaction between the independent variables and the response through the creation of a response surface. The type of response variable also impacts the number of data points required. As Coleman and Montgomery [7] discuss, the preferred forms are continuous response variables.

Variability in the experiment must be considered. Variability in the regression model coefficients or in system response predictions is a function of the response data collected (in the form of the estimated error) and the experimental design. Since the planning process controls the design, the planning process team might consider designs that achieve variance properties that improve either the variance of the response function coefficients or the variance of any system performance predictions. Computer packages will usually provide the means to generate designs that are optimal with respect to regression model coefficients (i.e., D-optimal) or are optimal with respect to the prediction variance (i.e., I-optimal). Montgomery [5] provides a nice discussion of these and the other categories of alphabetic optimality conditions and how these conditions affect the experimental design.

10.4.2.6   Define the Test Execution Strategy

The final step is to define the test execution strategy. In practice this includes all the logistical details of conducting the test. For UAS testing this may include safety of flight reviews and human subject experimentation reviews. Our concern in this discussion are details for the strategy to facilitate success from the statistical and analytical perspective.

The **schedule** for the experiment provides the experimental run order. This will include any amount of randomization of the design, in-test preparation for subsequent tests (e.g., repositioning the path of the UAS for the next phase of the experiment), and the swapping of human assets, if needed (e.g., swapping out operator teams).

For UAS testing, article failure is a risk. Thus, the schedule might want to be constructed so that should a failure occur, canceling the remainder of the test, preliminary results can be gleaned from the test runs successfully completed. Running experiments in successive blocks of one complete unreplicated design is a strategy toward this end.

Since the human operator is prominent in UAS operations, the test execution strategy should consider human operator behavior controls. Whether or not they are a focus of the experimental objective, human operators will have specific tasks to accomplish during the test. The test strategy needs to define those tasks, provide assurances to limit deviation from those tasks and clarify why deviations by the operator are discouraged. An operator altering test protocol because "a different way is easier" can cause changes to the basic experimental conditions that could potentially invalidate the results derived from the complete set of data collected.

10.5   Applications of the UAS Planning Guidelines

As the use of Small Unmanned Aerial System (SUAS) vehicles in the DoD increases, an important area of research involves improving the capability of the aircraft to operate autonomously. One method of adding autonomous capability to an SUAS system is the inclusion of state-based logic in the navigation software. State-based logic, often used to create objects modeled as finite state machines, involves code that changes the state of a machine based on inputs from the environment [15].

The next section describes the proposed methodology employed to add state-based logic to the open-source autopilot software, ARDUPLANE, used to fly SUAS platforms typically used by model airplane hobbyists. The UAS test planning process was used in both the development and testing phases of the process to create the state-based logic in the SUAS. The description of the testing process is divided into the testing process steps described in the previous section.

10.5.1   Determine the Specific Research Questions

There were two research questions for this phase of the software development: In which SUAS states is it advantageous to change system parameter settings in order to improve the aircraft autopilot performance as it follows a vehicle, and which system parameters should be changed? The legacy software code was designed to receive a GPS waypoint, navigate to within a specified radius of that waypoint, and then enter a loiter about the waypoint. "Follow me" mode was an option that allowed the SUAS to continuously accept updated waypoints from a GPS transmitter located on a vehicle, thereby allowing the SUAS to autonomously follow the vehicle.

A central assumption to this research was that environmental factors such as wind direction and speed would affect the ability of the SUAS to follow a vehicle. The code written to execute the "Follow me" mode, referred to in the following section as the legacy code, did not take environmental factors into account. It was proposed that by modifying software parameters based on the state of the SUAS – as defined by environmental factors – the ability of the SUAS to follow a vehicle would improve.

10.5.2   Determining the Role of Human Operators

The goal of this software development was to improve the capability of an SUAS to navigate autonomously, so there were limited roles for human operators. The majority of testing was conducted using flight simulator software written to train hobbyists to fly model aircraft. An operator role during the simulation testing was to ensure test parameters were set appropriately. A more crucial task was to reduce variability in the response data by ensuring the initial test conditions in the simulation were as homogeneous as possible.

There were two key roles for human operators during real-world experimentation. First, a safety pilot was present at the range to take over operation of the aircraft from the autopilot in the event of unexpected or dangerous aircraft activity. Second, the test engineer was respon-sible for uploading the test parameter settings to the autopilot. The test engineer guided the target vehicle with a laptop computer running software that transmitted waypoints to the SUAS. As with the simulation environment, it was essential that the test engineer help to reduce variability in the response caused by differences in the initial conditions at the start of each experimental run. This was accomplished by placing the SUAS in a loiter about the target vehicle prior to the start of the test. The test run was initiated when the SUAS was at a speci-fied point in the loiter path; this starting point was used for all test runs.

10.5.3   Determine the Response Data

There were several response variables that could be used to measure the ability of the aircraft to follow a vehicle in autopilot mode. Candidate response variables were the average distance between the SUAS and the vehicle, the minimum distance, the maximum distance, and the variance of the distance. In the context of vehicle surveillance, it is often important for the SUAS to maintain a constant distance from the target vehicle. It is therefore beneficial to min-imize the variance of the following distance in these circumstances, even if the overall average distance is not reduced as much as possible. It was decided to make the primary response var-iable the variance of the following distance, and the secondary response variable the average following distance.

10.5.4   Define the Experimental Factors

Factors in this research fell into two categories: environmental factors that affected the state of the SUAS, and factors related to parameters in the autopilot software that determined the performance of the aircraft. An Ishikawa (fishbone) diagram, shown in Figure 10.1, was used as a tool to identify potential factors.

Major categories of factors that might affect the response variables were parameters affecting aircraft responsiveness, parameters related to navigation settings, ground vehicle maneuvers, and environmental conditions. After reviewing the software documentation for the ARDUPLANE, throttle slew rate, maximum bank angle, and roll time constant appeared to be the most relevant factors related to aircraft responsiveness. Waypoint radius, waypoint loiter radius, and target airspeed were considered to be relevant parameters for navigation. Vehicle speed and the type of vehicle maneuver were suggested as factors related to the ground vehicle activity. Finally, wind direction and wind speed were chosen as the environmental factors of interest.

While all of the factors were considered important, there were practical limitations to inves-tigating the effects of every factor. First, the ARDUPLANE was not capable of determining vehicle speed. The SUAS simply attempted to reach the last known waypoint until an updated waypoint was transmitted. It was determined to be outside the scope of this research to write software that would compute vehicle speed and modify the aircraft speed accordingly. A sim-plifying solution was to design test maneuvers where the vehicle maintained a constant speed and the SUAS target airspeed was set to the same constant speed. Using this methodology, the factors target airspeed and vehicle speed were set to the same constant value and removed as test factors. Although this was a simplification of the pursuit paradigm, it was a reasonable

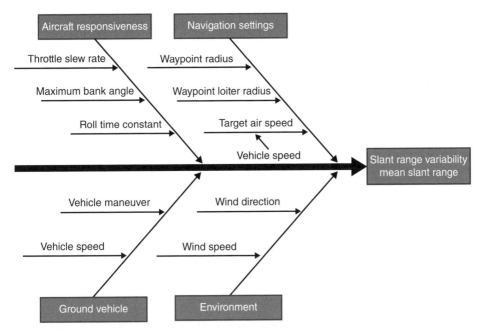

Figure 10.1 Ishikawa diagram for factor identification.

approximation to the operation of an SUAS that was programmed to estimate and match a target vehicle's speed.

The lower branches of the Ishikawa diagram contain factors that are unrelated to the autopilot code, but affect the performance of the SUAS. These factors define an operating condition, referred to as a state, for the SUAS. For example, an aircraft being blown by a 5 mph crosswind from the left and following a vehicle driving in a straight line is in a different state than when the aircraft is flying into a 10 mph headwind following a vehicle that makes a U-turn.

The upper branches of the Ishikawa diagram contained code parameters that were set to constant values in the legacy software. It was suspected that the affect of these factors on the response variable (slant range variance) depended on the vehicle and environmental factors. That is to say there are interactions between the parameter factors and the environmental factors. If indeed there were interactions between software parameters and the state defining parameters, then it might be advantageous to change the system parameters based on the state of the SUAS.

10.5.5   Establishing the Experimental Protocol

In general, establishing the experimental protocol is crucial to the success of any test. Every test participant should understand the procedure for their part of the test. Following an exact test procedure for every experimental run helps to minimize the variance in the response variable due to random noise.

The role of human operators in this test was limited so establishing the experimental protocol was relatively simple. The crucial part of the process to standardize was at the start of the test. As mentioned above, the SUAS was placed in a loiter about the target vehicle and the test run was commenced when the SUAS was at the same point in the loiter path.

It is also important to ensure that all test participants understand the run order for the experiment. As previously discussed, randomization is a deliberate method of spreading out the effect of unknown nuisance factors. It will be tempting for some test personnel to rearrange the experiments in a more convenient order; changing factor levels after every test run can be inconvenient and time consuming. These personnel must be educated as to why the experiment should be conducted in the order specified by the test matrix.

10.5.6   Select the Appropriate Design

Using the philosophy of sequential experimentation (an experimental campaign), several test events were planned that involved both simulation and real-world testing. Different types of designs were used throughout the campaign. This section discusses the details of each testing stage in the process of adding state machine logic in the ARDUPLANE software.

10.5.6.1   Verifying Feasibility and Practicality of Factor Levels

The purpose of the first test event in the simulation environment was to confirm the feasibility and practicality of the factor levels. Being unfamiliar with the ARDUPLANE code, it was not certain that the factor levels for the software parameters were chosen appropriately. To avoid spending excessive time on this preliminary experiment, a 12-run Plackett–Burman design for six factors was selected. All non-zero correlations between ME or 2FI effects are $\pm 1/3$; this design provides reasonable estimates of the MEs in a small number of runs.

The preliminary experiment turned out to be a prudent decision, as the minimum level for maximum bank angle was initially set too low. With an excessively low maximum bank angle the aircraft could not properly turn to follow a vehicle, resulting in a very high variance and mean for the following distance. This poor performance due to the bank angle masked any effect of the other factors. Fortunately this was discovered before too much experimentation had occurred. Based on the results of the Plackett–Burman experiments, the low level for the maximum bank angle was increased.

10.5.6.2   Factorial Experimentation

The goal of the next experiment, conducted using the flight simulator, was to determine how the response variables changed based on what state the SUAS was in, and what factors affected the responses. The state of the aircraft was defined using four levels of wind direction (north, south, east, and west) and three levels of vehicle maneuver (straight route, right turn, and U-turn) for a total of 12 states. For each state, an experiment was run with six factors: throttle slew rate, maximum bank angle, roll time constant, waypoint radius, waypoint loiter radius, and wind speed.

Since each state required a separate experiment, a resolution VI 2^{6-1} design with 32 runs was too large. Running such a design for each of the possible 12 states would have resulted in $12 \times 32 = 384$ total experiments. Two alternative designs were considered: a 24-run no-confounding design and a 17-run DSD [16]. There were positive and negative aspects related to each design.

The 24-run no-confounding design has orthogonal MEs which are also orthogonal to 2FI. There is also no complete confounding between any pair of 2FI effects, with all non-zero correlations at a value of 1/3. Empirical analysis of this design in Stone [17] has shown it provides accurate regression coefficient estimates for models of up to six MEs and up to 2FI.

The other option, the DSD, has many advantages as well. This design has orthogonal ME, which are also orthogonal to 2FI and quadratic effects. Each column of the design has three center factor levels, allowing the estimation of full quadratic models with three or fewer factors. For six factors, DSDs have 13 runs, which can result in high correlations between 2FI and quadratic effects. Using an eight-factor DSD with two columns removed (a 17-run design) provides additional runs which lower the correlations between the effects somewhat.

In an effort to avoid further experimentation the 24-run no-confounding design was tried first. With more runs and a smaller maximum correlation than the 17-run DSD, it was thought this design offered the potential for the highest quality ME + 2FI model. However, if quadratic effects were necessary to model curvature in the response surface, the 17-run DSD was the better design, given the goal of avoiding further experimentation.

Center runs (all factors set to a zero level) can be used for a global test for curvature. Four center runs were included with the 24-run no-confounding design to determine if there was a lack of fit when using an ME and 2FI model. A positive lack-of-fit test will indicate that a quadratic effect is necessary to model curvature in the response function, but it will not determine which specific factor requires a squared term in the regression model.

Selecting the state with a right cross wind and the vehicle turning to the right, the 6-factor 24-run no-confounding design with 4 center runs was conducted. Before running the experiments for the remaining 11 states, the response data was analyzed using regression analysis to determine if the 24-run design was appropriate for the experimental conditions. The regression analysis and lack-of-fit test showed that the quadratic effect for one of the factors was significant. This unexpected result was evidence that the DSD, which is capable of estimating quadratic effects, was a better design choice. The DSD had the added benefit of reducing the overall experimental runs to $12 \times 17 = 204$ experiments.

10.5.6.3 The First Validation Experiment

A validation experiment was run using real-world testing to determine the accuracy of the simulation results. Time constraints had prevented the use of real-world testing to obtain the 204 experimental data points and the flight simulator was the only feasible option to obtain the modeling data. However, there was time to conduct 12 real-world experiments.

Twelve experiments using the SUAS would not generate enough data to conduct a proper analysis and draw conclusions, but it was enough to validate the simulation results. If the simulation flight paths differed significantly from the real-world SUAS flight path, this would indicate the results of the simulation should be considered much less accurate. Since the state variables for wind direction and wind velocity could not be controlled, the only true independent variable in the live test event was vehicle maneuver.

Four experiments were run for each of the three vehicle maneuvers: right turn, U-turn, and a straight route. GPS coordinates for the SUAS were recorded as the SUAS attempted to follow a vehicle driving each maneuver for 60 s. The wind direction and wind velocity were determined and recorded by the SUAS software so that a similar environment could be created in the simulation. After the live test, 12 simulation runs using identical wind conditions and vehicle maneuver were conducted and the simulated GPS coordinates for the SUAS were recorded.

The flight paths of the SUAS in both the live test and simulation were compared to validate the simulation accuracy. The results showed very similar flight paths in both the simulation and real-world test as long as the initial conditions at the start of the test run were the same. That is, the SUAS had to be in the same relative position to the vehicle when the vehicle began its maneuver. This process demonstrated the sensitivity of the test run results to the initial conditions at the start of the run. With the understanding that relatively small changes in the location of the SUAS in relation to the vehicle could result in different following routes, great care was taken to begin each test run when the SUAS was at the same point in the loiter path.

10.5.6.4   Analysis: Developing a Regression Model

Using the simulation results from the 204-run experiment, a linear regression model was developed to predict the response variables for various settings of the investigated factors. Recall that the response variables were the average following distance and the variance of the following distance. The following distance was recorded approximately once per second, resulting in autocorrelated time series data.

Examining the correlation at various lags showed that samples every 7 s produced samples with autocorrelation of around 0.5. This was deemed acceptable to satisfy the independence assumption for the linear regression model. Interestingly the data came from a weakly stationary time series process, meaning that the overall average and variance were approximately equal to the average and variance of almost any equally spaced sample points.

For given combinations of vehicle maneuver and wind direction, a multiple linear regression model was fit using wind speed and the five independent variables related to the software parameter settings. The software used for the analysis was capable of reporting the levels of the independent variables that would optimize the response variable. Since wind speed, along with wind direction and vehicle maneuver defined a state, the wind speed variable was set to one of two levels: 3 and 11 mph. The remaining variables were then set to their optimal levels to minimize the variance of the following distance. Depending on the levels of the state-defining variables, not all of the software parameter variables affected the response variable. If a parameter variable was not significant in the regression model, the variable was set to its default level.

The optimal levels of the parameter variables were used to develop the state-based logic for the SUAS. For a given wind direction, wind level, and vehicle maneuver, the code set the five software parameters to their optimal levels as determined by the regression model. This allowed the SUAS to adapt to conditions of the environment, as well as changes in the behavior of the target vehicle.

10.5.6.5   Software Comparison

The last simulation experiment compared the performance of the SUAS running the legacy software, as measured by the response variables, to the performance using the updated software. For each of the 24 possible states, as defined by four wind directions, two wind speeds, and three vehicle maneuvers, the legacy software was simulated for two replications and the updated software was simulated for two replications. This test facilitated a state-by-state comparison of the two software versions.

The final real-world experiment was also a comparison of the legacy software to the modified software, and served to validate the comparison test conducted in the simulation environment. Again, the time available for live testing limited the experiment to 12 runs. Since the wind direction and wind speed could not be controlled, the only independent variable was vehicle maneuver. For each of the three maneuvers (straight route, right turn, and U-turn) the legacy software was run twice and the updated software was run twice. Although the power to detect a difference in the variance of the following distance was relatively low, there was enough data from this test to determine if the simulation results were not representative of the real-world performance of the SUAS.

10.6   Conclusion

The explosion of UAS technology is a testament to engineering excellence. The full use of this technology, as currently realized and as envisioned, will require changes in policies, procedures, and even laws. Such changes will not occur unless the engineering and test data decisively support the changes. Statistically rigorous test results emanate from statistically rigorous test planning. This chapter prescribes such a planning approach for the UAS community along with a case study employing that approach.

Acknowledgments

This research was supported by the Office of the Secretary of Defense, Director of Operational Test and Evaluation (OSD DOT&E) and the Test Resource Management Center (TRMC) within the Science of Test research consortium.

Disclaimer

The views expressed in this article are those of the authors and do not reflect the official policy or position of the United States Air Force, Department of Defense, or the US Government.

References

1. Keane JF and Carr SS 2013 A brief history of early unmanned aircraft. John Hopkins APL Technical Digest 32(3), 558–571.
2. Blom JD 2010 Unmanned Aerial Systems: A Historical Perspective. Occasional Paper 37. Combat Studies Institute Press.
3. Tetrault C 2009 A short history of unmanned aerial vehicles (uavs). http://www.draganfly.com/news/category/uavnews/.
4. Welshans JS 2014 Much more than an insect pest – the kettering bug. The ITEA Journal 35(4), 311–313.
5. Montgomery DC 2013 Design and Analysis of Experiments, Eighth Edition. A Wiley-Interscience Publication, John Wiley and Sons, Inc.
6. Hahn GJ 1977 Some things engineers should know about experimental design. Journal of Quality Technology 9(1), 13–20.

7. Coleman DE and Montgomery DC 1993 A systematic approach to planning for a designed experiment. Technometrics 35(1), 1–12.
8. Cohen ML, Rolph JE, and Steffey DL (Eds.) 1998 Statistics, Testing and Defense Acquisition: New Approaches and Methodological Improvements. National Research Council.
9. Johnson RT, Hutto GT, Simpson JR, and Montgomery DC 2012 Designed experiments for the defense community. Quality Engineering 24(1), 60–79.
10. Freeman LJ, Ryan AG, Kensler JL, Dickenson RM, and Vining GG 2013 A tutorial on the planning of experiments. Quality Engineering 25(4), 315–332.
11. Carr LK, Lambrecht S, Shaw G, Whittier W, and Warner C 2003 Unmanned Aerial Vehicle Operational Test and Evaluation Lessons Learned. IDA Paper P-3821. Institute for Defense Analysis.
12. Warner C 2011 Continuing the emphasis on scientific rigor in test and evaluation. ITEA Journal 32(1), 15–17.
13. Drury JL and Scott SD 2008 Awareness in unmanned aerial vehicle operations. The International C2 Journal 2(1), 1–28.
14. Hodson DD and Hill RR 2014 The art and science of live, virtual, and constructive simulation for test and analysis. The Journal of Defense Modeling and Simulation: Applications, Methodology, Technology 11(2), 77–89.
15. Black PE 2014 Dictionary of algorithms and data structures, nist. http://www.nist.gov/dads/HTML/determ FinitStateMach.html.
16. Jones B and Nachtsheim C 2011 A class of three-level designs for definitive screening in the presence of second order effects. Journal of Quality Technology 43(1), 1–15.
17. Stone BB 2013 No-confounding Designs of 20 and 24 Runs for Screening Experiments and a Design Selection Methodology. Ph.D. thesis, Arizona State University, Tempe, AZ.

11

Total Cost of Ownership (TOC): *An Approach for Estimating UMAS Costs*

Ricardo Valerdi[1] and Thomas R. Ryan, Jr.[2]

[1] *Department of Systems and Industrial Engineering, University of Arizona, Tucson, AZ, USA*

[2] *Department of Systems Engineering, United States Military Academy, West Point, NY, USA*

11.1 Introduction

Cost, schedule, and quality may not drive a technology, but they shape the chances of that technology becoming actualized. In recent years one of the leading customers of unmanned systems, the US Department of Defense (DoD), has continued to struggle with management of cost and schedule causing programs to deliver products that are "good enough," delayed months to years, or even worse decommissioned. Cost estimation techniques in use today are vast and based on techniques unrelated to emergent systems. One of the most prevalent requirements in the unmanned systems arena is autonomy. The acquisition community will need to adopt new methods for estimating the total cost of ownership of this new breed of systems. Singularly applying traditional software and hardware cost models do not provide this capability because the systems that were used to create and calibrate these models were not [1] Unmanned Autonomous Systems (UMAS). Autonomy, although not new, will redefine the entire way in which estimates are derived. The goal of this chapter is to provide a method that attempts to account for how cost estimating for autonomy is different than current methodologies and suggest ways it can be addressed through the integration and adaptation of existing cost models.

Operations Research for Unmanned Systems, First Edition. Edited by
Jeffrey R. Cares and John Q. Dickmann, Jr.
© 2016 John Wiley & Sons, Ltd. Published 2016 by John Wiley & Sons, Ltd.

11.2   Life Cycle Models

When designing a product the recommended practice is to consider design decisions and their impact throughout the entire life cycle. This is a holistic approach that allows the engineer to examine all phases, and ensure that the stakeholders' (e.g., operators, testers, maintainers) needs are met [2]. This is the same approach that should be taken when identifying product costs, thinking holistically throughout the life cycle. For purposes of discussing the realm of UMAS we will focus on two life cycle standards: DoD 5000 [3, 4] and ISO/IEC 15288 Systems Engineering–System Life Cycle Processes [5].

Both product life cycle standards are organized into discrete phases. Each phase has a distinct role in the life cycle and helps separate major milestones throughout the life cycle of a product. These life cycle stages help answer the "when" and are useful in identifying costs such as development, production, and operational.

11.2.1   DoD 5000 Acquisition Life Cycle

Although there are many commercial customers being identified and pursued within the UMAS arena, the largest acquirer of autonomous systems is the U.S. Defense Department. The DoD 5000 is a useful framework to apply to a product as it forces engineers to produce specific sub-products in each of the five phases [3].

1. In the first phase, Materiel Solution Analysis, the DoD requires an initial capabilities document and an analysis of alternatives study.
2. During the second phase, Technology Development, the goals are to produce a demonstrable prototype that will allow the customer to make decisions in the risk, technology, and design.
3. The third phase, Engineering and Manufacturing Development, forces the engineer to again demonstrate prototype articles, conduct integrated testing (Developmental, Operational, and Live Fire Test and Evaluation), prepare for both the Critical Design Review and the proposal for product continuation.
4. During the fourth phase, Production and Deployment, engineers are now preparing low-rate and full-scale production.
5. The final phase, Operations and Support, consists of activities such as maintaining capabilities, logistical support, upgrades, customer satisfaction, and prepare for proper disposal.

The five phases and major milestones are shown in Figure 11.1.

11.2.2   ISO 15288 Life Cycle

A definition of the system life cycle phases is needed to help define the boundaries between engineering activities. A useful standard is ISO/IEC 15288 Systems Engineering–System Life Cycle Processes (ISO/IEC 15288). However, the phases established by ISO/IEC 15288 were slightly modified to reflect the influence that ANSI/EIA 632 Processes for Engineering a System has on the Constructive Systems Engineering Cost Model's (COSYSMO's) System Life Cycle Phases, and are shown in Figure 11.2.

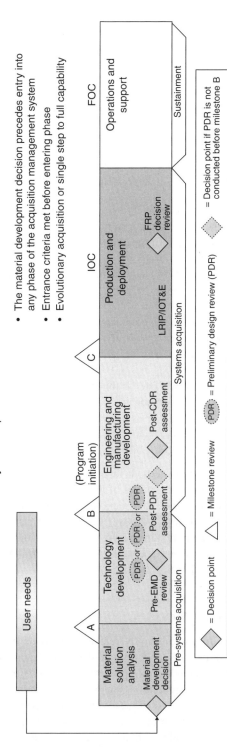

Figure 11.1 DoD 5000 acquisition framework. *Source*: Spainhower, K. (2003). Life Cycle Framework. Retrieved June 10, 2014, from https://dap. dau.mil/aphome/das/Pages/Default.aspx

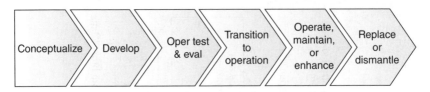

Figure 11.2   COSYSMO system life cycle phases.

Life cycle models vary according to the nature, purpose, use, and prevailing circumstances of the product. Despite an infinite variety in system life cycle models, there is an essential set of characteristic life cycle phases that exists for use in the systems engineering domain.

1. The Conceptualize stage focuses on identifying stakeholder needs, exploring different solution concepts, and proposing candidate solutions.
2. The Development stage involves refining the system requirements, creating a solution description, and building a system.
3. The Operational Test and Evaluation stage involves verifying/validating the system and performing the appropriate inspections before it is delivered to the user.
4. Other users' needs.
5. The Operate, Maintain, or Enhance stage involves the actual operation and maintenance of the system required to sustain system capability.
6. The Replace or Dismantle stage involves the retirement, storage, or disposal of the system.

We will revisit these life cycle models later in this chapter and we decompose various types of costs into their respective phases to demonstrate Total Cost of Ownership.

11.3   Cost Estimation Methods

The exploration of new cost modeling methods involves the understanding of the cost metrics relevant to UMAS as well as an understanding of their sensitivity to cost from a production and operational standpoint. In this light, this section provides an overview of different cost estimation approaches used in industry and government. Significant work has been done to understand the costs of aircraft manufacturing [7–9] but these studies only deal with manned commercial and military aircraft. Nevertheless, they provide useful insight on how one could approach the estimation of UMAS life cycle cost.

11.3.1   Case Study and Analogy

Recognizing that companies do not constantly reinvent the wheel every time a new project comes along, there is an approach that capitalizes on the institutional memory of an organization to develop cost estimates. Case studies represent an inductive process, whereby estimators and planners try to learn useful general lessons by extrapolation from specific examples. They examine in detail elaborate studies describing the environmental conditions and constraints that were present during the development of previous projects, the technical and managerial decisions that were made, and the final successes or failures that resulted.

They then determine the underlying links between cause and effect that can be applied in other contexts. Ideally, they look for cases describing projects similar to the project for which they will be attempting to develop estimates and apply the rule of analogy that assumes previous performance is an indicator of future performance. The sources of case studies may be either internal or external to the estimator's own organization. Home-grown cases are likely to be more relevant for the purposes of estimation because they reflect the specific engineering and business practices likely to be applied to an organization's projects in the future. Well-documented case studies from other organizations doing similar kinds of work can also prove very useful so long as their differences are identified.

11.3.2   Bottom-Up and Activity Based

Bottom-up estimating begins with the lowest level cost component and rolls it up to the highest level for its estimate. The main advantage is that the lower level estimates are typically provided by the people who will be responsible for doing the work. This work is typically represented in the form of subsystem components, which makes this estimate easily justifiable because of their close relationship to the activities required by each of the system components. This approach also allows for different levels of detail for each component. For example, the costs of an airplane can be broken down into seven main components: center-body, wing, landing gear, propulsion, systems, payloads, and assembly. Each of these components, such as the wing, can be decomposed into subcomponents such as winglet, outer wing, and inner wing. This decomposition is illustrated in more detail in Figure 11.3. This can translate to a fairly accurate estimate at the lower level components. The disadvantages are that this process is labor intensive and is typically not uniform across products. In addition, every level introduces another layer of conservative management reserve which can result in an overestimate at the end.

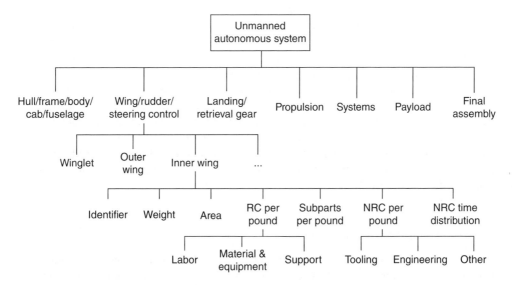

Figure 11.3   Product breakdown structure of a typical UMAS.

11.3.3   Parametric Modeling

This method is the most sophisticated and most time consuming to develop but often provides the most accurate result. Parametric models generate cost estimates based on mathematical relationships between independent variables (i.e., requirements) and dependent variables (i.e., effort or cost). The inputs characterize the nature of the work to be done, plus the environmental conditions under which the work will be performed and delivered. The definition of the mathematical relationships between the independent and dependent variables is the heart of parametric modeling. These relationships are commonly referred to as the Cost Estimating Relationships (CERs) and are usually based upon statistical analyses of large amounts of data. Regression models are used to validate the CERs and operationalize them in linear or non-linear equations. The main advantage of using parametric models is that, once validated, they are fast and easy to use. They do not require a lot of information and can provide fairly accurate estimates. Parametric models can also be tailored to a specific organization's characteristics such as productivity rates, salary structures, and Work Breakdown Structures (WBSs). The major disadvantage of parametric models is that they are difficult and time consuming to develop and require a lot of clean, complete, and recent data to be properly validated. Despite the wide range of estimation approaches available for commercial and military aircraft, no parametric models have been created specifically for UMAS. This could be attributed to the fact that UMAS have not been around for very long and, as a result, there are insufficient data available to validate such models. Before proposing a framework for such a model, unique issues pertaining to the UMAS life cycle are discussed.

11.4   UMAS Product Breakdown Structure

It is widely recognized that creating a WBS or Product Breakdown Structure (PBS) is the most complete way to describe a project [10]. The level of detail required to properly utilize, or manage with, the PBS such as the one shown in Figure 11.3 is a crucial component to assigning costs to a product's subcomponents. In this section we will discuss some of the commonalities and shared considerations of designing a WBS/PBS within an unmanned system at the system level. Budgeted amounts for various unmanned and autonomous systems are shown in Tables 11.1–11.4 at the second or third level of a WBS/PBS.

One observation from the UMAS examples provided in Tables 11.1–11.4 is the range of unit costs. On the high end, the Flyaway Unit Cost of the Global Hawk Unmanned Aircraft System is $92.87 million [11, p. 177]. On the low end, the Modular Unmanned Scouting Craft Littoral is $700,000 [13]. Another observation from these examples is the wide range of units purchased, as few as four Commercial Off-The-Shelf (COTS)/Government Off-The-Shelf (GOTS) packages for converting manned systems to unmanned and as many as 311 Small Unmanned Ground Systems [12].

11.4.1   Special Considerations

The unique physical and operational characteristics of UMAS require special consideration when exploring cost modeling approaches. In Figure 11.4, the DoD has laid out its desires for UMAS over the next 30 years. It has organized its requirements by air, ground, and maritime

Table 11.1 Air system (Unmanned Air System (UAS)).

Unmanned Aerial Vehicle – Global Hawk	Unit cost ($M)	Number of units	Total cost ($M)	Program allocation[a] (%)
Aerial vehicle	69.84	45	3143.16	66.60
Ground control station	21.82	10	218.21	4.62
Support element	n/a	n/a	1357.84	28.77
Projected total cost	n/a	n/a	4719.21	100.00

[a] Since the program allocation was only available for the Global Hawk, we applied the same ratios to other unmanned programs.

Source: Department of Defense Fiscal Year (FY) 2015 Budget Estimates. (2014, March 1). Retrieved January 7, 2015, from http://www.saffm.hq.af.mil/shared/media/document/AFD-140310-041.pdf

Table 11.2 Ground system (Unmanned Ground System (UGS)).

UGS COTS/GOTS	Unit cost ($M)	Number of units	Total cost ($M)
Ground vehicle	*3.39*	4	*13.56*
Ground control station	*0.23*	4	*0.94*
Support element	n/a	n/a	5.86
Projected total costs	n/a	n/a	20.36

Ground Control Stations are the user controls (i.e., the video game-like interface to maneuver vehicle).
Italicized numbers = extrapolation based on RQ-4 Global Hawk program ratio.
Unaltered numbers are from the Exhibit P-40 Presidential Budget FY2015 or equivalent cost data.
Source: Department of Defense Fiscal Year (FY) 2015 Budget Estimates. (2014, March 1). Retrieved January 7, 2015, from http://asafm.army.mil/Documents/OfficeDocuments/Budget/budgetmaterials/fy15/pforms//opa34.pdf

Table 11.3 Ground system (UGS).

Small Unmanned Ground Vehicle (SUGV)	Unit cost ($M)	Number of units	Total cost ($M)
Ground vehicle	*0.180*	311	*55.90*
Ground control station	*0.012*	311	*03.88*
Support element	n/a	n/a	24.15
Projected total costs	n/a	n/a	83.93

Ground Control Stations are the user controls (i.e., the video game-like interface to maneuver vehicle).
Italicized numbers = extrapolation based on RQ-4 Global Hawk program ratio.
Unaltered numbers are from the Exhibit P-40 Presidential Budget FY2015 or equivalent cost data.
Source: Department of Defense Fiscal Year (FY) 2015 Budget Estimates. (2014, March 1). Retrieved January 7, 2015, from http://asafm.army.mil/Documents/OfficeDocuments/Budget/budgetmaterials/fy15/pforms//opa34.pdf

Table 11.4 Marine system (UGS).

Modular Unmanned Scouting Craft Littoral (MUSCL)	Unit cost ($M)	Number of units	Unit cost ($M)
Maritime vehicle	0.700	*13*	*9.03*
Surface control station	0.048	*13*	*0.62*
Support element	n/a	n/a	3.90
Projected total costs	n/a	n/a	13.56

Ground Control Stations are the user controls (i.e., the video game-like interface to maneuver vehicle).
Italicized numbers = extrapolation based off of RQ-4 Global Hawk program ratio.
Unaltered numbers are from the Exhibit P-40 Presidential Budget FY2015 or equivalent cost data.
Source: Department of Defense Fiscal Year (FY) 2015 Budget Estimates. (2014, March 1). Retrieved January 7, 2015, from http://www.finance.hq.navy.mil/fmb/15pres/OPN_BA_5-7_Book.pdf

Goals		2013	2014	2015	2016	2017	2018	2019	2020	2021	2022	2023	2030+
		Near Term					**Mid Term**				**Far Term**		
Technology projects	UAS	Secure C2 links. Certified GBSAA. Certified displays. Improved sensors. Interoperable payload					Certified ABSAA and separation algorithms. Integrated equipment				Integrated SAA. Evolution with NextGen		
	UGS	Expand physical architectures. Increase autonomation for specific tasks. V2V comms					Expanded autonomy systems and avoidance algorithms				Autonomous architecture		
	UMS	Improved power, comm, and sensor systems					Effective autonomy systems and avoidance algorithms. Security architectures						
Desired capability	UAS	Incremental access to the NAS. Effective information fusion					Routine access to the NAS. Due regard capability. Effective exploitation				Increased safety and efficiency for flight in NAS and worldwide. Effective forensics		
	UGS	Robust physical capabilities					Effective manned-unmanned teaming				Adaptable systems		
	UMS	Autonomy for specialized missions in localized areas. Increasingly networked systems					Increased missions in expanded geographical areas				Autonomous missions worldwide		

Figure 11.4 Operating environment technology development timeline (2013–2030). *Source*: Unmanned Systems Integrated Roadmap FY2013-2038. (2013). Washington, DC: Department of Defense.

operational environments, as well as projected the types of exploration initiatives that should allow for success of these autonomous systems. This figure is not meant to be totally exhaustive, but guide the general direction of the military's UMAS vision.

11.4.1.1 Mission Requirements

The mission requirements are specified tasks to which the UMAS must comply in order to perform [14]. These requirements are shaped by the Operational Environment (OE), or venue by which the UMAS will perform its intended functions or capabilities – this can be physical and situational. The physical environment can consist of Air, Ground (surface and subsurface), and Marine (surface and submersible.)

11.4.2 System Capabilities

In essence, what will the UMAS do for the customer? These functions must also include current capabilities such as Attack, Logistical, and Reconnaissance. This area also includes any of the "-ilities" that a UMAS might need to adhere to that are not specified in its mission requirements. These may include manufacturability, reliability, interoperability, survivability, and maintainability.

11.4.3 Payloads

A final consideration for UMAS is its payload. This could also be categorized as special equipment. For example, a logistical UMAS (or cargo transportation system like the Squad Mission Support System (SMSS™)) needs to have a tow system or recovery package in addition to the ground vehicle; or if it is an attack/reconnaissance system – it needs to support munitions, missiles, or gun platforms.

Although many more areas can be identified for consideration when engineering a system for autonomy, this section was meant to highlight the WBS/PBS in more detail rather than the technical capabilities of the UMAS itself. The cost to build and produce a system is a bottom line decision for the producer (and the engineer), but the DoD needs and expects that a WBS represent all phases of the life cycle. By accurately representing the system in a more complete WBS/PBS the cost estimates will have more fidelity and a higher confidence, because estimators will be able to link the lowest level of that structure to a group of cost drivers within a cost model.

11.5   Cost Drivers and Parametric Cost Models

Cost drivers are characteristics of projects that best capture the effort, typically measured in Person-Months, required to complete them [15]. As mentioned in Section 11.3.3 of this chapter, developing these characteristics, or drivers, is data and labor intensive. The developer of the model must establish a strong mathematical relationship, usually a form of regression, between an identified characteristic and its impact on the project. The number of cost drivers for each type of estimate will vary according to the type of component (hardware, software, etc.).

Each cost driver has a scale, usually five levels, which allows the user of the model to best represent characteristics of the product. For example, a cost driver can be described using Very Low, Low, Nominal, High, Very High; and each one of these choices has a value that will either increase or decrease cost [16]. Each level is clearly defined so the user can estimate the complexity of a system as realistically as possible. The key for success with utilizing parametric modeling and its drivers is to fully understand and be realistic with assignment of scale values.

11.5.1   Cost Drivers for Estimating Development Costs

Our proposed method for system level estimation is to combine five different parametric models that best represent the amount of effort required to successfully build, test, produce, and operate a UMAS. These include: (i) Hardware, (ii) Software, (iii) Systems Engineering and Program Management, (iv) Performance-Based characteristics, and (v) Weight-Based characteristics.

Each of the five models is described below and should be considered when developing a complete life cycle estimate; however, it is not mandatory to utilize all five since each UMAS will have unique cost and performance considerations.

11.5.1.1   Hardware

SEER-H is a hybrid model that utilizes analogous estimates, as well as harnessing parametric mathematical cost estimation relationships specific to hardware products. SEER-H aids in the estimation of hardware development, production, and operations costs [17]. Unlike the other estimation tools available, SEER-H has an exhaustive suite and could be used to estimate

Table 11.5 Material composition rating scale [17].

Material	Key property
Aluminum/malleable metals	Metal alloy, easily manufactured Example: aluminum, magnesium, copper, aluminum–lithium
Steel	Hard, rigid metal alloy, resistant to rust Example: steel, stainless steel
Commercially available exotic	Commodity available exotic materials Example: titanium, precious metals, boron, higher end composites
Other exotic	Requires very complex metallurgical processes, available only through special orders Example: metal matrix composites, particulate strengthened composite materials, research materials
Composite	Commodity available, continuous filament, or particulate strengthened composite materials Example: graphite or boron epoxy, fiber glass
Polymer	Nonmetallic compound, easily molded, may be hardened or pliable Example: plastics, thermoplastics, elastomers
Ceramic	Very strong, brittle Example: ceramic, clay, glass, tile, porcelain

many technical areas. The number of cost drivers in SEER-H is extensive, therefore we will focus on only three within their Mechanical/Structural Work Elements category:

- *Material Composition:* the material that will dominate the system and its difficulty to acquire
- *Certification Level:* the amount of Test and Evaluation with demonstration required for the materials utilized
- *Production Tools and Practices:* how ready the materials are for production.

11.5.1.1.1 Material Composition
This SEER-H driver is categorized by the predominant material used to build the system, subsystem, or their components as shown in Table 11.5. The estimator should also consider some of the materials that may not dominate, but are identified as critical. The total cost may be a combination of critical and dominant materials.

11.5.1.1.2 Certification Level
Certification level represents the requirements imposed on the manufacturer by the customer as shown in Table 11.6. This parameter quantifies the additional cost associated with the customer's certification requirements; therefore, any extra certification, inspections, or intangible property security controls, and so on, will increase cost.

11.5.1.1.3 Production Tools and Practices
This parameter describes the extent to which efficient fabrication methodologies and processes are used, and the automation of labor-intensive operations. The rating should reflect the state of production tools that are in place and already being used by the time hardware production begins (Table 11.7).

Table 11.6 Certification level rating scale [17].

Rating	Description
Very high	Very high level of qualification testing including fatigue, fracture mechanics, burst, temperature extremes, and vibration testing. Example: manned space product
High	High level of qualification testing including fatigue, fracture mechanics, burst, temperature extremes, and vibration testing. Example: space product
Nominal+	Qualification testing for mission requirements including static and dynamic load testing, wind tunnel testing, and all other tests required for military aircraft. Example: military airborne/aircraft product.
Nominal	Qualification testing in accordance with FAA requirements, as specified for commercial or general aviation aircraft. Example: airborne/aircraft product
Nominal−	Qualification testing in accordance with U.S. Army Mobility requirements, or U.S. Navy specifications. Testing includes meeting shock, vibration, temperature, and humidity requirements. Example: military ground-mobile or sea product
Low	Nominal qualification testing for mission requirements covering equipment located in controlled environments (temperature, humidity). Example: military ground system
Very low	Minimal testing required (functional check-out). Example: commercial grade product

Table 11.7 Production tools and practices rating scale [17].

Rating	Description
Very high	Production tooling associated normally with large-scale production (20 000 units or above). Highly sophisticated tools, die casts, molds. High degree of mechanization, robotics manufacture, assembly, and testing. High degree of integration between computer-aided manufacturing and design. Example: die casting, multi-cavity molds, progressive dies, and other sophisticated tools.
High	Production tooling normally applied to medium scale, averaging 20 000 unit production. Tools are custom designed with simple dies. Some degree of mechanization, numerically controlled machine tools, some integration with computer aided design. Example: simple die casting, complex investment casting, custom die sets for sheet metal fabrication
Nominal	Production tooling facilitates production of 1000–2000 units. Complex tools, simple dies, and castings. Little mechanization, few numerically controlled machining operations. Some automated links with computer aided design. Example: complex sand castings, investment castings of some complexity, and simple custom die sets are included in the tooling category
Low	Tooling designed for the production of up to 1000 units. Standard tools, casts, dies, and fixtures are supplemented with some custom tools and jigs. Occasional or experimental use of automated links with computer aided design. Example: sand castings, investment castings, and simple custom die sets. Many aerospace/DoD programs are in this category
Very low	Minimum tooling required to produce up to about 50–100 units. Many operations of manufacture, assembly, and test are by skilled labor. The use of standard tools and fixtures is predominant. No automated links. Example: simple sand castings

11.5.1.2   Software

The recommended parametric estimation tool for UMAS software aspects is the Constructive Cost Model (COCOMO II). This model has 30 years of refinement, and is an industry and academic standard for parametric modeling [15]. The number of cost drivers in COCOMO II vary from 7 to 17 depending on the life cycle phase of the project in which the estimate is being performed [15]. Since less information is known at the beginning of the project, the COCOMO II model provides less parameters to rate. As more information is known about the software project, the number of parameters increases. This section is not meant to replace the COCOMO II User's Manual,[1] but rather to provide relevant details about the relevant cost drivers. Three drivers are relevant for UMAS software estimation:

- *Size:* measured by number of lines of code
- *Team Cohesion:* weighted average of four characteristics
- *Programmer Capability:* how efficient programmers are as a whole.

11.5.1.2.1   Size
Size, in units of thousands of source lines of code (KSLOC), is derived from estimating the size of software modules that will constitute the application program. It can also be estimated using Unadjusted Function Points (UFPs), converted to SLOC, then divided by 1000. Equation (11.1) is the basic COCOMO II algorithm which includes Size as the central component to calculating effort in Person-Months (PM).

$$PM = A \times \left(SIZE \right)^{E} \times \prod_{i=1}^{n} EM_{i}$$ (11.1)

11.5.1.2.2   Team Cohesion
This parameter accounts for the human component in software design. These elements are not limited to but contain differences in multiple stakeholder objectives, cultural backgrounds, team resiliency, and team familiarity. The focus is how the design team interacts externally within the project (Table 11.8).

11.5.1.2.3   Programmer Capability
This parameter also deals with a human aspect of software engineering, however, it differs from team cohesion in the direction of the focus. In this parameter the assessment is on the internal workings of the team's capability as it relates to their efficiency, thoroughness, internal communication, and cooperation (Table 11.9).

11.5.1.3   Systems Engineering and Project Management

To estimate the Systems Engineering and Project Management required effort for a UMAS, we use the COSYSMO. This parametric models output accounts for integrating system components and will quantify intangible efforts such as requirements, architecting, design,

---

[1] http://csse.usc.edu/csse/research/COCOMOII/cocomo_main.html.

Table 11.8 Team cohesion rating scale.

Characteristic	Very low	Low	Nominal	High	Very high	Extra high
Consistency of stakeholder objectives and cultures	Little	Some	Basic	Considerable	Strong	Full
Ability, willingness of stakeholders to accommodate other stakeholders' objectives	Little	Some	Basic	Considerable	Strong	Full
Experience of stakeholders in operating as a team	None	Little	Little	Basic	Considerable	Extensive
Stakeholder teambuilding to achieve shared vision and commitments	None	Little	Little	Basic	Considerable	Extensive

Table 11.9 Programmer capability rating scale.

Programmer Capability (PCAP) descriptors	15th percentile	35th percentile	55th percentile	75th percentile	90th percentile	
Rating levels	Very low	Low	Nominal	High	Very high	Extra high
Effort multipliers	1.34	1.15	1.00	0.88	0.76	n/a

verification, and validation [16]. This model also depends on 18 size and cost drivers.[2] By introducing some of the most important drivers we capture the most important cost considerations of a UMAS. The three most relevant systems engineering cost drivers are:

- *Number of System Requirements:* number of specified functions a system must perform to meet the user's needs
- *Technology Risk:* how mature or demonstrated the technologies are
- *Process Capability:* how well/consistent the team/organization perform in terms of the Capability Maturity Model Integration (CMMI).

11.5.1.3.1 Number of Requirements

This parameter asks the estimator to count the number of requirements for the UMAS at a specific level of design. These requirements may deal with number of system interfaces, system specific algorithms, and operational scenarios. Requirements are not limited to but may include functional, performance, feature, or service-oriented in nature depending on the methodology used for specification. Of note, requirement statements usually contain the words "Shall, Will, Should, or May" (Table 11.10).

---

[2] http://cosysmo.mit.edu.

Table 11.10 Number of requirements rating scale.

Easy	Nominal	Difficult
Simple to implement	Familiar	Complex to implement
Traceable to source	Can be traced to source with some effort	Hard to trace to source
Little requirements overlap	Some overlap	High degree of requirement overlap

11.5.1.3.2 Technology Risk

This parameter asks you to evaluate a UMAS's subsystem's maturity, readiness, and obsolescence of the technologies being implemented. Immature or obsolescent technologies will require more systems engineering effort (Table 11.11).

11.5.1.3.3 Process Capability

Like some of the COCOMO II parameters, this COSYSMO example focuses on the consistency and effectiveness of the project team performing the systems engineering processes. The assessment of this driver may be based on ratings from a published process model (e.g., CMMI [18], EIA-731 [19], SE-CMM [15, 20], ISO/IEC15504 [21]). It can alternatively be based on project team behavioral characteristics, if no previous external assessments have occurred (Table 11.12).

11.5.1.4 Performance-Based Cost Estimating Relationship

One important consideration of every product is its ability to perform the specified requirements well. The model that best captures the performance characteristics of a product was created by the U.S. Army for Unmanned Aerial Vehicle Systems, but can be modified to fit other autonomous systems [22]. The methodology for estimating performance are not restricted to this list, but should fit in similar categories for air, land, sea, or space (Table 11.13).

The cost drivers that are recommended for performance measurement are based on an aerial platform, but are modified in this chapter to provide ideas on what areas to consider (Table 11.14).

The U.S. Army's performance-based CER is shown in Eq. (11.2):

$$\text{UAV T1R1}(\text{FY03\$K}) = 118.75 * (\text{Endurance} * \text{Payload-Wt.})^{0.587}$$
$$* e^{-0.010(\text{FF\_Year-1900})} * e^{-0.921(\text{Prod1/0})} \tag{11.2}$$

where:

UAV T1R 1	=	Theoretical first unit cost of UAV air vehicle hardware normalized for learning (95% slope) and rate (95% slope), via unit theory. In FY03 $K.
Endurance	=	UAV air vehicle endurance in flight hours.
Payload-Wt.	=	Weight of total payload in pounds. Total payload includes all equipment other than the equipment that is necessary to fly and excludes fuel and weapons.
FF-Year	=	Year of first flight.
Prod 1/0	=	1 if air vehicle is a production unit.
	=	0 if air vehicle is a development or demonstration unit.

Table 11.11 Technology risk rating scale.

	Very low	Low	Nominal	High	Very high
Lack of maturity	Technology proven and widely used throughout industry	Proven through actual use and ready for widespread adoption	Proven on pilot projects and ready to roll-out for production jobs	Ready for pilot use	Still in the laboratory
Lack of readiness	Mission proven (TRL 9)	Concept qualified (TRL 8)	Concept has been demonstrated (TRL 7)	Proof of concept validated (TRL 5 and 6)	Concept defined (TRL 3 and 4)
Obsolescence			Technology is the state-of-the-practice	Technology is stale	Technology is outdated and should be avoided in new systems
			Emerging technology could compete in future	New and better technology is ready for pilot use	Spare parts supply is scarce

Table 11.12 Process capability rating scale.

	Very low	Low	Nominal	High	Very high	Extra high
CMMI assessment rating	Level 0 (if continuous model)	Level 1	Level 2	Level 3	Level 4	Level 5
Project team behavioral characteristics	Ad hoc approach to process performance	Performed system engineering process, activities driven only by immediate contractual or customer requirements, system engineering focus limited	Managed system engineering process, activities driven by customer and stakeholder needs in a suitable manner, system engineering focus is requirements through design, project-centric approach – not driven by organizational processes	Defined system engineering process, activities driven by benefit to project, system engineering focus is through operation, process approach driven by organizational processes tailored for the project	Quantitatively managed system engineering process, activities driven by system engineering benefit, system engineering focus on all phases of the life cycle	Optimizing system engineering process, continuous improvement, activities driven by system engineering and organizational benefit, system engineeringfocus is product life cycle and strategic applications
Systems Engineering Management Plan (SEMP) sophistication	Management judgment is used	SEMP is used in an ad hoc manner only on portions of the project that require it	Project uses a SEMP with some customization	Highly customized SEMP exists and is used throughout the organization	The SEMP is thorough and consistently used; organizational rewards are in place for those that improve it	Organization develop best practices for SEMP; all aspects of the project are included in the SEMP; organizational reward exists for those that improve it

Table 11.13 Performance-based characteristics rating scale.

Performance-based categories	Descriptions
Vehicle or body of UMAS	Define and measure how well the vehicle or body of the UMAS performs its intended requirements
Sensors	Define and measure how well the UMAS can interact and react with its intended (or unintended) environment
Control system	Define and measure how efficiently the command and control system interacts with UMAS

Source: Cherwonik, J. and A. Wehrley (2003). Unmanned Aerial Vehicle System Acquisition Cost Research: Estimating Methodology and Database. Washington D.C., USA: The Office of the Deputy Assistant of the Army for Cost and Economics.

Table 11.14 Performance cost drivers.

Performance drivers	Description/use of driver
Operational environment constraints	Define and measure the physical boundaries guiding the UMAS
Endurance	Define and measure the amount of time or distance the UMAS can perform its intended task prior to needing human interaction
Sensor resolution	Define and measure the sensitivity, accuracy, resiliency, and efficiency of the UMAS sensors
Base of operations	Define and measure how constrained the UMAS is by its logistical requirements and the resources required for effective operations

Source: Cherwonik, J. and A. Wehrley (2003). Unmanned Aerial Vehicle System Acquisition Cost Research: Estimating Methodology and Database. Washington D.C., USA: The Office of the Deputy Assistant of the Army for Cost and Economics.

Table 11.15 Weight-based cost drivers.

Weight-based drivers	Description/use of driver
Weight of total system	Define and measure the weight of total system as it relates to its intended objectives (this does not include ordnance or other attachable options)
Payload weight	Define and measure the amount and type of ordnance or any additional attachable option that is deemed mission critical
Sling load or recovery operation capacity[a]	Define and measure the amount of weight the UMAS can support as a sling load or in a tow capacity, in addition to its nominal capacity

[a] If applicable.
Source: Cherwonik, J. and A. Wehrley (2003). Unmanned Aerial Vehicle System Acquisition Cost Research: Estimating Methodology and Database. Washington D.C., USA: The Office of the Deputy Assistant of the Army for Cost and Economics.

11.5.1.5 Weight-Based Cost Estimating Relationship

A final consideration for estimating the cost of UMAS is its weight. Weight may already exist as an important cost driver in other estimation models such as hardware and performance; however, we feel that this particular estimation relationship is strong enough to also be a

standalone component. When operational implementation is considered for a given autonomous system, weight plays a critical role in the success or failure. Some drivers, modified
from source to apply to UMAS, are shown in Table 11.15.

The U.S. Army's weight-based CER is shown in Eq. (11.3):

$$\text{UAV T1R1}(\text{FY03\$K}) = 12.55 * (\text{MGTOW})^{0.749} * e^{-0.371(\text{Prod}1/0)} \qquad (11.3)$$

where:

UAV T1R 1	=	Theoretical first unit cost of UAV air vehicle hardware normalized for learning (95% slope) and rate (95% slope). In FY03 \$K.
MGTOW	=	UAV air vehicle maximum take-off weight in pounds.
Prod 1/0	=	1 if air vehicle is a production system
	=	0 if air vehicle is a development or demonstration model.

11.5.2   Proposed Cost Drivers for DoD 5000.02 Phase Operations and Support

11.5.2.1   Logistics – Transition from Contractor Life Support (CLS) to Organic Capabilities

Managing logistic support is complex and not easy to summarize into a single parameter.
However, all systems require maintenance, which can be described within the range provided
in Table 11.16. The goal of this parameter is to allow life cycle planners to nest their system
engineering plan into DoD requirements and minimize Contractor Life Support (CLS).

11.5.2.2   Training

The development costs for a UMAS can be significant, but one area of consideration is how
quickly and efficiently users can be trained to employ the system. With increasing levels of
autonomy this warrants its own cost driver.

The planning for and implementation of such training considerations shown in Table 11.17
will be challenging. The DoD acknowledges these challenges and offers a perspective of
expectation management displayed in Figure 11.5. The training objectives attempt to lay out
how UMAS and other emergent systems will be inculcated into the existing training system.

Table 11.16 Logistics cost driver.

Uniformed servicemen only	>2 years transition	2–5 year transition	<5 year transition	CLS only
System was designed in a manner that current life support is sufficient for operational use	Very few contractors (1–5) needed at Colonel (0–6) level command units to ensure proper life support	Few contractors (6–10) needed at Colonel (0–6) level command units to ensure proper life support	Contractors are needed at every level of command; Captain (0–3) through Colonel (0–6). Minimum 1 at each level	System is so technologically advanced that operational use will require a permanent contractor presence

Table 11.17 Training cost driver considerations.

Minimal impact	Medium impact	High impact	Extreme impact	Unknown impact
Training fits current TRADOC[a] throughput and requires minimal certification (e.g., system is a modified version of a previously integrated system – autonomous raven)	Training program is similar to a current DoD method; however, needs to be a standalone block of instruction or course. Can use existing facilities and infrastructure currently provided	Training program is not similar to any current DoD method. Needs to be a standalone course. Needs facilities and infrastructure not currently provided	Training program is not similar to any current DoD method. Needs to be a standalone course. Needs facilities and infrastructure not currently available	Training systems are still being developed and will require extensive integration

[a] U.S. Army Training and Doctrine Command.

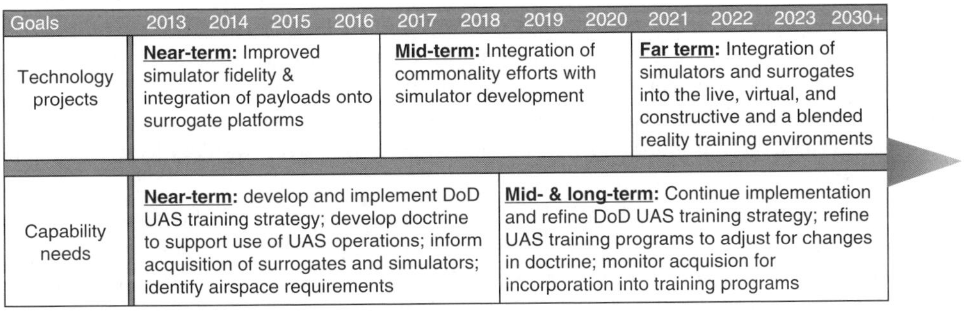

Goals	2013 2014 2015 2016	2017 2018 2019 2020	2021 2022 2023 2030+
Technology projects	**Near-term:** Improved simulator fidelity & integration of payloads onto surrogate platforms	**Mid-term:** Integration of commonality efforts with simulator development	**Far term:** Integration of simulators and surrogates into the live, virtual, and constructive and a blended reality training environments
Capability needs	**Near-term:** develop and implement DoD UAS training strategy; develop doctrine to support use of UAS operations; inform acquisition of surrogates and simulators; identify airspace requirements	**Mid- & long-term:** Continue implementation and refine DoD UAS training strategy; refine UAS training programs to adjust for changes in doctrine; monitor acquision for incorporation into training programs	

Figure 11.5 UMAS training objectives (2013–2030). *Source*: Unmanned Systems Integrated Roadmap FY2013-2038. (2013). Washington, DC: Department of Defense.

As engineers build their systems, understanding these strategies will help with system implementation in areas that are not implicit in the system being procured.

11.5.2.3 Operations – Manned Unmanned Systems Teaming (MUM-T)

The goal of the DoD's investment in UMAS is to enhance warfighters' capability while reducing risk to human life, maintaining tactical advantage, and performing tasks that can be dull, dirty, or dangerous [14]. However, all of the systems will require some level of manned-with-unmanned cooperation. The more these two worlds efficiently work together the better the operational outcome (Table 11.18).

11.6 Considerations for Estimating Unmanned Ground Vehicle Costs

For a large-scale project that requires the integration of multiple engineering disciplines, specifically in the field of UMAS, no single estimation tool can completely capture total life cycle costs. By applying the proper estimation models, or a combination of these models,

Table 11.18 Manned unmanned systems teaming cost driver.

Very low teaming	Low teaming	Nominal teaming	High teaming	Very high teaming
Meets no joint interoperability requirements, and generates data that need to be transferred to a common operating picture	Meets minimum branch-specific interoperability requirements, but is not compatible with all systems employed by its home branch	Meets branch-specific interoperability requirements, and shares information with manned systems, branch-specific	Meets all branch-specific interoperability requirements, as well as one or more joint requirements. Also shares information with manned systems	Meets all joint interoperability requirements, and shares a common operating picture other manned and unmanned systems can utilize

the estimator can ensure complete coverage of each program element and their relative cost impact across the UMAS project life cycle.

The example used to illustrate the cost estimating process is the Lockheed Martin Unmanned Autonomous Ground System, SMSS™. By utilizing the Product Work Breakdown Structure (P-WBS) cost experts can then apply an estimation tool at the appropriate level. The sum of each sub-estimate is then integrated into the overall project level estimation. Considerations for which levels within the P-WBS require estimates is unique to each UMAS project. Contractual requirements will be the determining factor on how detailed the estimate needs to be.

In response to the critical need for lightening the Soldier and Marine Infantryman's load in combat as well as providing the utility and availability of equipment that could not otherwise be transported by dismounted troops, the SMSS™ is being developed by Lockheed Martin. The SMSS™ can address the requirements of Light Infantry, Marine, and Special Operating Forces to maneuver in complex terrain and harsh environments, carrying all types of gear, material, and Mission Equipment Packages (MEPs).

The SMSS™ is a squad-sized Unmanned Ground Vehicle (UGV) platform as shown in Figure 11.6, about the size of a compact car, capable of carrying up to 1500 lb of payload. Designed to serve as a utility and cargo transport for dismounted small unit operations, it possesses excellent mobility in most terrains. The SMSS™/Transport lightens the load of a 9–13 man team by carrying their extended mission equipment, food, weapons, and ammunition on unimproved roads, in urban environments, and on cross-country terrain. Control modes include tethered, radio control, teleoperation (Non-Line of Sight (NLOS) and Beyond Line of Sight (BLOS)), supervised autonomy, and voice command. Technology Readiness Level TRL level is 7–9.

As shown in Table 11.19, the five proposed cost models adequately capture all of the P-WBS elements of the SMSS™. In some cases, the cost of individual elements can be captured by more than one cost model. To ensure that costs are not double-counted, the estimator should decide which of the cost models will be used for each WBS element. This decision could be based on the amount of fidelity provided by each cost model or the ability of the cost model to capture the WBS element's characteristics that influence cost.

Figure 11.6 Squad Mission Support System (SMSS™). *Source*: http://www.lockheedmartin.com/us/products/smss.html

Once the appropriate cost models are determined for each WBS element, the cost can be calculated as the sum of the outputs of the five cost models as shown in Eq. (11.4).

$$Cost\,(convert\,all\,individual\,outputs\,to\,\$K) = (Hardware) + (COCOMO\,II) + (COSYSMO)$$
$$+ (Weight\,Based\,CER) + (Perfomance\,Based\,CER)$$
$$(11.4)$$

The expected unit cost would be in the range of $1 million to $100 million depending on the capabilities and complexities of the UMAS. This is based on the historical results from the unit cost of the Global Hawk Unmanned Aircraft System ($92.87 million) and Modular Unmanned Scouting Craft Littoral ($700,000). If the estimated cost falls outside of this range, careful analysis should be done to ensure that the capabilities of the UMAS being estimated are truly beyond the scope of the historical data.

Another basis of comparison could be the two CERs described in this chapter which consider flight hours and maximum takeoff weight. While these cost drivers would only be relevant for Unmanned Aerial Vehicles, they can serve as sanity checks when performance and weight are important considerations.

For the purposes of this chapter, we are unable to provide a comparison of actual costs versus estimated costs to validate our proposed cost modeling approach. One reason is the proprietary nature of the data. Another is the lack of fidelity that is available to compare UMAS costs using the same cost elements, namely: vehicle, ground control station, and support elements.

Table 11.19 Types of estimates needed per product breakdown structure element.

Type of model recommended

Ref. #	WBS element [4]	Hardware (SEER-H)	Software (COCOMO II)	Systems engineering (COSYSMO)	Weight-based CER	Performance-based CER
1	Squad Mission Support System (SMSS™)					
1.1	Common Mobility Platform Vehicle					x
1.1.1	Vehicle Integration, Assembly, Test, and Checkout			x		
1.1.2	Hull/Frame/Body/Cab				x	x
1.1.2.1	Main Chassis Structure	x			x	
1.1.2.1.1	Frame and Hull				x	x
1.1.2.1.2	Hood				x	
1.1.2.1.3	Deck Panels				x	
1.1.2.1.4	Skid Plate				x	
1.1.2.2	Electronics Box Structure				x	
1.1.2.3	Front Brush Guard				x	
1.1.2.4	Rear Brush Guard				x	
1.1.2.5	Front Sensor/Component Mount				x	x
1.1.2.6	Rear Sensor/Component Mount				x	x
1.1.2.7	Equipment Rack				x	
1.1.2.8	Pack Racks/Tail Gate				x	
1.1.3	System Survivability			x		x
1.1.4	Turret Assembly		x	x		x
1.1.5	Suspension/Steering					x
1.1.6	Vehicle Electronics		x	x		x
1.1.7	Power Package/Drive Train	x				
1.1.8	Auxiliary Automotive	x		x	x	x
1.1.9	Fire Control					x
1.1.10	Armament				x	x

No.	Item				
1.1.11	Automatic Ammunition Handling		x		x
1.1.12	Navigation and Remote Piloting				x
1.1.12.1	Navigation Unit	x	x		
1.1.12.2	Robotics Subsystem	x			x
1.1.12.3	Autonomy Subsystem	x			x
1.1.13	Special Equipment			x	
1.1.14	Communications	x	x		x
1.1.15	Vehicle Software Release	x			
1.1.16	Other Vehicle Subsystems			x	
1.2	Remote Control System		x		
1.2.1	Remote Control System Integration, Assembly, Test, and Checkout			x	x
1.2.2	Ground Control Center Subsystem			x	
1.2.3	Operator Control Unit Subsystem		x		
1.2.4	Remote Control System Software Release	x			

11.7   Additional Considerations for UMAS Cost Estimation

11.7.1   Test and Evaluation

Many systems engineering and project management experts advise concurrent planning of Test and Evaluation (T&E) during the earliest phases of a project [2]. In a similar fashion, estimating the cost of these activities should also begin earlier rather than later. As budget are allocated and costs are estimated, some key considerations on how UMAS may be tested might be: analytical testing, prototyping, production sampling, demonstration, and modification [2]. The current practice in many organizations is to focus most on the cost of product development and when the project reaches the T&E phases use the remaining funding. This often leads to reduced testing and schedule slippages.

11.7.2   Demonstration

Demonstration is one of the unique aspects of T&E because there are many categories or subsets of demonstrating a product's capability. The two that are most important are demonstrating systems integration and demonstrating full operational capability. The costs associated with these are very different, and will also vary by type of UMAS. Some questions to consider when estimating the costs of demonstrating UMAS systems include:

1. *Level of Autonomy:*
   a. At what level of autonomy is the UMAS designed to operate?
   b. How will the level of autonomy influence safety, reliability, and integration to other systems?
2. *Systems Integration:*
   a. Will these demonstrations coincide with the design reviews or be separate events?
   b. What key system capabilities will your team want to demonstrate?
   c. Will you focus only on risky technology or demonstrate solutions to previously developed concepts?
3. *Full Operational Capability:*
   a. Who is your audience? Depending on whether it is government or commercial, this will play a huge factor in where and how you demonstrate.
   b. Will you need to create an operational scenario to show how the UMAS integrates into the current paradigm of its intended field? For example, will you need to have a mock battle, or create a queuing backlog at a distribution plant or border crossing?

11.8   Conclusion

In this chapter we described unique considerations of UMAS. In particular, life cycle models that help structure cost estimates, existing cost estimation methods, P-WBS, and parametric models. These led to a case study that described an Army Unmanned Vehicle and a recommended approach for estimating the per unit life cycle cost. We concluded by discussing two unique considerations of estimating the cost of UMAS – levels of autonomy, test and evaluation, and demonstration – that have the potential to significantly influence the complexities involved with transitioning a UMAS into operation.

As UMAS continue to be developed and deployed into operation we anticipate the maturity and accuracy of estimating their costs will similarly increase. At the moment, reliance on complete WBSs, comparisons with historical data, and utilization of existing parametric cost models can provide a reliable estimation process that can be used to develop realistic cost targets.

Acknowledgments

This material is based upon work supported by the Naval Postgraduate School Acquisition Research Program under Grant No. N00244-15-1-0008. The views expressed in written materials or publications, and/or made by speakers, moderators, and presenters, do not necessarily reflect the official policies of the Naval Postgraduate School nor does mention of trade names, commercial practices, or organizations imply endorsement by the U.S. Government.

References

1. Valerdi, R., Merrill, J., and Maloney, P. (2013). "Cost metrics for unmanned aerial vehicles." AIAA 16th Lighter-Than-Air Systems Technology Conference and Balloon Systems Conference, Atlanta, GA.
2. Blanchard, B. and Fabrycky, W. (2010). Systems Engineering and Analysis (5th ed.). Englewood Cliffs, NJ: Prentice-Hall.
3. Hagan, G. (2011, May 4). "Overview of the DoD systems acquisition process." DARPA Webinar. Lecture Conducted from DARPA.
4. Mills, M.E. (2014). Product Work Break Down Structure for SMSS™ provided by Lockheed Martin Missiles and Fire Control.
5. ISO/IEC (2002). ISO/IEC 15288:2002(E). Systems Engineering – System Life Cycle. Geneva, Switzerland: International Organization for Standardization.
6. Spainhower, K. (2003). Life Cycle Framework. Retrieved June 10, 2014, from https://dap.dau.mil/aphome/das/Pages/Default.aspx.
7. Cook, C.R. and Grasner, J.C. (2001). Military Airframe Acquisition Costs: The Effects of Lean Manufacturing. Santa Monica, CA: Project Air Force RAND.
8. Markish, J. (2002). "Valuation Techniques for Commercial Aircraft Program Design," S.M. Thesis, Aeronautics and Astronautics Department, MIT, Cambridge, MA.
9. Martin, R. and Evans, D. (2000). "Reducing Costs in Aircraft: The Metals Affordability Initiative Consortium," JOM, Vol. 52, Issue 3, pp. 24–28. http://www.tms.org/pubs/journals/JOM/0003/Martin-0003.html.
10. Larson, E.W. (1952). Project Management: The Managerial Process / Erik W. Larson, Clifford F. Gray (5th ed p. cm). New York: McGraw-Hill.
11. Department of Defense Fiscal Year (FY) 2015 Budget Estimates. (2014a, March 1). Aircraft Procurement, Air Force, Retrieved January 7, 2015, from http://www.saffm.hq.af.mil/shared/media/document/AFD-140310-041.pdf
12. Department of Defense Fiscal Year (FY) 2015 Budget Estimates. (2014b, March 1). Retrieved January 7, 2015, from http://asafm.army.mil/Documents/OfficeDocuments/Budget/budgetmaterials/fy15/pforms/opa34.pdf
13. Department of Defense Fiscal Year (FY) 2015 Budget Estimates. (2014c, March 1). Retrieved January 7, 2015, from http://www.finance.hq.navy.mil/fmb/15pres/OPN_BA_5-7_Book.pdf
14. Department of Defense. (2013). Unmanned Systems Integrated Roadmap FY2013-2038. Washington, DC: Department of Defense.
15. Boehm, B.W. (2000). Software Cost Estimation with Cocomo II. Upper Saddle River, NJ: Prentice Hall.
16. Valerdi, R. (2008). The Constructive Systems Engineering Cost Model (COSYSMO): Quantifying the Costs of Systems Engineering Effort in Complex Systems. Saarbrucken, Germany: VDM Verlag Dr. Muller.
17. SEER-H® Documentation Team: MC, WL, JT, KM. (2014). SEER for Hardware Detailed Reference – User's Manual. El Segundo, CA: Galorath Incorporated.

18. CMMI (2002). Capability Maturity Model Integration – CMMI-SE/SW/IPPD/SS, V1.1. Pittsburg, PA: Carnegie Mellon – Software Engineering Institute.
19. ANSI/EIA (2002). EIA-731.1. Systems Engineering Capability Model. Philadelphia, PA: American National Standards Institute (ANSI)/Electronic Industries Association (EIA).
20. Clark, B.K. (1997). "The Effects of Software Process Maturity on Software Development Effort," Unpublished Dissertation, Computer Science Department, University of Southern California.
21. ISO/IEC (2012). ISO/IEC 15504:2003. Information Technology — Process Assessment, Parts 1–10. Geneva, Switzerland: International Organization for Standardization
22. Cherwonik, J. and Wehrley, A. (2003). Unmanned Aerial Vehicle System Acquisition Cost Research: Estimating Methodology and Database. Washington, DC: The Office of the Deputy Assistant of the Army for Cost and Economics.

12

Logistics Support for Unmanned Systems

Keirin Joyce
University of New South Wales at the Australian Defence Force Academy
(UNSW Canberra), Australia

12.1   Introduction

Operations research focuses on optimizing real-world tasks [1] and is directly relatable to logistics, as the constituent components of a capabilities support system can be modeled, analyzed, and optimized with readily accessible, modern software. Advances in autonomy will result in both a change in mission and an increase in mission variety for unmanned systems. It follows that the logistic support of those systems must also adapt with the introduction of new technologies and mission sets. Indeed, some of the new missions of unmanned systems will be in the field of logistics support delivery.

This chapter aims are to address logistics support modeling, to discuss specific system adaptations due to the introduction of unmanned systems, and to pose areas for consideration when modeling a new support system.

12.2   Appreciating Logistics Support for Unmanned Systems

Due mainly to the speed at which the current generation of unmanned systems has been fielded, support of these systems has been limited to and constrained by traditional logistics models. This has been reasonably easy to achieve due to the well-funded cooperation of enthusiastic original equipment manufacturers (OEMs) and that the equipment is constructed from standard contemporary components that are already being supported. This will change, however, as unmanned systems that are currently in research, development, and prototype will radically alter operating scenarios. Mission sets (including logistics missions) will change and expand, the logistics support framework will improve and adapt to challenging new missions,

Operations Research for Unmanned Systems, First Edition. Edited by
Jeffrey R. Cares and John Q. Dickmann, Jr.
© 2016 John Wiley & Sons, Ltd. Published 2016 by John Wiley & Sons, Ltd.

and operations in turn will be able to accomplish more – a virtuous cycle for operations courtesy of new logistics technologies. We will cover these changes later in the chapter, but for now will start with the foundations of the current arrangements.

12.2.1   Logistics

Within the private sector, logistics is defined as that part of the supply chain process that plans, implements, and controls the efficient and effective transportation and storage of goods including services, and related information from the point of origin to the point of consumption for the purpose of conforming to requirements [2]. Business logistics is very much based on efficiency to attain profitability [3] and is structured along the following activities [4] that generally form the make-up of all business logistics systems:

- Customer service
- Demand forecasting
- Inventory management
- Logistic communications
- Materials handling
- Order processing
- Packaging
- Parts and service support
- Plant and warehouse site selection
- Procurement
- Reverse logistics
- Traffic and transportation
- Warehousing and storage.

The study of business logistics began in earnest in the 1950s and 1960s when concepts of total cost analysis and cross-activity management (integration) were identified as areas to squeeze profitability out of an organization. The military had, of course, been executing logistics for two millennia prior to this, but not necessarily with profitability as a focus. The key challenge to military logistics is (and always has been) the provision of support through an extended supply chain in austere environmental and threat conditions with the objective of ensuring the fighting capability of the fielded force.

In general terms, in the barracks or in peace-time, military logistics is driven by similar considerations to business [3], and aims to achieve as high a readiness as possible, aspiring to 100% availability and priming of a supply chain of spares and consumables within pre-paredness lead times. In the field, in operational usage, it is generally accepted that there will be some degraded capability, and hence as high availability as possible (functions maximized) is aimed for (sometimes by over priming logistics in the fielded force, referred to as "iron mountains" [5]). Availability is all about equipment maintenance being enabled by a supply chain that has minimized operational interruptions and distribution to resources [6]. Thus, the private sector plans and allocates resources to operations in order to achieve financial outcomes, whereas the defense apparatus plans and budgets operations for operational outcomes [7].

Table 12.1 Comparison of US, Australian, and UK military principles of logistics.

US Department of Defense	US Army	Australian Defence Force	Australian Army	UK Ministry of Defence
Simplicity	Simplicity	Simplicity	Simplicity	Simplicity
Economy	Economy	Economy	Economy	Efficiency
Flexibility	Improvisation	Flexibility	Flexibility	Agility
Responsiveness	Responsiveness	Responsiveness	Responsiveness	
Survivability	Survivability	Survivability	Survivability	
Sustainability	Continuity	Sustainability	Sustainability	
	Anticipation	Foresight	Foresight	Foresight
		Balance	Balance	
	Integration	Cooperation		Cooperation
Attainability				

Military logistics doctrine is grounded in sets of principles that are taught to and practiced by all logistics managers throughout their military training regimes. As this field of study has evolved, the US, Australia, and the UK have developed remarkably similar doctrines, as shown in Table 12.1.

Of note, no business text provides a list of principles, keywords, or guiding terms for the development of a civilian logistics system, its application or its execution. The civilian systems look instead to functional descriptions of the sub-components of their logistics with an overarching management structure that performs the function of integrating the activities. Intrinsically, however, the military principles are easily translatable to business logistics as keywords or principles. The exception would be the principle of "survivable." For the military, this might be considered the most important, for if a logistics unit is destroyed by the enemy, there is no point to the remaining principles [8]. Survivability is generally not a consideration for the civilian sector in the same way, however, threats from competitors, poor performance of sub-contractors and suppliers, changes to legislation, or community pressure could all be seen as less violent equivalents to military operations. The civilian sector might consider "endure (the business cycle)" as a principle.

Information technology developments enabled equipment project and capability managers to fuse whole-of-lifecycle approaches in the late 1980s and early 1990s. Pioneered by the US Department of Defense, and quickly adopted by the UK and Australia, the integration of a logistics support (ILS) system allowed "an integrated and iterative process for developing material and a support strategy that optimizes functional support, leverages existing resources, and guides the system engineering process to quantify and lower life cycle cost and decrease the logistics footprint" [9] (demand for logistics), making the system easier to support. This process, although originally developed for military purposes, is now widely used in commercial product support or customer service organizations. From the principles of logistics, and in support of a functional development of a logistics support system, a set of 10 ILS elements are used to design an ILS system. These elements encapsulate the constituent "stovepipes" of logistics function and are as follows:

- Engineering support (or design interface)
- Maintenance planning

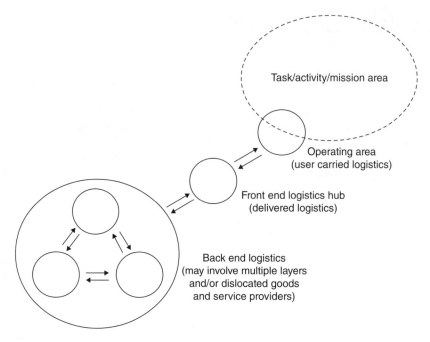

Figure 12.1   Simplified logistics network diagram.

- Personnel (and manpower)
- Supply (warehousing) support
- Support and test equipment
- Training and training support
- Technical data
- Computer (resources) support
- Facilities
- Packaging, handling, storage, and transport.

In an attempt to simplify the definitions of logistics as a practice, Figure 12.1 is presented to graphically convey "logistics" as a simplified network diagram. Such networks lend themselves to modeling and present great cost benefits to be reaped from optimizing the systems. Operations researchers have been doing such for more than 45 years.

12.2.2   Operations Research and Logistics

Computer-based discrete-event simulation (DES) has long been a tool for analysis of logistics and supply chain systems [10]. Before that, and still for analyzing more simple components of the system, data analysis has been used in strategic and operational decision making by taking data from actual/predicted orders and providing management and decision makers with stochastic sets for linear programming or simulation. Linear programming, while once the most widely used planning tool in strategic and operational logistics management, has been overtaken in recent times by simulation. The earliest simulations occurred in the 1960s and were characterized by the Logistics Composite Model (L-COM) developed by the US Air Force Project RAND and cycled by the large mainframes of defense computing power. The L-COM model, as an example, simulated operations and support functions at an air force base via network analysis, and comprised three main programs: pre-processing (translating data and generating sortie requirements from the flying program); simulation (flight and base support processes including weapon loading, refueling, and malfunctions); and post-processing (generation of reports and statistical graphical plots) [11]. The very basic nature of the computers available to run this model meant that the probabilities input were simple, non-variable percentage chances determined by the research team as they wrote the lines of code to describe the 1500 daily tasks. Validation of the model against an independent data collection project (Project PACER SORT) indicated satisfaction of L-COM's "broad validation objectives," and enabled L-COM to serve as the stepping stone for further US Air Force Project RAND model developments over the decades.

Monte Carlo simulation is the technique generally used in the generation of system models. First used by Fermi and Ulam in 1930 to calculate the properties of the neutron, the basis for these simulations is the generation of random numbers and assigning ranges of numbers to values that fit a probability distribution function (PDF) [1]. A typical example, used by Valles-Rosales and Fuqua [1], to show this technique is the case of rolling a die: Generate a uniform random number series from zero to one (U(0,1)) and allocate an equal range to each die roll outcome as follows:

0.001–0.166: die roll of 1
0.167–0.332: die roll of 2
0.333–0.498: die roll of 3
0.499–0.665: die roll of 4
0.666–0.832: die roll of 5
0.833–0.999: die roll of 6.

The real situation of a die roll, and successive die rolls, can then be modeled by using a pseudo-random number generator (PRNG) from a computer. This simplistic example can be escalated to simulate any fixed range of uniform numbers, but can also have functions substituted to more closely replicate non-linear results. In this way models can be created from the simplest dice roll up to the most complex of systems such as nuclear power plant processing.

With the advent of cheaper and more powerful computer software and hardware, system simulation models have evolved to allow management to evaluate and optimize various strategies, inclusive of logistics systems. This is done by manipulating the logistics system variables, which can be manually input (akin to L-COM), or randomly generated through the background selection and programming of PDFs. These developments occurred primarily in the 1990s, first with the manufacturing industry who had simple linear processes to model and were under pressure to harness operational efficiencies [12], but then developed into "material handling systems, civil engineering, the automotive industry, transportation, health, military, service industries, communication and computer systems, activity scheduling, personnel allocation, business process re-engineering, and human systems management, among others." There are now dozens of commercially available software (CAS) products available to choose from and almost all start with the generation of a basic queuing model of a bank teller, with customers coming and going, and tellers coming on and off shift. Da Silva and Botter [12] have developed a method to assist managers to choose a product from the dozens available.

In the field of system simulation, a DES models the operation of a system as a discrete sequence of events in time, where the state variables change only at those discrete points in time at which events occur. Each event occurs at a particular instant in time, as a consequence of activity times and delays, and marks a change of state in the system. Entities may compete for system resources, possibly joining queues while waiting for an available resource. Activity and delay times may "hold" entities for durations of time. DES lends itself perfectly to logistics, as every output of a logistics system analysis is a function of time. A DES model is conducted over time ("run") by a mechanism that moves simulated time forward [13].

Computerized modeling used to follow four categories: planar models; warehousing models (internal and external); network models; and discrete models [4]. However, with modern discrete event simulators being able to combine service attributes, cost data, and networks across multiple (almost infinite) locations, discrete models are now favored as they can do all of Stock and Lambert's [4] old categories. Many simulation programs are now available that use graphical interfaces or visual interactive modeling systems (VIMSs) [14]. VIMSs assist even a non-trained modeler in modeling a system [1]. Programs include ProModel, Simul8, Arena, and ExtendSim, among others.

Contemporary software packages rely at their core on DES as applied to the modeling of operations systems: "a configuration of resources combined for the provision of goods or services" [15]. This definition goes on to identify four specific functions of operations systems as: manufacture; transport; supply; and service (which are four transactional elements that constitute "logistics" as discussed in the previous section). Computer-based DES enhances our understanding of logistics and supply chain systems by offering the flexibility to understand changes in system behavior under dynamic cost parameters and policies [16] and by utilizing compressed time [17]. Logistics systems lend themselves to simulation because of the network and linkage nature of infrastructure placement, and the ability to generate relatively quantifiable data. In addition, the size and complexity of logistics systems, their stochastic nature, data detail necessary for investigation, and the relationships between system components, make simulation modeling an appropriate modeling approach to understand such systems [10].

Key to the accuracy of a stochastic model is the selection of the correct statistical distribution to approximate the behavior of each model block as a reflection of what happens in the real world, noting that the models are probabilistic and dependent on the accuracy of the modeled "world." An exact science it is not. The choice of one distribution over another is dependent on the type and extent of the data gathered, the detail required, and informed guesswork [18]. Typical distributions utilized in the modeling of logistics systems include Beta (items in a shipment), Exponential (electrical component/system time between failure (TBF) or time to repair (TTR)), Normal (the bell curve–for events of natural causes), Logistic (growth models, particularly where exceptional cases play a large role), Uniform (activity durations) and Weibull (mechanical component/systems TBF or TTR). Texts such as Krahl *et al.* [18] and O'Connor [19] provide advice as to selection, and software is available to assist in fitting data to a distribution.

Models of current systems are tested, verified, and validated using known facts and data sets, and then can be optimized by careful application of optimization techniques developed by operational researchers [4]. However, verification and validation of a discrete event model is difficult to perform when modeling a system that does not yet exist, where real-world nuances are not known by the modeler, or when investing in wholesale change for the future of a system [20]. As such, an accepted approach has been to model the current (or a representative) system, verify and validate that model against data obtained from that current system, and then begin to modify the model for the future based upon the same data, distribution functions, and principles as the validated current model was built upon. In an effort to address this "heuristic" verification and validation gap, Manuj *et al.* [10] presented an eight-step process, called the simulation model development process (SMDP), for the design, implementation, and evaluation of logistics and supply chain simulation models. The SMDP also identifies rigor criteria for each step. Further research does exist, such as [21], to guide verification and validation activities through Generic Methodologies for Verification and Validation (GM-VV). Such methodologies assist in the design and acceptance of new models and simulations.

Optimization is the most important activity when analyzing both existing and developing logistics support systems, but it is even more so in the development of logistics support systems that will support new capabilities. Indeed, most of the decisions that affect the logistics support of a product occur well before it is manufactured [22]. These decisions are the result of design, manufacturing, and cost trade-offs among performance measures such as reliability, maintainability, and availability, which then affect decisions regarding the logistics system framework. Chapman *et al.* [23] state that the decisions (design and support) that are made initially, during the design and development of a product, will determine 80% of the total lifecycle costs. Such costs, depending on operational concept and length of life of type, will be in the order of two to five times the cost of the acquisition.

Simulation optimization must have a structured approach to determine the optimal input parameter values of a simulation model that will yield system improvement. Optimization routines need to efficiently and robustly calculate the system performance for different input factors, which can be quantitative (numerical) or qualitative (structural) [24]. In models of logistic systems the goal of the optimization is to improve the efficiency of the support blocks to increase the utilization of the supported systems.

A typical optimization involves two phases: a search phase and an iteration phase. Techniques for optimization can be classified into two general groups that support these phases. The first is search methods, which evaluate the simulation at different points and use different rules to find the optimal. The second is gradient-based methods where the gradient information from running the model is used to find the optimal.

Search methods are most suitable for models with mostly discrete input parameters, that is, the parameters take on only a finite set of values. Search methods include Complete Enumeration, Heuristic Search Technique, Response Surface Method, Pattern Search, and Random Search. Gradient-based approaches use the simulation to find the gradient, which is then used to speed up the search for the optimum. Gradient estimation techniques include Finite Difference, Steepest Descent, Meta-modeling, Frequency Domain Method, Simulated Annealing, Gradient Surface Method, Perturbation Analysis, and Likelihood Ratio.

Many optimization routines are now provided with DES software, in graphical user interface programs available as an optimization block to "plug-in" [25]. These routines seek improved performance measures by resetting the input parameters using heuristics. The optimizer runs the model many times using different values for selected parameters, searching the solution space until it is satisfied that it has found an acceptable solution, then populates the model with the optimized parameter values [18].

Pertinent examples in the academic literature of logistics system components and networks being optimized, including facility locations and routing [26], maintenance resources [27, 28], support to amphibious operations [29], spares components, supply chains [30, 31], and airlift planning [32, 33] provide access to robust mathematics, distribution fitting, and techniques for verification and validation, sensitivity analysis, and optimization.

12.2.3   Unmanned Systems

Injecting unmanned systems into the "Task/Activity/Mission Area "Operating Environment" in Figure 12.1 changes little in a simplistic contemporary setting. The manned equipment that previously undertook a mission has simply been exchanged for unmanned equipment to either replace or augment the humans performing the task. However, it still relies on humans to operate it. The difference comes in the near future when unmanned systems technologies and capabilities start expanding the range, endurance, and scope of missions: operational distances go from electronic line-of-sight to satellite-enabled intercontinental ranges; sortie durations go from the limitations of combustion engine fuel tank capacities to hydrogen/hybrid/solar-electric propulsion systems that power equipment for days/weeks/ months; and roles and tasks scope changes with the enhancement of sensor systems, autonomy, the range of payloads that can be installed onto the platforms, and overcoming various ethics challenges.

Unmanned Systems are currently defined as follows:

- The Unmanned Aircraft System (UAS) is a system whose components include the necessary equipment, network, and personnel to control an unmanned aircraft [34]. This system boundary includes a ground segment to launch, recover, and control (typically called a Ground Control Station (GCS)) the air vehicle. UAS is interchangeable within this chapter with Unmanned Aerial Vehicle (UAV), Remotely Piloted Aircraft (RPA), Remotely Piloted Aircraft System (RPAS), and Remotely Piloted Vehicle (RPV). UASs are generally orga- nized together into five groups (generally) based on size and weight, with nano/micro sitting outside the groups at the far left, and Unmanned Combat Air Vehicles (UCAVs) taking an outside place on the far right [35, 36].
- The Unmanned Ground Systems (UGS) is a powered physical system with (optionally) no human operator aboard the principal platform, which can act remotely to accomplish assigned tasks [36]. UGSs may be mobile or stationary and include all associated supporting components such as operator control units (OCUs).
- The Unmanned Maritime System (UMS) comprises both Unmanned Surface Vehicles (USVs) and Unmanned Undersea Vehicles (UUVs), all necessary support components, and the fully integrated sensors and payloads required to accomplish the required missions [36]. UUVs fall into two categories: tethered Remotely Operated Vehicles (ROVs) and un-tethered, free-swimming vehicles called Autonomous Underwater Vehicles (AUVs) [37].

However, noting the increasing cross-dimensional nature of technologies (autopilots, autonomy control systems, and artificial intelligence (AI) networks are transferable across all three dimensions), and in a military context, the continued development of "jointness," many are adopting the more generic term "Unmanned Systems" abbreviated most commonly as "UxS." This is particularly noteworthy among organizations like the United States Marine Corps (USMC), which operates systems across all three domains, and among small defense forces such as in Australia. It is further evident in corporate unmanned technology organizations that acknowledge the cross-domain nature of unmanned systems in their proprietary designs, operation, and maintenance.

Manned-Unmanned Teaming (MUM-T) is an operating concept where an unmanned system provides services for a third party on the ground, while pioneered by the military (MUM-T in the guise of the provision of surveillance imagery or target prosecution), is spreading its wings with expanding utility, and in the future unmanned systems in all domains will provide products/services to third parties. Research and development teams are also investing heavily in hybrid unmanned capabilities such as UUV launching UAS, USV launching UAS (the Riverwatch Nacra unmanned catamaran that launches and recovers a Virtualrobotix R Brain 4 hexacopter), and UAS air-dropping UUV and UGS.

While many chapters of this book cover some of the increased scope of range, endurance, and mission that unmanned systems may undertake, it is useful to restate some of that here. A good reference for this is the US Department of Defense Unmanned Systems Integrated Roadmap FY2013-2038 [36], which articulates a UxS vision across all domains over 25 years. Their relevant points are summarized in Table 12.2.

As human endurance is discounted further and further as a design consideration in unmanned systems, particularly as autonomy and AI advancements continue, availability of the unmanned system asset/hardware becomes the key performance indicator for the support of those capabilities. In a military sense, availability must be maximized at all times (aiming

Table 12.2 Mission expansion considerations from US Department of Defense [36].

UAS	Bug (nano) reconnaissance
	Bird (micro) reconnaissance
	Group 2/3/4 vertical take off and landing (VTOL) surveillance/small yield strike
	Group 3/4 VTOL resupply
	Group 3/4 VTOL casualty evacuation
	Joint Strike Fighter (JSF) robot wingmen
	Penetrating strike aircraft (UCAV)
	Strategic bomber (unmanned B-52/B-1/B-2 replacement)
UGS	Man-packable explosive ordnance robots
	Area detection and clearance systems (mine hunting/clearing) vehicles
	Squad equipment transport vehicles/robots
	Squad robot wingmen
	Convoy autonomy vehicles
	Ultralight reconnaissance/nano/micro robots
UMS	Surface patrol boats (unmanned surveillance and port/littoral support)
	Hydrographic reconnaissance (unmanned hydrographic mapping/sensing)
	Underwater battlespace sensors (mobile sound surveillance network (SOSUS))
	Underwater battlespace clearance (surface and underwater mine clearance)
	Large displacement underwater vehicles (unmanned submarines)

for 100% readiness to undertake a military mission) and for a public/private sector unmanned system, availability must be 100% at a pre-planned times and places, which raises the question, what are the challenges to supporting these systems?

12.3 Challenges to Logistics Support for Unmanned Systems

The unmanned systems support community faces two sets of challenges, linked to chronologic proximity: immediate and future.

12.3.1 Immediate Challenges

The first generation of fielded unmanned systems, spanning the first 90 years of their evolution, predominantly was focused on the rapid delivery of immediate capability to the warfighters. Examples range from the Curtiss-Sperry Aerial Torpedo and Kettering Bug of World War I development, Interstate DR-2 World War II bomber, Ryan BQM-34 Firebee, and Gyrodyne QH-50 DASH of Vietnam, IAI Scout of the Israeli Armed Forces fighting the conflicts of the 1960s and 1970s, to the first RQ-1 Predator variants flying in Kosovo [38, 39]. Because of the need to rapidly develop and field these initial capabilities, long-term sustainability planning has often occurred late in the development cycle [36], or not at all in the case of systems that became obsolescent and were superseded/replaced.

Fleet managers find themselves with a raft of immediate logistics support challenges brought about from the Global War On Terror's (GWOT) rapid advancement and fielding of technology, low rate initial production (LRIP) fielding, and research and developmental fielding. This is evident in sustaining rapid development/acquisition systems that have an OEM monopoly for the provision of support, limited reliability, availability, maintainability (RAM) and/or intellectual property (IP) data to enable taking ownership of support back from the OEM, and multiple configuration baselines, which are difficult to manage and sustain.

New buyers of these systems/services have often found themselves confronted with the decision to contract heavily with the OEM for their logistics support, or to invest in large (unaffordable) overheads if they wish to take a greater ownership of the logistics support to that system.

12.3.2 Future Challenges

The logistic legacy challenges of the GWOT described above will be compelled to change due to financial pressure–they cost too much!

Current fleet sizes in the military are large, but with the advancement of technology and unmanned systems missions, they will continue to grow, particularly if the research concepts being invested in swarms of miniature systems come to fruition. Systems owners cannot afford to accept the current status quo of poorly planned logistics support, especially as their fleets, or the fleets that they are joining, continue to expand globally. Whether it be the US military, supporting a large systems fleet that is deployed around the globe, or whether it be a small to medium enterprise tapping into a global network of a flexible payload platform like the Insitu ScanEagle, the tyranny of distance will remain a challenge in supporting the system. However, with every challenge comes opportunities and advances in global supply chain/distribution, 3D printing of components, and so on can be modeled to offset and/or alleviate these challenges.

Table 12.3 Grouping unmanned systems for the purposes of analyzing logistics systems.

	UAS	UGS	UMS
Group A	Nano/micro/mini Group 1	Man-packable systems	Robots/tethered
Group B	Group 2 Group 3 Group 4 UCAV	Man-portable systems	
Group C	Group 5 ULE/ULR	Vehicle integrated systems	Un-tethered

12.4 Grouping the Logistics Challenges for Analysis and Development

Despite the varied sizes, roles, domains, and missions of unmanned systems, when it comes to logistics support operations analysis and their challenges now, and in the future, they can be grouped into three areas (Table 12.3).

12.4.1 Group A – No Change to Logistics Support

In the first group of UxS to be discussed we have equipment that is used by single persons and small teams. The group is characterized by portability, operators conducting (or not) their own (organic) front end maintenance with spares and consumables that they carry with them, and comparatively low up-front cost equipment (sometimes described as "throw away") that frequently results in "repair by replacement." Specifically:

- *UAS:* Nano/micro/mini and Tier 1 systems. Current military examples include the Prox Dynamics Black Hornet nanocopter in use by the British Army, and the AeroVironment Wasp, and Raven small UAS that are in use primarily by US forces, but also used in small quantities by many other national militaries/paramilitaries around the world. These systems are acquired for costs in the order of US\$10 000 to US\$100 000, but many commercial systems such as the Parrot AR.Drone are being sold to remote control enthusiasts, photographers, and the film and sports industries over the internet in costs of US\$100. Due to the relatively low cost of systems and components, research and development efforts in these fields are heavily contributed by start-ups, small to medium enterprises, and universities. One commercial application being keenly pursued by many companies is UAS delivery (DHL and microdrones packages via Paketkopter, Amazon books, and "ambulance drone" which delivers a defibrillator). This effort will yield technologies in the future that will see further miniaturization of current equipment (easyJet is testing a hexacopter carrying a miniaturized laser scanner for aircraft inspection tasks), increased endurance due to power source advances (the AeroVironment Puma AE has flown for more than 48 h on a cryogenic liquid hydrogen fuel cell and LaserMotive has used ground-based laser power beaming to keep a Lockheed Martin Stalker airborne for greater than 48 h), AI taking a greater role in command and control of the equipment, and practical implementation of swarming.

- *UGS:* Man-packable equipment. The military, paramilitary, and emergency services currently utilize Unattended Ground Sensors (UGS) for detection using movement, infrared, or seismic triggers, such as the ARA eUGS or the Textron Systems MicroObserver, tactical persistent surveillance systems (cameras on sticks) for monitoring of high-value equipment and installations, disaster robots (Remotec ANDROS Wolverine), and "throwbots" (ReconRobotics Throwbot, QinetiQ Dragon Runner, and iRobot FirstLook) to support entry operations. This class also includes household robots (vacuums and lawnmowers). There is not a large or emerging civilian industry for these tools as yet, although UGS have been tested for wildlife monitoring applications and disaster robots have been called upon increasingly within the last decade to assist at disaster sites [37]. To support the continued military/paramilitary markets, future systems will continue to be miniaturized and run for longer, leveraging off the same research and development efforts being undertaken in the UAS space.

Applications of these technologies in the future could be widespread, and only limited by the extents of imagination, from real estate, to tourism, the film industry, search and rescue agencies, wildlife protection, resource companies, utility companies, and, of course, the military.

While the expansion of applications for these small UxS will continue, their characteristics and very nature will result in little change to support methodology. As the equipment is carried on the person and used by one person or a small team working remotely, the person/small team also needs to continue to undertake all support activities at the operating location. Battery charging, consumable replenishment, minor repairs, and repair by replacement must be undertaken by supplies carried with the equipment/person/small team, and only replenished by a larger hub, depot, warehouse, or operations center. Support at that hub/depot/warehouse/operations center is dealt with in the same manner as the Group B systems.

12.4.2   Group B – Unmanned Systems Replacing Manned Systems and Their Logistics Support Frameworks

Group B equipment has been the largest area of UxS acquisition investment over the past decade, with most acquisitions being a result of supporting the GWOT. In this context, unmanned systems have been utilized as a counter-insurgency technique to remove combatant humans from danger and replace them with equipment that is unmanned (although they are not truly unmanned, but tele-operated from a safe distance). In some extremes, UAS missions in Afghanistan are under command and control from operating bases in the US and the UK. Examples are as follows:

- *UAS:* Systems include UAS Groups 2–4 and UCAV, exemplified in current systems by the Insitu ScanEagle and AeroVironment RQ-20 Puma (Group 2), Insitu RQ-21 Blackjack (Integrator), and AAI Corp RQ-7 Shadow 200 (Group 3), Elbit Hermes 450, Thales IAI Watchkeeper 450, IAI Heron, Northrop Grumman MQ-8B Fire Scout and GA-ASI MQ-1 Predator and MQ-9 Reaper (Group 4), and Northrop Grumman X-47B UCLASS. In the future, designs such as rotary wing Groups 2–4 (Schiebel Camcopter, Boeing A160 Hummingbird, Kaman K-MAX, Northrop Grumman MQ-8C Fire Scout) will prove efficient and robust enough to field, and provide much more tactical launch and recovery sites than the current systems. Rotary wing platforms will not only take over surveillance roles but also increase mission scope to include unmanned aerial resupply and medical evacuations (already tested over the past three years in Afghanistan). Group 4 long endurance (LE) systems (currently 12–30 h) will rise into the ultra long endurance (ULE) category due to advances in propulsion efficiency and electronics reliability. UCAV systems will come to the fore to supplement/replace manned combat platforms (UCLASS, BAE Taranis, EADS Barracuda, Dassault nEUROn). One further field of future commercialization is the air cargo sector – military and commercial – which would be exemplified by un-crewed 747s, 777s, A380s, C-5s, C-17s, and the like, flying cargo across the globe in 5–20 h sortie durations.

- *UGS:* Man-portable systems, able to be moved and emplaced by manpower alone, but requiring vehicles to transport the equipment from operating site to operating site. Current examples include attended/unattended Ground Surveillance Radars (GSR) such as those used for forward operating base (FOB) and patrol long halt protection in Afghanistan (Rockwell Collins Patrol Persistent Surveillance System, Northrop Grumman ExTASS), and unexploded ordnance (UXO) robots used by military/paramilitary forces in dealing with bomb and improvised explosive device (IED) threats (examples include the iRobot PackBot and QinetiQ TALON). In the future, systems such as radars will miniaturize somewhat, however, the power sources required will continue to see the equipment look much the same as it currently does. Equipment currently utilizing humans to tele-operate robotic arms and provide navigational guidance will see significant advances due to the efforts of universities and other robotic market suppliers (medicine, space, etc.) – these advances should see people removed from the operation of these systems as AI takes over the guidance and operation of robotic elements.
- *UMS:* Robots/tethered systems operated from a mother ship. Current examples include datalinked and tethered systems such as the Saab Double Eagle, Seaeye Falcon, and Leopard. Not much is anticipated to change in this field of systems as the nature of underwater operations dictates datalink restrictions. AI advances may allow some of these systems to cut away their links and rise into the third group of autonomous operating systems allowing them to go further/deeper (as component technology dictates/allows).

Before the GWOT, government organizations acquired the equipment, technical packages (in many cases the military provided the research and development funding to meet the performance specification of the system, and as such owned all or a large portion of the IP), initial spares and training, and provided this to government operating units in government facilities. Government accountability and processing dictates that this takes a good measure of time to undertake, usually measured in years. Examples here included the UK Ministry of Defence Phoenix UAS program and the US Department of Defense RQ-2 Pioneer UAS program.

The GWOT, however, initiated a new and rapid development of these more recent systems over the past decade, defined by rapid fielding and spiral development cycles that government acquisition has not been able to match with current practices. This has resulted in governments needing to outsource much of their support framework to the OEM or the OEM's sub-contractors. Operations now see an increased presence of civilians in operating units. For example, there have been RQ-7B Shadow 200 units operating in Iraq that have been government-owned, contractor-operated (GOCO) (approximately 20 contractors), and all Shadow 200 units (military and GOCO) are supported by civilian contractor Field Service

Representatives (FSRs) in the operating locations. Forward repair activities of Shadow 200, including warehousing and maintenance rearward of the operating teams, are also located in Afghanistan. Another example is USAF MQ-1 Predator units in Afghanistan supported by a USAF maintenance crew of fewer than 10, and augmented by a contractor maintenance crew of ~20 operating from an airfield in theater.

In the military context, this group of systems operate a support system that sees the military run all, or almost all, support operations at the tactical location. The small exception to this is the embedding of an FSR from the OEM with the operating unit so as to provide in depth engineering/technical support at the operating location. This presence of civilians, however, is only a pseudo-change to traditional support frameworks as the civilians are undertaking the roles that military personnel would normally undertake if not for the rapid fielding of equipment. The military has, however, always utilized civilian service providers from as far back as Alexander the Great. Nonetheless, there are advantages to the utilization of OEM and OEM sub-contracted logistics support framework components, and as such, current logistics support frameworks for Group B fall into two classes:

- Organic (traditional) – run by the operator's logistics organization
- Contractor Logistic Support (CLS) – OEM (generally) contracted by the operator to provide logistics support through private means (private warehousing, freight, and maintenance).

From an operations research perspective, these two classes of support can be easily analyzed against each other. This is because, essentially, they utilize exactly the same processes and discrete events for modeling. A discrete event modeler can easily insert the different times, rates (usage, failures, etc.), and, importantly, costs to achieve system support simulation results that can be weighed against each other by operational and business decision makers.

There are examples in current reporting and media that discuss and show the relative effectiveness of these different support frameworks, which are now being modified to achieve future support efficiency targets. These new key performance indicators have been theoretically (through operations research and modeling) optimized from the last decade's results. Some of these include:

- USAF MQ-1/9 Predator/Reaper units training USAF maintenance personnel to take over FSR maintenance tasks in order to reduce the reliance on (and cost of) FSR.
- US Army RQ-7B Shadow 200 units (and other foreign operators) training army maintenance personnel to take over FSR maintenance tasks in order to reduce the reliance on (and cost of) FSR.
- US Army incorporating (FAA/US Army Aviation Regulation) design acceptance requirements into future modifications/upgrades to ensure aviation engineering compliance and supportability of systems.
- US Navy requiring regulatory compliance on MQ-4C Triton designs, which is in contrast to USAF RQ-4A/B Global Hawk fielding experience.

Group B definitely creates the greatest propensity of current and immediate logistics challenges due to the quantity of GWOT-acquired UAS/UGS. Adapting the logistics support systems of these challenges will be a valuable learning experience for the support framework designers of the future systems.

12.4.3 Group C – Major Changes to Unmanned Systems Logistics

By far the most interesting capabilities, from a logistics support perspective, Group C equipment is the subject of the highest levels of research and development effort and funding. Examples are as follows:

- *UAS:* Systems include UAS Group 5 and the developmental Ultra-Long Endurance/Range (ULE/R) systems (currently characterized by aircraft of very light weight structures), exemplified in current systems by the Northrop Grumman Global Hawk (Group 5 – size and weight of business jets and smaller wide-body passenger aircraft) and solar electrically powered QinetiQ Zephyr and Titan Solara. In the future, designs such as hybrid propulsion Group 5 (Boeing Phantom Eye) will prove efficient and robust enough to field, and provide great increases in range and endurance that will thrust them into the ULE/R category. Another contributor to ULE/R will be stratospheric geostationary blimps/airships like the Thales Alenia Space StratoBus.
- *UGS:* Vehicle integrated systems utilizing chassis and power plants found in military and commercial vehicle fleets, ranging in size from robot infantry squadmen, quad-bikes, and up to small and medium trucks. Current military examples include infantry patrol support vehicles like the Lockheed Martin MULE and SMSS, and General Dynamics MUTT, which carry a patrol's heavier equipment while they are on foot; sentry robots utilized for base/facility perimeter patrolling such as the NREC Gladiator, LSA Autonomy RAP, Northrop Grumman CaMEL, and HDT Protector; and robot convoy vehicles like those displayed at the US Army Robotics Rodeo (held in 2009, 2010, 2012, and 2013) including the Lockheed Martin AMAS. However, the civilian automotive sector is also investing in the technology to make cars driverless and provides examples to the annual AUVSI Driverless Car Summit such as the Google Self-Driving Car (which is a kit able to be installed onto contemporary vehicles), among many other manufacturers pursuing these concepts. In the future these systems will mature and be fielded/commercialized through having their technologies proven. This will see cars that can drive for 1000+ km (on current engine system endurance) without vehicle human operators, or convoys of vehicles with only one or two human operators at the wheel. Robot squadmen is also a field of UGS that will most likely mature in the middle future, where robot infantrymen like the DARPA ATLAS robot may accompany a manned infantry organization.
- *UMS:* Autonomous systems that are operating away from ships or independently within harbors or along coastlines are already in prototype and initial fielding. Current underwater examples include long-range un-tethered hydrographic scientific data capturing systems utilizing buoyancy ("gliders") such as the Slocum Glider (3 months and 1800 km in a single mission) and LE battery systems such as Bluefin-21 (in use with the Malaysia Airlines Flight 370 search in 2014), Kongsberg Remus 100, 600, or 6000 (used in the search for Air France 447 in 2010) and WHOI Nereus. Above the waterline examples include harbor patrol systems such as the Textron CUSV and aerRobotix CatOne, and the Danish Navy's Mine Countermeasure Missions mine hunters and sweepers (100-ton displacement ships). Not much is anticipated to change in this field of systems as the nature of these operations dictates data collection and range restrictions, however, as the datalink or propulsion systems improve, the operational ranges may be expanded; that is, the scientific systems may stay out for longer/go deeper, and harbor patrol may increase in scope to coastline patrolling. There is also scope here to apply remote/autonomous technology to sea freight ships, which can be at sea for 6–10 weeks taking their cargo between continents.

Currently, in both the military and civilian context, as these aerial systems are still in test and initial fielding, they are currently being logistically supported as per Group B. The constraint here is that operational cycles are either kept short due to current endurance restrictions (a Global Hawk engine will currently keep going for 33 h) or operational restrictions (systems are generally returned to the same place they are launched due to limited operating infrastructure (Global Hawk) or due to test requirements (the small team of testers/developers are in one place – such as Zephyr's endurance record (14 days) and UAE proof of concept)).

UGS, provided they maintain a design and support similarity with manned vehicles into the future, should not change support structure markedly as a vehicle servicing center will remain essentially unchanged in structure and location. The changes will be required if the future UGS uses a unique fuel system (such as adapting UAS fuel systems into the UGS space to increase endurance) or if servicing/repair is required in a very remote location. In this case a Group C logistics support model should be developed for analysis, however, it may simply prove that the UGS vehicle may have to be backloaded to a service center as per current support doctrine.

For UAS or Unmanned Undersea Sytems (UUS), however, in the future, a logistics support framework will be required that is essentially global. Global Hawk or Zephyr could conceivably launch from one continent on an operational sortie that takes days, weeks, or months and be recovered for replenishment or maintenance on another continent. UUS may also launch from one harbor on a collection/patrol mission that lasts months, and end its mission in a harbor or rendezvous with a support ship in another ocean. This would also apply to the remote/autonomous operation of sea freight ships.

Little operationalization of these concepts has yet been done. One exception is the five-hub model that has been developed by the US Navy in operating and supporting its Global Hawk (Triton) fleet. In this model, the five hubs of Triton can cover a significant percentage of the earth's ocean surface simultaneously and can be supported by flight corridors in between the hubs to move aircraft in between locations. At each of these hubs a logistics node of supply and maintenance can be positioned that daily supports the operational endurance of these aircraft. However, if one thinks forward into a future where the Global Hawk/Triton (or successor system), now equipped with a ULE/R propulsion system, stays airborne for a week, or a month, it is then resource inefficient and costly to maintain a support team at each node.

The Group C space is a true challenge for adapting both capabilities and logistics support. The following section looks at some ways that the support framework may be adjusted to account for these missions and the technology advances that are involved.

12.5 Further Considerations

One of the advantages of modeling logistics system over some other types of military systems is that almost all components of a logistic system can be stochastically modeled for quantitative outcomes when bounded by a robust set of approximations and appropriate distributions. To that end, logistics for any future unmanned system can be modeled and for cost and operational efficiency and effectiveness. Considerations would include the following:

- *FIFO Maintenance:* For ULE UAS, would a single maintenance team conduct fly-in/fly-out (FIFO) servicing, hopping around the global nodes in their own jet as each ULE aircraft comes back to its hub for replenishment? Heuristics tell us that this would be more cost effective than positioning five (in the case of US Navy Global Hawk/Triton) permanent maintenance teams around the globe at the nodes/hubs.

- *Military or Civilian Maintainers:* Given the assumed safety provided by a hub (usually a major airhead or logistics base), would these teams still be military or would they be civilianized? There are advantages to the utilization of OEM and OEM sub-contracted logistics support framework components if there is a negligible security risk.
- *Centralized Maintenance:* For ultra-long range (ULR) UAS, does the new system utilize global flight corridors to operate from a single hub in the user country, and simply fly itself to wherever it is required in the world, returning to the single hub when required for replenishment? This may go against contemporary logic of "wasting" sortie time on a transit leg, but when you have dozens of days worth of sortie hours available, these operating costs to transit may be a good investment when weighed with globally pre-positioning, or globe-trotting, maintenance teams/spares.
- *Amalgamation of Trades:* UxS share cross-domain technologies in their control systems, datalinks, and electro-mechanical components. Married with order of magnitude increases in component reliability and safety, will there still need to be trade division in the mainte-nance workforce where aircraft/avionics mechanics only work on UAS, ground mechanics work on UGS, and marine mechanics work on UMS? Commonality of componentry and control systems suggests that the maintenance of all UxS could be undertaken by a combined trade: the robotics technician. The one exception would be combustion engines/power plants as these large sub-systems, currently piston and turbofan technology, require a specialized skill set in an engine mechanic that is aside from the fitter/electrician/electronics/optics/computers that is the robotics technician. There are obvious efficiencies here in centralizing workforce at nodes/hubs where this team of robotics technicians could maintain all UxS.
- *Maintenance Robots:* Will technicians at nodes/hubs be required at all? Advances being pioneered in the space, medical, and automotive industries see node/hub maintenance being performed remotely where a qualified technician could be rearwards in a central location instructing a trades assistant or the equipment operator via video streaming datalinks (as is now done in some highly specialized surgery and dubbed "telemedicine"). Even further, a technician could control a maintenance drone/robot on a hangar/workshop floor with a maintenance jig akin to factory robots or industrial manipulators [37]. That maintenance drone/robot could even be fully programmable and AI equipped to undertake most of the maintenance operations autonomously, only going back to a technician if fault diagnosis is beyond their programmed means. This "tele-robotic" or "teleoperative" maintenance would be enabled by enhanced built-in test/diagnostics resident within the equipment itself.
- *Anticipatory Spares Ordering:* Built-in test/diagnostics could enable the pre-provisioning of spares. Imagine a scenario where the UxS is on station for weeks and detects a deterio-rating system. Its datalink could enable the platform to self-diagnose, identify, and order that component in order to shorten the delivery time and have it waiting for its service when it returns to the node/hub from its mission.
- *Distribution:* Contemporary logistics is already utilizing cost analysis to deem that third-party logistics distributors provide profitable economies. These include freighting individual items through FedEx/DHL, up to complete outsourcing of distribution to companies such as Linfox and Toll and their fleets of trucks/rail carriages. Unmanned technology can take this a step further. Unmanned trucks can run continuously without concern for driver rest periods, creating even greater efficiencies. Unmanned cargo ships and aircraft will also provide cost

efficiencies when operating crews are removed. At the small end of distribution, FedEx and DHL couriers may also be removed from the equation by utilizing small UAS to deliver parcels direct to a door from a distribution center. These concepts and technologies are as easily applied to civilian or military logistics systems, and are easily modeled to establish optimization yields.

- *Technical Support Network:* Dealing with difficult diagnosis will become easier if customers share their experiences though a global data technical support network. Such a data library of experiences could be created by uploading and sharing their reports and returns: An environmentally induced fault solved by an Australian maintenance team operating in the Simpson Desert could be shared with US operator/maintenance personnel operating in the Mojave Desert to shorten diagnosis, and subsequently, repair times.
- *Organic Spares Manufacture:* 3D printing/manufacturing technology continues to advance, miniaturize, and become more affordable. This could enable a drastic reduction in spares holdings on shelves at nodes/hubs, instead allowing the node/hub to manufacture the component on the spot, when required. Such technology will bring true meaning to the term "just-in-time logistics." At a minimum, the technology should see large cost savings to a fleet manager, as he will not have to pay a manufacturer for these components, instead manufacturing them organically, when required. Of course, this does require IP rights/engineering support to be sourced from the equipment manufacturer during the acquisition phase.
- *EA/SD/TR:* The GWOT has generated new acquisition methodologies within the forces of the allied coalition as new technologies were quickly harnessed and fielded as fast as possible–many of these were unmanned systems technologies. Evolutionary Acquisition (EA) has been adopted by the US Department of Defense as a process for defense system development, in which a system is developed in stages as part of a single acquisition program. The different stages can be additional hardware and software capabilities, or performance gains due to advances in technological maturity and reliability growth [40]. The stages are delivered by "Spiral Development" (SD) defined by Farkas and Thurston [41] as a process where a desired capability is identified, but the end-state requirements are not known at program initiation. Those requirements are refined through demonstration and risk management; there is continuous user feedback; and each increment provides the user the best possible capability. Parallel to SD is Technology Refresh (TR), a sustainment methodology to keep the equipment as "sharp" as possible and avoid component obsolescence: It involves the periodic replacement of Commercial Off-The-Shelf (COTS) and Military Off-The-Shelf (MOTS) components; for example, processors, displays, computer operating systems, and CAS, within larger systems to assure continued supportability of that system through an indefinite service life [42]. TR is already being adopted in EA (for example, the Joint Strike Fighter (JSF) program) in acknowledgment of the rapid obsolescence of computing/processors, and is supported from an engineering perspective by the essential design requirement for Open Systems Architecture (vendor independent, non-proprietary systems, and components).

All three of these strategies can be modeled, essentially only altering the time to upgrade systems, however, each has its own set of logistics issues. The first issue concerns support contracting: having different planned lots, blocks, or configurations makes contracting for long-range support difficult [43]. The second can be described as the compounding effect of multiple partial configurations: where logistics support can become highly inefficient if SD/TR is only partially applied to fleets. Consider the next SD to a fleet of equipment and that there might only be a budget to upgrade half of the fleet: the capability user obviously demands the enhancement, and the modification is approved to half the fleet causing there to now be two configuration baselines in the fleet (that have to be logistically supported with (potentially) double the spares, double the pages in the maintenance manual and the buying power for those spares now halved). On the next SD there is budget for two-thirds the fleet... now there are three baselines. On the next SD there is budget for three-quarters, and now there are four configuration baselines. TR has the same effect, where there may be budget to refresh only half the fleet, so that is done (allowing half the old components to be cannibalized to support the half that were not refreshed)–two configurations. Then the next TR occurs with two-thirds funding–three configurations, and so on. Combine SD and TR configurations and the fleet manager has an exponentially increasing permutation of baselines to manage. EA/SD/TR are all the rage for the modern capability user [44], and are supportable, but only if properly resourced by funds, efficient schedules, and/or a very robust configuration management program combined with a disciplined and flexible capability management planning system.

12.6 Conclusions

The forthcoming technologies will create new equipment task employment on a scale not yet dreamed, and will remove the human from a great portion of them. To support the exciting expansions of the front end of activities/operations/missions, the back end will be forced to adapt creatively also, embracing many of the technologies itself as a means to that end. The logistics framework of production, distribution, and maintenance will remain extant from a first principles perspective, but how these functions are executed will require analysis of a much broader range of technology enhances options. DES will remain, for the foreseeable future, the best way that an operations researcher will be able to analyze these options to help engineer the most optimized system for logistics support. This chapter has provided the reader with a range of considerations and future-casts of enabling technology in designing a logistics system to support future unmanned systems.

References

1. Valles-Rosales, D.J. and Fuqua, D.O. (2007). 'Optimizing Logistics Through Operations Research,' in *Army Logistician* Jan/Feb 2007, 39:1:49–51.
2. Council of Supply Chain Management Professionals (CSCMP) (2014). www.clm1.org, accessed 29 Nov 14.
3. Gallasch, G.E., Lilith, N., Billington, J., Zhang, L., Bender, A. and Francis, B (2008). 'Modeling Defense Logistics Networks,' in *International Journal of Software Tools and Technology Transfer*, (2008), 10:75–93.
4. Stock, J.R. and Lambert, D.M. (2001). *Strategic Logistics Management*, 4th Edition, McGraw-Hill, New York.
5. Pagonis, W.G. (1992). *Moving Mountains: Lessons in Leadership and Logistics from the Gulf War*, Harvard Business School Press, Cambridge, MA.
6. Miner, N.E., Welch, K.M., Handy, S.M., and Andrade, L. (2010). 'Logistics Modeling, Simulation and Analysis for Lifecycle Decision Making,' in *Logistics Spectrum*, Apr–Jun 2010, 44:2:4–10.
7. Yoho, K.D., Rietjens, S., and Tatham, P. (2013). 'Defense Logistics: An Important Research Field in Need of Researchers,' in the *International Journal of Physical Distribution and Logistics Management*, (2013), 43:2:80–96.
8. Clark, D. (2014). 'Only the Strong Survive–CSS in the Disaggregated Battlespace,' in the *Australian Army Journal*, Winter Edition 2014, 11:1:21–33.
9. Defense Acquisitions University (2014). *Integrated Logistic Support (ILS) Elements*, http://acqnotes.com/acqnote/careerfields/integrated-logistics-support-ils, accessed 24 Nov 14.

10. Manuj, I., Mentzer, J.T., and Bowers, M.R. (2009). 'Improving the Rigor of Discrete-Event Simulation in Logistics and Supply Chain Research,' in the *International Journal of Physical Distribution and Logistics Management*, (2009), 39:3:172–201.

11. RAND Corporation (1968). *The Logistics Composite Model: An Overall View*, Memorandum RM-5544-PR May 1968, The RAND Corporation for United States Air Force Project RAND, Washington, DC.

12. Da Silva, A.K. and Botter, R.C. (2009). 'Method for Assessing and Selecting Discrete Event Simulation Software Applied to the Analysis of Logistic Systems,' in *Journal of Simulation*, (2009), 3:95–106.

13. Banks, J. (1999). 'Discrete Event Simulation,' in *Proceedings of the 1999 Winter Simulation Conference*, 7–13.

14. Robinson, S. (2005). 'Discrete-Event Simulation: From the Pioneers to the Present, What Next?,' in *Journal of the Operational Research Society*, (2005), 56:619–629.

15. Wild, R. (2002). *Operations Management*, 6th Edition, Continuum, London.

16. Rosenfield, D.B., Copacino, W.C., and Payne, E.C. (1985). 'Logistics Planning and Wvaluation When Using 'What-If' Simulation,' in the *Journal of Business Logistics*, (1985), 6:2:89–119.

17. Chang, Y. and Makatsoris, H. (2001). 'Supply Chain Modeling Using Simulation,' in the *International Journal of Simulation*, (2001), 2:1:24–30.

18. Krahl, D., Diamond, B., Lamperti, S., Nastasi, A., and Damiron, C. (2007). *ExtendSim 7*, Imagine That Inc.San Jose, CA.

19. O'Connor, A.N. (2011). *Probability Distributions Used in Reliability Engineering*, Reliability Information Analysis Center (RIAC), University of Maryland, Baltimore, MD.

20. Amouzegar, M., Drew, J.G., and Tripp, R.S. (2010). 'A Simulation Model for the Analysis of End-to-End Support of Unmanned Aerial Vehicles,' in the *International Journal of Applied Decision Sciences*, (2010), 3:3:239–258.

21. Voogd, Rosa M., and Sebalj, D. (2012). 'The Generic Methodology for Verification and Validation to Support Acceptance of Models, Simulation and Data,' *Journal of Defense Modeling and Simulation: Applications, Methodology, Technology*, (2013), 10:347.

22. Hatch, M.L. and Badinelli, R.D. (1997). 'Concurrent Optimization in Designing for Logistics Support,' in the *European Journal of Operational Research*, (1997), 115:77–97.

23. Chapman, W.L., Bahill, A.T., and Wymore, A.W. (1992). *Engineering Modeling and Design*, CRC Press, Boca Raton, Florida.

24. O'Rorke, M. and Burke, A. (2001). *Optimization Problems in Discrete Event Simulation*, University College Dublin, Dublin, Ireland.

25. Hagendorf, O. (2009). *Simulation Based Parameter and Structure Optimization of Discrete Event Systems*, Liverpool John Moores University, Liverpool, UK.

26. Ghanmi, A. (2011). 'Canadian Forces Global Reach Support Hubs: Facility Location and Aircraft Routing Models,' in *Journal of the Operational Research Society*, (2011), 62:638–650.

27. Guarnieri, J., Johnson, A.W., and Swartz, S.M. (2006). 'A Maintenance Resources Capacity Estimator,' in *Journal of the Operational Research Society*, (2006), 57:1188–1196.

28. Fan, C.-Y., Fan, P.-S., and Chang, P.C. (2010). 'A System Dynamics Modeling Approach for a Military Weapon Maintenance Supply System,' in *International Journal of Production Economics*, (2010), 128:457–469.

29. Lenhardt, T.A. (2006). 'Evaluation of a USMC Combat Service Support Logistics Concept,' in *Mathematical and Computer Modeling*, (2006), 44:368–376.

30. Julka, N., Thirunavukkarasu, A., Lendermann, P., Gan, B.P., Schirrmann, A., Fromm, H., and Wong, E. (2011). 'Making Use of Prognostics Health Management Information for Aerospace Spare Components Logistics Network Optimization,' in *Computers in Industry*, (2011), 62:613–622.

31. McGee, J.B., Rossetti, M.D., and Mason, S.J. (2005). 'Quantifying the Effect of Transportation Practices in Military Supply Chains,' in *Journal of Defense Modeling and Simulation*, April 2005, 2:2:87–100.

32. Baker, S.F., Morton, D.P., Rosenthal, R.E., and Williams, L.M. (2002). 'Optimizing Military Airlift,' in *Operations Research*, Jul/Aug 2002, 50:4:582–602.

33. Ciarallo, F.W., Hill, R.R., Mahadevan, S., Chopra, V., Vincent, P.J., and Allen, C.S. (2005). 'Building the Mobility Aircraft Availability Forecasting (MAAF) Simulation Model and Decision Support System,' in *Journal of Defense Modeling and Simulation*, April 2005, 2:2:57–69.

34. United States Department of Defense (2010). *Joint Publication (JP) 3-52, Joint Airspace Control*, US DoD, Washington, DC.

35. United States Army (2010). *Eyes of the Army: US Army Roadmap for Unmanned Aircraft Systems 2010-2035*, US Army UAS Center of Excellence, Fort Rucker, Alabama.

36. United States Department of Defense (2013). *Unmanned Systems Integrated Roadmap, FY2013-2038*, US DoD, Washington, DC.

37. Murphy, R.R. (2014). *Disaster Robotics*, MIT Press, Cambridge, Massachusetts.

38. Newcombe, L.R. (2004). *Unmanned Aviation: A Brief History of Unmanned Aerial Vehicles*, American Institute of Aeronautics and Astronautics Inc., Reston, Virginia.

39. Keane, J.F. and Carr, S.S. (2013). 'A Brief History of Early Unmanned Aircraft,' in *Johns Hopkins APL Technical Digest*, (2013), 32:3:558–571.

40. US Army, 2011.

41. Farkas, K. and Thurston, P. (2003). 'Evolutionary Acquisition Strategies and Spiral Development Processes: Delivering Affordable, Sustainable Capability to the Warfighters,' in *Defense Acquisition University PM Magazine*, July-August 2003, 10–14.

42. Haines, L. (2001). 'Technology Refreshment in the DoD: Proactive Technology Refreshment Plan Offers DoD Programs Significant Performance, Cost, Schedule Benefits,' in *Defense Acquisition University PM Magazine*, March-April 2001, 22–27.

43. Drew, J.G., Shaver, R., Lynch, K.F., Amouzegar, M.A., and Snyder, D. (2005). *Unmanned Aerial Vehicle End-to-End Support Considerations*, Project Air Force - RAND Corporation, Santa Monica, California.

44. LTCOL B and C (2014). 'An Essay on Technical Refreshment and Evolutionary Acquisition of Land Materiel,' in *The Link: Australian Defence Logistics Magazine*, Issue 11: June 13, 17–21.

13

Organizing for Improved Effectiveness in Networked Operations*

Sean Deller[1], Ghaith Rabadi[2], Andreas Tolk[3], and Shannon R. Bowling[4]

[1]*Joint and Coalition Warfighting (J7), Washington, DC, USA*
[2]*Department of Engineering Management & Systems Engineering (EMSE),*
Old Dominion University, Norfolk, VA, USA
[3]*The MITRE Corporation, USA*
[4]*College of Engineering Technology and Computer Science, Bluefield State College,*
Bluefield, WV, USA

13.1 Introduction

The effectiveness of a networked force depends on many factors, one of which is how we choose to organize it. But how should we measure organizational effectiveness? There are few quantifiable metrics that can discriminate between various networked forces that differ solely in their arrangement, and fewer still that can consistently discriminate between configurations that differ by a single link, regardless of the significance of that link. The research presented in this chapter aimed to gain insight into how an Information Age combat force should be organized by presenting an initial attempt to determine the utility of the Perron–Frobenius Eigenvalue (λ_{PFE}) as a measure of the effectiveness of the organization of a networked force.

The organizations of military forces throughout history have largely depended on the capabilities of the weapons of that age. As technological advances led to increases in lethality and speed, military organizations were forced to adapt. Early warfare from the era of the phalanx to the line and column formations of the eighteenth century focused on massed manpower.

*Contribution originally published in MORS Journal, Vol. 17, No. 1, 2012. Reproduced with permission from Military Operations Research Society.

Operations Research for Unmanned Systems, First Edition. Edited by
Jeffrey R. Cares and John Q. Dickmann, Jr.
© 2016 John Wiley & Sons, Ltd. Published 2016 by John Wiley & Sons, Ltd.

The expanded lethality of firearms and artillery, coupled with the advent of the machine gun and indirect fire capabilities, shifted that focus from massed manpower to massed firepower. Increased speed led to the emergence of maneuver warfare and the employment of nonlinear tactics. The leap from Industrial Age to Information Age warfare will require an evolutionary step in military organization of equal magnitude. How then, should an Information Age combat force be organized in order to optimize its effectiveness?

Jain and Krishna [1] introduced the relationship between the Perron–Frobenius Eigenvalue (λ_{PFE}) of a graph and its autocatalytic sets, and used graph topology to study various network dynamics. Cares [2] employed a similar approach in developing the Information Age Combat Model (IACM) to describe combat (or competition) between distributed, networked forces, or organizations. The IACM focuses on the λ_{PFE} as a measure of the ability of a network to produce combat power. Cares proposed that the greater the value of the λ_{PFE}, the greater the effectiveness of the organization of that networked force. The research presented in this chapter is an initial attempt to determine the utility of the λ_{PFE} through the development of an agent-based model based on the IACM.

The results of this work indicate that the value of the λ_{PFE} is a significant measurement of the performance of a networked force, but that this significance depends on the existence of unique λ_{PFE} values for the configurations under consideration. However, the utility of the λ_{PFE} can be enhanced by adding a measurement we introduce in this paper called *robustness* to improve the effectiveness of the λ_{PFE} value as a quantifiable metric of network performance. It is important to reinforce that the intent of this work is not to propose a completed model for simulating combat between networked forces, but only to gain a first-order understanding of the value, if any, of the λ_{PFE} as a quantitative measure of the effectiveness of the organization of a networked force.

13.2   Understanding the IACM

The basic objects of the IACM are not platforms or other entities capable of independent action, but rather nodes that can perform elementary tasks and links that connect these nodes. *Sensor* nodes receive signals about observable phenomena and send that information to *Decider* nodes, which direct the actions of the other nodes. *Influencer* nodes receive those directions and interact with other nodes to affect the state of those nodes. *Targets* are nodes that have military value but are not Sensors, Deciders, or Influencers. Information flow between the nodes is generally necessary for any useful activity to occur. All nodes belong to one of two opposing sides, conventionally termed BLUE (depicted in black in the figures) and RED (depicted in gray).

The basic combat network shown in Figure 13.1 represents the simplest situation in which one side can influence another, commonly referred to as a *combat cycle*. The BLUE Sensor (S) detects the RED Target (T) and informs the BLUE Decider (D) of the contact. The Decider then instructs the BLUE Influencer (I) to engage the Target. The Influencer initiates effects, such as exerting physical force, psychological or social influence, or other forms of influence on the target. The process may be repeated until the Decider determines that the desired effect has been achieved. It should be noted that the effect assessment requires sensing, which means that this will be conducted in a new cycle.

The simplest complete combat network represents all the ways in which Sensors, Deciders, Influencers, and Targets interact meaningfully with each other. There are 18 types of links in this network, each of which appears twice due to the symmetry between BLUE and RED. These are listed in Table 13.1, where the nodes are identified as in Figure 13.2. Because

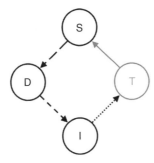

Figure 13.1   The basic combat network represents the simplest situation in which one side can influence another. *Source*: Deller, Sean; Rabadi, Ghaith; Tolk, Andreas; Bowling, Shannon R., "Organizing for Improved Effectiveness in Networked Operations," *MORS Journal*, Vol. 17, No. 1, 2012, John Wiley and Sons, LTD.

Table 13.1   Types of links available in the IACM.

Link type	From	To	Interpretation	Link type	From	To	Interpretation
1	S_{BLUE} S_{RED}	S_{BLUE} S_{RED}	S detecting own S, or S coordinating with own S	10	I_{BLUE} I_{RED}	D_{BLUE} D_{RED}	I attacking own D, or I reporting to own D
2	S_{BLUE} S_{RED}	D_{BLUE} D_{RED}	S reporting to own D	11	I_{BLUE} I_{RED}	I_{BLUE} I_{RED}	I attacking own I, or I coordinating with own I
3	S_{BLUE} S_{RED}	S_{RED} S_{BLUE}	S detecting adversary S	12	I_{BLUE} I_{RED}	T_{BLUE} T_{RED}	I attacking own T
4	D_{BLUE} D_{RED}	S_{BLUE} S_{RED}	S detecting own D, or D commanding own S	13	I_{BLUE} I_{RED}	S_{RED} S_{BLUE}	I attacking adversary S, or S detecting adversary I
5	D_{BLUE} D_{RED}	D_{BLUE} D_{RED}	D commanding own D	14	I_{BLUE} I_{RED}	D_{RED} D_{BLUE}	I attacking adversary D
6	D_{BLUE} D_{RED}	I_{BLUE} I_{RED}	D commanding own I	15	I_{BLUE} I_{RED}	I_{RED} I_{BLUE}	I attacking adversary I
7	D_{BLUE} D_{RED}	T_{BLUE} T_{RED}	D commanding own T	16	I_{BLUE} I_{RED}	T_{RED} T_{BLUE}	I attacking adversary T
8	D_{BLUE} D_{RED}	S_{RED} S_{BLUE}	S detecting adversary D	17	T_{BLUE} T_{RED}	S_{BLUE} S_{RED}	S detecting own T
9	I_{BLUE} I_{RED}	S_{BLUE} S_{RED}	I attacking own S, or S detecting own I	18	T_{BLUE} T_{RED}	S_{RED} S_{BLUE}	S detecting adversary T

Targets are passive and Deciders and Sensors do not exert influence, 28 other link types have been excluded. Links between a node and itself in Figure 13.2 have been interpreted as connecting two different nodes of the same type and side. Note that some of the links (types 1, 4, 9, 10, 11, and 13 in Table 13.1) have ambiguous interpretations; both interpretations of this link will be used, but the context of the model always makes clear which is intended.

Once the IACM has been defined in terms of a network of nodes and links, the language and tools of graph theory (see, for example, [3]) can be used for both description and analysis. A concise description of any graph is provided by the adjacency matrix **A**, in which the row and column indices represent the nodes, and the matrix elements are either one or zero according

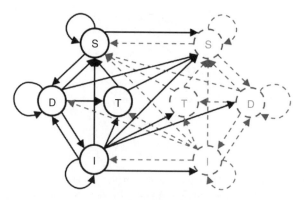

Figure 13.2 The simplest complete combat network represents all the ways in which Sensors, Deciders, Influencers, and Targets interact meaningfully with each other. *Source*: Deller, Sean; Rabadi, Ghaith; Tolk, Andreas; Bowling, Shannon R., "Organizing for Improved Effectiveness in Networked Operations," *MORS Journal*, Vol. 17, No. 1, 2012, John Wiley and Sons, LTD.

				To				
	S	D	I	T	S	D	I	T
S	1	1	0	0	1	0	0	0
D	1	1	1	1	1	0	0	0
I	1	1	1	1	1	1	1	1
T	1	0	0	0	1	0	0	0
S	1	0	0	0	1	1	0	0
D	1	0	0	0	1	1	1	1
I	1	1	1	1	1	1	1	1
T	1	0	0	0	1	0	0	0

From (label along left side)

Figure 13.3 An adjacency matrix for the simplest complete combat network. *Source*: Deller, Sean; Rabadi, Ghaith; Tolk, Andreas; Bowling, Shannon R., "Organizing for Improved Effectiveness in Networked Operations," *MORS Journal*, Vol. 17, No. 1, 2012, John Wiley and Sons, LTD.

to the rule: $A_{ij}=1$, if there exists a link **from** node i **to** node j and $A_{ij}=0$, otherwise. Figure 13.3 is an adjacency matrix representation of the simplest complete combat network depicted in Figure 13.2. Note that the links in both the graphical depiction and the adjacency matrix are directional, in that their meanings differ depending on which nodes they go "from" (left column) and "to" (top row). Given that the number of different subnetworks for any N × N matrix is $2^{(N \times N)}$, it is obvious that attempts to analyze the effectiveness of various arrangements of nodes and links for any but the simplest combat networks quickly becomes impossible.

One method used in studying the evolution of complex adaptive systems (chemical, biological, social, and economic) is calculation of the principal (maximum) eigenvalue of the adjacency matrix [4]. The Perron–Frobenius theorem guarantees the existence of a real, positive principal eigenvalue of A_{ij} if A_{ij} is an irreducible nonnegative matrix. Because the IACM focuses on combat cycles, all potential Target nodes are connected to all opposing Sensor and Influencer nodes. The graphical depictions of the various configurations in this study use a single, representative Target node to highlight the differences in the configurations, but they remain, in effect, strongly connected and, therefore, irreducible.

This eigenvalue, λ_{PFE}, is a measure of the selective connectivity within the network (i.e., networks with the same number of links may have different λ_{PFE} values depending on the placement of the links). Jain and Krishna [1] noted when studying population dynamics that networks with autocatalytic sets (ACSs) always outperform networks without them as "a consequence of the infinite walks provided by the positive feedback inherent in the ACS structure, while non-ACS structures have no feedbacks and only finite walks." Consequently, Cares [2] proposed that it represents the ability of a network to produce feedback effects in general and combat power specifically in the case of the IACM.

13.3 An Agent-Based Simulation Representation of the IACM

The structure of the IACM makes it clear that the λ_{PFE} is a quantifiable metric with which to measure the organization of a networked force, but is it an indicator of effectiveness? To determine this we constructed an agent-based simulation representation of the IACM and conducted a series of force-on-force engagements using opposing forces of equal assets and capabilities, but differing in their connectivity arrangements or configurations. These differences in connectivity often, but not necessarily, lead to unequal λ_{PFE} values.

The agent-based paradigm was utilized for this purpose because the resulting models provide both the ability to account for small unit organization and the autonomy of action that is necessary for our investigation. An additional advantage of utilizing an agent-based simulation was the ability to work around the ambiguities of link interpretation in the IACM. For example, instead of a mutually exclusive choice between defining a directional link from a BLUE Influencer to a RED Sensor (type 13 in Table 13.1) as either the Influencer "targeting" the Sensor or as the Sensor "sensing" the Influencer, both abilities can be represented in the agent-based simulation.

The first challenge in modeling the IACM concerned the adjacency matrix representation of the network. The IACM as originally described by Cares [2] uses a single adjacency matrix to reflect the collective organization of both BLUE and RED forces. In this approach, the λ_{PFE} value is dependent on the configurations of both the BLUE and RED forces and might well represent the extent to which feedback effects occur in the engagement. Obviously BLUE and RED each seek separately to maximize their own organizational effectiveness while minimizing the organizational effectiveness of the opposing force. This cannot be represented by a single λ_{PFE} value, so we calculate separate values (λ_{BLUE} and λ_{RED}) to measure the potential effectiveness of each opposing configuration. These calculations required the adjacency matrices include a single Target node representative of all the enemy forces capable of being targeted. In other words, the values of λ_{BLUE} and λ_{RED} are determined solely by the arrangement of their respective assets, independent of the asset arrangement of the opposing force.

The agent-based simulation environment utilized for this research was NetLogo [5]. The code of the agent-based model closely follows the logic of the IACM, with a few notable exceptions. Agents served as Sensors, Deciders, and Influencers, but Targets were not included as they served no purpose other than to absorb losses. Including Target agents with no detect, direct, or influence capabilities would only serve to clutter the results. Additionally, Deciders cannot be destroyed in the present model. This was done in recognition of their unique role in connecting multiple Sensors and Influencers. Destruction of a Decider typically renders a number of other nodes useless (effectively destroyed), making it a particularly high value target. Because targets are detected and engaged in random order in our model, we wished to give all targets equal value to avoid generating atypical engagements that might bias the results.

The agent rules sets function in accordance with the IACM. Sensors detect enemy nodes within the sensing range parameter, and communicate that information to their assigned (connected) Deciders. Deciders communicate the sensing information to their assigned Influencers. Influencers destroy the nearest enemy node that is both "sensed" by a Sensor connected to that Influencer's Decider and within the influencing range parameter. Deciders direct Sensor movement toward areas of suspected enemy nodes. Deciders direct Influencers to move toward the nearest "sensed" enemy node. All nodes are assumed to perform their functions perfectly and instantaneously. Agent interactions are assumed to be deterministic, that is, the probabilities of detect, communicate, and kill are all "1." A stochastic dimension to the model can be built once a better understanding of the research questions is gained, and this new dimension can be used to model errors and delays representing technological and human performance factors. Most importantly, the rules sets and parameter values for both BLUE and RED agents were identical.

Because this effort focuses on gaining insight into the relationship between the λ_{PFE} value and the effectiveness of a networked force, the agent-based simulation rules of engagement were quite simple. The battlespace (i.e., "world") within the model is deliberately featureless in order to focus on the configurations themselves. The agents are randomly distributed across the two-dimensional battlespace at the beginning of each engagement. This random set-up avoids the need for defining at-start locations and how the forces should close and engage with each other and, most importantly, the potential effects this might have on the results of the engagement. Each engagement continued until either one force was victorious (i.e., all of the Sensors and Influencers of the opposing force were annihilated), or a stalemate emerged (i.e., both forces were incapable of continued combat because neither side had a Decider with a functioning Sensor and Influencer). Multiple runs of each engagement will result in a probability of a win for each particular configuration.

13.4   Structure of the Experiment

To best associate any difference in force effectiveness to the difference in connectivity, the opposing forces consisted of the same number of Sensors, Deciders, and Influencers, differing only in how they were arranged (i.e., linked). Because the potential value of a Sensor may not equal the potential value of an Influencer, the composition of each configuration considered in this work contained an equal number of Sensors and Influencers to preclude any bias toward those configurations that have more of one or the other. Additionally, both types of nodes had identical performance capabilities (i.e., the sensing range was chosen equal to the influencing range, and the speeds of movement of the two types of node were equal). Consequently, the composition of both forces followed an X-Y-X-1 (Sensor-Decider-Influencer-Target) template.

For any particular values of X and Y, there is a finite number of ways to arrange those assets. To gain a "first order" understanding of the IACM, we made two key scoping decisions. First, each Sensor and Influencer would only be connected to one Decider (but any given Decider could be connected to multiple Sensors and Influencers). Second, the connectivity within any X-Y-X-1 force was limited to only those links necessary to create combat cycles (i.e., link types 2, 3, 6, 13, and 15 in Table 13.1), which are the essence of the λ_{PFE} (the most basic element of the IACM). Whereas the other link types can significantly enhance both the λ_{PFE}

value and the performance of any given network configuration, the present model provides a baseline to assess what those potential effects may be.

Although the X-Y-X-1 template significantly scoped the focus of this chapter, the number of possible configurations for a given force still becomes large very quickly. For example, there are a total of nine possible ways to distribute four Sensors and four Influencers across three Deciders (see Table 13.2). No matter how you distribute them, one Decider will have two Sensors linked to it, and one Decider (which may or may not be the same Decider) will have two Influencers assigned to it. Fortunately, because the nodes of the IACM are generic it is possible to reduce this set by eliminating those configurations that are, in effect, isomorphic. The only *meaningful* difference between the nine possible configurations of a 4-3-4-1 networked force is whether the Decider that is linked to two Sensors is the same Decider that is linked to two Influencers (see Figure 13.4). The remaining seven possible configurations are all modeled identically to these two configurations in the IACM (and are shaded gray in Table 13.2).

Adding a single Sensor and Influencer yields a 5-3-5-1 networked force, which can be organized in 36 different ways. By applying this same logic, we reduce those 36 possible configurations to only eight meaningfully different configurations. Even with these most basic of examples, the difference between the number of possible configurations and the number of meaningfully different configurations becomes quite apparent.

Identifying the meaningfully different configurations is crucial for the purpose of scoping the problem. Whereas a 7-3-7-1 networked has 225 possible configurations, applying this same logic reduces this to a much more manageable number of only 42 meaningfully different configurations. Testing each of the 225 possible combinations of a 7-3-7-1 networked force against all 225 possible configurations of an opposing 7-3-7-1 networked force would require 50 625 (i.e., 225^2) unique engagements, but 42 combinations would only require 1764 (i.e., 42^2) unique engagements. Because the number of meaningfully different combinations for any given set of nodes is a function of the number of unique values of the allocation combinations of X across Y, we attempted to define the function in order to automatically generate the combinations. This was not a simple task. Although the allocation resembles a partition problem, the exact numerical sequence of the numbers of meaningful combinations was difficult to establish. Because determining what this function might be is not the purpose of this research,

Table 13.2 All possible configurations of a 4-3-4-1 network.

Configuration number	Number of sensors linked to Decider 1	Number of sensors linked to Decider 2	Number of sensors linked to Decider 3	Number of influencers linked to Decider 1	Number of influencers linked to Decider 2	Number of influencers linked to Decider 3
1	2	1	1	2	1	1
2	2	1	1	1	2	1
3	2	1	1	1	1	2
4	1	2	1	2	1	1
5	1	2	1	1	2	1
6	1	2	1	1	1	2
7	1	1	2	2	1	1
8	1	1	2	1	2	1
9	1	1	2	1	1	2

Table 13.3 The numbers of meaningfully different configurations of all X-Y-X-1 networked forces where X < 11 and Y < 8.

		Number of Deciders (Y)				
		3	4	5	6	7
Numbers of Sensors (X)	3	1				
and Influencers (X)	4	2	1			
	5	8	2	1		
	6	19	9	2	1	
	7	42	27	9	2	1
	8	78	74	30	9	2
	9	139	168	95	31	9
	10	224	363	248	105	31

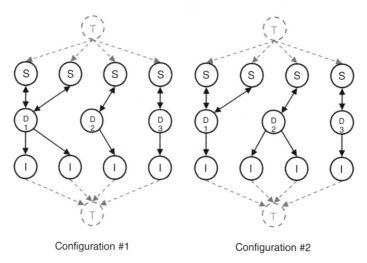

Configuration #1   Configuration #2

Figure 13.4 The two meaningfully different configurations of a 4-3-4-1 network. *Source*: Deller, Sean; Rabadi, Ghaith; Tolk, Andreas; Bowling, Shannon R., "Organizing for Improved Effectiveness in Networked Operations," *MORS Journal*, Vol. 17, No. 1, 2012, John Wiley and Sons, LTD.

we calculated the numbers of meaningfully different configurations for all X-Y-X-1 forces where X < 11 and Y < 8 using a simple algorithm based on the numbers of unique values for the distributions of Sensors and Influencers across the Deciders. The resulting totals are summarized in Table 13.3.

Each of these configurations has a unique adjacency matrix representative of the connectivity of its nodes. The adjacency matrices for all configurations will only differ in two sections (see the un-shaded sections of an example adjacency matrix in Figure 13.5), regardless of the total number of Sensors, Deciders, or Influencers. These un-shaded sections reflect the connectivity of each Sensor and Influencer to and from a particular Decider, and vary by configuration based on the allocation of Sensors and Influencers across the Deciders. The shaded areas represent the absolute absence of any links between those types of nodes, or the absolute existence of links between those types of nodes. Because 14 of 16 sections of the adjacency

To

From	S	S	S	S	S	S	S	D	D	D	I	I	I	I	I	I	I	T
S	0	0	0	0	0	0	0	1	0	0	0	0	0	0	0	0	0	0
S	0	0	0	0	0	0	0	1	0	0	0	0	0	0	0	0	0	0
S	0	0	0	0	0	0	0	1	0	0	0	0	0	0	0	0	0	0
S	0	0	0	0	0	0	0	1	0	0	0	0	0	0	0	0	0	0
S	0	0	0	0	0	0	0	1	0	0	0	0	0	0	0	0	0	0
S	0	0	0	0	0	0	0	0	1	0	0	0	0	0	0	0	0	0
S	0	0	0	0	0	0	0	0	0	1	0	0	0	0	0	0	0	0
D	0	0	0	0	0	0	0	0	0	0	1	1	1	1	1	0	0	0
D	0	0	0	0	0	0	0	0	0	0	0	0	0	0	0	1	0	0
D	0	0	0	0	0	0	0	0	0	0	0	0	0	0	0	0	1	0
I	0	0	0	0	0	0	0	0	0	0	0	0	0	0	0	0	0	1
I	0	0	0	0	0	0	0	0	0	0	0	0	0	0	0	0	0	1
I	0	0	0	0	0	0	0	0	0	0	0	0	0	0	0	0	0	1
I	0	0	0	0	0	0	0	0	0	0	0	0	0	0	0	0	0	1
I	0	0	0	0	0	0	0	0	0	0	0	0	0	0	0	0	0	1
I	0	0	0	0	0	0	0	0	0	0	0	0	0	0	0	0	0	1
I	0	0	0	0	0	0	0	0	0	0	0	0	0	0	0	0	0	1
T	1	1	1	1	1	1	1	0	0	0	0	0	0	0	0	0	0	0

Figure 13.5 An adjacency matrix for one of the 42 meaningfully different configurations of a 7-3-7-1 network. *Source*: Deller, Sean; Rabadi, Ghaith; Tolk, Andreas; Bowling, Shannon R., "Organizing for Improved Effectiveness in Networked Operations," *MORS Journal*, Vol. 17, No. 1, 2012, John Wiley and Sons, LTD.

matrices for each of the 42 configurations are identical, the variance between the λ_{PFE} values is greatly reduced.

Identical configurations always have the same λ_{PFE} value; however, it is possible for meaningfully different configurations to share the same λ_{PFE} value. In the case of a 7-3-7-1 networked force, the 42 meaningfully different configurations had 13 unique λ_{PFE} values ranging from 1.821 to 2.280. When this occurs, the λ_{PFE} loses its utility as an indicator of potential performance between these configurations. The number of unique λ_{PFE} values for the meaningful configurations for all X-Y-X-1 forces where $X<11$ and $Y<8$ are depicted in Table 13.4. Note that the numbers of unique λ_{PFE} values are not directly proportional to the numbers of meaningfully different configurations. For example, while an 8-3-8-1 networked force has 78 meaningfully different configurations with 20 unique λ_{PFE} values, the 95 meaningfully different configurations of a 9-5-9-1 networked force only have 13 unique λ_{PFE} values. This reduction will have a significant impact on the analysis of the modeling results presented later in this research.

The full range of mathematical values for a λ_{PFE} of an adjacency matrix containing 18 nodes is from 0 (for a network with no links at all) to 18 (for a maximally connected network). Note that the range of λ_{PFE} values for the 42 meaningfully different combinations of a 7-3-7-1 force is only a small segment (1.821–2.280) of the full range of possible values due to the relatively small differences of the links within any two of those configurations. Although the variation between the λ_{PFE} values is small, it is of significant utility because the values of other common statistical measures of networked systems (such as link-to-node ratio, degree distribution, and others compiled from various studies by Cares [2]) do not vary at all. Consequently, none of those other metrics can measure any potential variation in the effectiveness of these 42 configurations.

Table 13.4   The numbers of unique λ_{PFE} values of all X-Y-X-1 networked forces where $X < 11$ and $Y < 8$.

		Number of Deciders (Y)				
		3	4	5	6	7
Numbers of Sensors (X)	3	1				
and Influencers (X)	4	2	1			
	5	4	2	1		
	6	8	4	2	1	
	7	13	8	4	2	1
	8	20	13	8	4	2
	9	27	20	13	8	4
	10	38	27	20	13	31

13.5   Initial Experiment

The initial experiment consisted of all possible force-on-force engagements of the 42 meaningfully different configurations of two 7-3-7-1 networked forces (BLUE and RED). Because each of these configurations contains the same numbers of Sensors, Deciders, and Influencers (each type with identical capabilities) differing only in their connectivity, it is most likely that any difference in performance would be a consequence of this connectivity difference. A comprehensive test of each of these 42 configurations against each other required 1764 different engagements. Each engagement was represented by 30 replications of the agent-based simulation, each with a random distribution of the BLUE and RED nodes across the battlespace. Each replication resulted in a BLUE win, a RED win, or an undecided result (i.e., neither side contained a functioning combat cycle). The number of replications yielding an undecided result was 2717 (5.13% of the 52 920 total). A graphical representation of the results is presented in Figure 13.6, with the shaded surface representing those engagements where the probability of a BLUE win was greater than 0.5 and the un-shaded surface representing those engagements where the probability of a BLUE win was less than 0.5.

These initial results indicate that as the BLUE force is organized with a greater λ_{PFE} value, its effectiveness generally increases. Although the resulting surface is far from smooth, a general trend does appear: the smaller the λ_{PFE} value, the smaller the probability of a win. This trend becomes more apparent in Figure 13.7, where the probability of a BLUE win for any particular configuration is averaged over all RED configurations. Note that many BLUE configurations had an identical λ_{PFE} value (there were 13 unique λ_{PFE} values for the 42 configurations).

Clearly, it appears that the probability of a BLUE win increases for those BLUE configurations with a greater λ_{PFE} value. A simple linear regression confirms this with a coefficient of determination (R^2) of 0.896 for the following equation:

$$y = 1.0162(x) - 1.5780 \tag{13.1}$$

where $y =$ the average probability of a BLUE win for that configuration and $x =$ the λ_{PFE} value of a configuration.

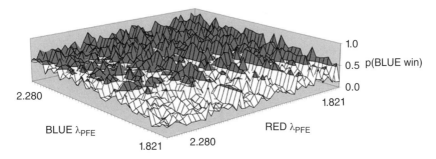

Figure 13.6 The probability of a BLUE win for each of the 42 BLUE configurations against each of the 42 RED configurations. *Source*: Deller, Sean; Rabadi, Ghaith; Tolk, Andreas; Bowling, Shannon R., "Organizing for Improved Effectiveness in Networked Operations," *MORS Journal*, Vol. 17, No. 1, 2012, John Wiley and Sons, LTD.

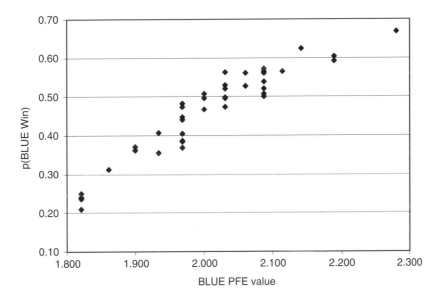

Figure 13.7 The average probability of a BLUE win by λ_{PFE} for 42 configurations of a 7-3-7-1 BLUE network. *Source*: Deller, Sean; Rabadi, Ghaith; Tolk, Andreas; Bowling, Shannon R., "Organizing for Improved Effectiveness in Networked Operations," *MORS Journal*, Vol. 17, No. 1, 2012, John Wiley and Sons, LTD.

13.6   Expanding the Experiment

Deller *et al.* reported on these findings and applied them in "Applying the IACM: Quantitative Analysis of Network Centric Operations" [6]. Since this paper was published, additional insights could be derived.

Adding just one additional Sensor and one additional Influencer to the 7-3-7-1 force (i.e., making the configuration 8-3-8-1) increased the number of meaningful combinations to 78. A comprehensive test of each of these 78 configurations against each other required 6084

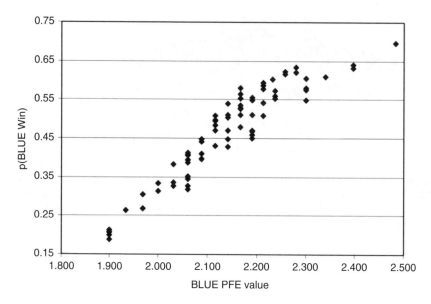

Figure 13.8   The average probability of a BLUE win by λ_{PFE} for 78 configurations of an 8-3-8-1 BLUE network. *Source*: Deller, Sean; Rabadi, Ghaith; Tolk, Andreas; Bowling, Shannon R., "Organizing for Improved Effectiveness in Networked Operations," *MORS Journal*, Vol. 17, No. 1, 2012, John Wiley and Sons, LTD.

different engagements. Each engagement was represented by 30 replications, each with a random distribution of the BLUE and RED nodes across the battlespace. Each replication resulted in a BLUE win, a RED win, or an undecided result (i.e., neither BLUE nor RED can complete the annihilation of the opposing force due to a lack of Sensors or Influencers). The number of replications yielding an undecided result was 8820 (4.83%). Figure 13.8 shows the average p(Win) value for each of these configurations. There were 24 unique λ_{PFE} values for the 78 configurations.

These results also indicate that as the BLUE force is organized with a greater λ_{PFE} value, its effectiveness generally increases. A simple linear regression confirms this with a coefficient of determination (R^2) equal to 0.876 for the following equation:

$$y = 0.9484(x) - 1.5633 \tag{13.2}$$

where y = the average probability of a BLUE win for that configuration and x = the λ_{PFE} value of a configuration.

Again, utilizing an ordinal scale, these BLUE configurations can be ranked from 1 to 78 based on their average probability of a BLUE win (subsequently ordered by their λ_{PFE} values for those configurations with equal p(Win) values, where possible).

Because increasing the number of Sensors and Influencers substantiated the initial results, the next logical step was to determine the impact of increasing the number of Deciders. A 9-5-9-1 force was selected because the number of meaningful combinations (95) was considerable, yet small enough to be modeled in a reasonable amount of time (it required approximately 78 h of agent-based simulation model runtime). A comprehensive test of each of these 95 configurations against each other required 9025 different engagements. Each

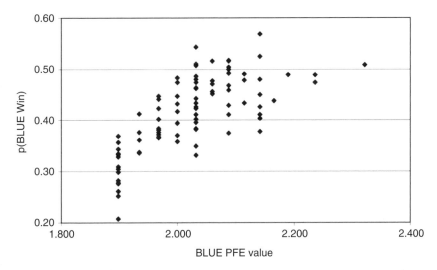

Figure 13.9   The average probability of a BLUE win by λ_{PFE} for 95 configurations of a 9-5-9-1 BLUE network. *Source*: Deller, Sean; Rabadi, Ghaith; Tolk, Andreas; Bowling, Shannon R., "Organizing for Improved Effectiveness in Networked Operations," *MORS Journal*, Vol. 17, No. 1, 2012, John Wiley and Sons, LTD.

engagement was represented by 30 replications, each with a random distribution of the BLUE and RED nodes across the battlespace. Each replication resulted in a BLUE win, a RED win, or an undecided result (i.e., neither BLUE nor RED can complete the annihilation of the opposing force due to a lack of Sensors or Influencers). The number of replications yielding an undecided result was 19 892 (7.35%). Figure 13.9 shows the average p(Win) value for each of these configurations. There were 13 unique λ_{PFE} values for the 95 configurations.

Note that the highest p(Win) value does not belong to the configuration with the highest λ_{PFE} value. This indicates that there is some other correlating factor, and the dramatic reduction in the coefficient of determination (R^2) to 0.519 for the resulting equation confirms that:

$$y = 0.5861(x) - 0.7736 \qquad (13.3)$$

where y = the average probability of a BLUE win for that configuration and x = the λ_{PFE} value of a configuration.

An unexpected result of the increase of the number of assets to a 9-5-9-1 force was the significant reduction in the number of unique λ_{PFE} values to 13, for a ratio of 13.68%. The first two experiments contained much larger ratios of unique λ_{PFE} values: 13 of 42 (30.95%) for the 7-3-7-1 force, and 24 of 78 (30.77%) for the 8-3-8-1 force. The impact of this reduction in 9-5-9-1 is a greater variety of p(Win) values for any particular unique λ_{PFE} value. Consequently, the coefficient of determination (R^2) value is reduced significantly.

The distribution of Sensors and Influencers for any particular Decider within a configuration can be either similar or disparate. Whether a configuration's disparity has a positive or negative impact on the p(Win) is determined by the balance of Sensors and Influencers for each Decider within that configuration. Likewise, this Sensor-Influencer balance affects the number of combat cycles contained within the network. This balance can also be easily measured, and we will use the term *robustness* (Barabasi [7] uses the term "robustness" to describe a network's resilience to failure due to the loss of some of its nodes). Robustness is defined as the minimum number of nodes lost that would render the entire configuration unable to destroy any more enemy nodes, mathematically expressed as:

$$Robustness = \left[\min\left(S_1, I_2\right)\right] + \left[\min\left(S_2, I_2\right)\right] + \cdots + \left[\min\left(S_n, I_n\right)\right] \qquad (13.4)$$

where S_n = the number of Sensors assigned to Decider n and I_n = the number of Influencers assigned to Decider n.

For example, Decider$_1$ of Configuration #3 has five Sensors but only one Influencer, while Decider$_2$ has one Sensor and four Influencers. This lack of balance has a negative impact on the configuration's performance as it reduces the minimum number of nodes that can be lost before a portion of the force is rendered combat ineffective. If the sole Influencer linked to Decider$_1$ is lost, then the five Sensors are combat ineffective as the information collected by the Sensors cannot be acted on. Consequently, the average probability of the BLUE Win for Configuration #3 was only 0.2365, which was the second-worst performance across all of the 7-3-7-1 configurations.

In essence, the robustness value reflects the rate of the reduction of the λ_{PFE} valuer over time. The greater the robustness value, the longer a configuration will maintain combat effectiveness. Configurations that were more robust have a greater p(Win) value, while less robust configurations had a lower p(Win) value. This is the point at which the robustness value becomes particularly useful. Once again, utilizing an ordinal scale, these BLUE configurations can be ranked from 1 to 95 based on their average probability of a BLUE win (subsequently ordered by their λ_{PFE} values for those configurations with equal p(Win) values, where possible). Figure 13.10 shows the resulting trend in robustness.

Because the λ_{PFE} values now have a reduced correlation to the values of p(Win), the robustness value becomes much more useful in discriminating between configurations. For example,

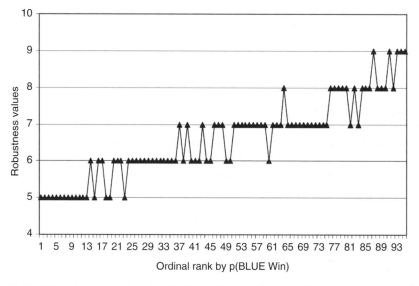

Figure 13.10 The robustness values of the 95 configurations of a 9-5-9-1 BLUE network. *Source*: Deller, Sean; Rabadi, Ghaith; Tolk, Andreas; Bowling, Shannon R., "Organizing for Improved Effectiveness in Networked Operations," *MORS Journal*, Vol. 17, No. 1, 2012, John Wiley and Sons, LTD.

20 of the 95 configurations share the λ_{PFE} value of 2.031. Which configuration should have a greater average p(Win) value? By looking at the robustness value for each configuration (which vary from 6 to 9), we see that the sole configuration with a robustness value of 9 (Configuration #93) has the highest average p(Win) value. This average p(Win) value of 0.5425 is substantially larger than any configurations with an equal λ_{PFE} value. Configuration #93 has the highest average p(Win) among the configurations with equal λ_{PFE} values regardless of whether or not ties are included (0.5710 without including ties). Note that the configuration with the highest average p(Win) (Configuration #43) is not the configuration with the highest λ_{PFE} value (although its λ_{PFE} value of 2.141 is quite high), but one of the configurations with the highest robustness value. A regression analysis of both the λ_{PFE} value and the robustness value yields a significant increase in the coefficient of determination (R^2) from a value of 0.621 to 0.850 and provides the following equation:

$$y = \left[0.0997\left(x_1\right) + 0.0613\left(x_2\right) \right] - 0.1617 \tag{13.5}$$

where y = the average probability of a BLUE win for that configuration, x_1 = the λ_{PFE} value of a configuration, and x_2 = the robustness value of a configuration.

The increase in the value of the coefficient of determination (R^2) remains significant even if we include the tie results again. In this case the value of R^2 is 0.805 and provides the following equation:

$$y = \left[\left(-0.0307\right)\left(x_1\right) + 0.0615\left(x_2\right)\right] + 0.0678 \tag{13.6}$$

where y = the average probability of a BLUE win for that configuration, x_1 = the λ_{PFE} value of a configuration, and x_2 = the robustness value of a configuration.

13.7   Conclusion

The purpose of this research was to gain insight into how an Information Age combat force should be organized to optimize its effectiveness. Given the lack of quantifiable metrics that can discriminate between various networked forces that differ solely in their arrangement, this research represents an initial attempt to determine the utility of the Perron–Frobenius Eigenvalue (λ_{PFE}) as a measure of the effectiveness of the organization of a networked force. The results of the agent-based simulation modeling indicate that the value of the λ_{PFE} is a significant measurement of the performance of a networked force. A force organized with a greater λ_{PFE} value will defeat a force with equal assets and capabilities, but organized in a less optimal manner, more often. The coefficient of determination (R^2) of both the 7-3-7-1 and 8-3-8-1 networked forces showed a strong degree of correlation between the λ_{PFE} value and the average probability of a Win.

Although the λ_{PFE} value alone was a sufficient indicator for networked forces with three Deciders, it was not sufficient for a networked force with five. The greater number of configurations with the same λ_{PFE} value reduced the effectiveness of the λ_{PFE} value in discriminating between those configurations. Consequently, another metric was required. The robustness factor was introduced in this research to improve the effectiveness of the λ_{PFE} value as a quantifiable metric of network performance, and can be utilized in other similar research as quantifying factors. By utilizing both the λ_{PFE} value and the robustness value, the coefficient of determination (R^2) for the 9-5-9-1 networked force showed a strong degree of correlation with the average probability of a Win. No other quantifiable network metrics can consistently discriminate between configurations that differ by a single link, regardless of the significance of that link.

The success of this research warrants further exploration in a number of areas, such as expanding the experiment to include additional link types (from Table 13.1) and larger networks. Replacing the deterministic links with stochastic values between 0 and 1 representing the probabilities of detect, communicate, and kill should be investigated. Additionally, the simulation can be developed to include specific capabilities of individual Sensors, Deciders, and Influencers such as rates of movement, sensing and influencing ranges, search patterns, and survivability. Will the λ_{PFE} remain a significant measurement of the performance of an Information Age combat force? Can we determine the value of a Sensor relative to an Influencer? Fundamentally, the answer we seek is to the question: how many assets can organizational optimization offset? Investigating the effects of engagements between forces without the same numbers of assets can determine whether a smaller, more optimally organized force can defeat a larger force.

Disclaimer

The views expressed in this chapter are those of the authors and do not reflect the official policy or position of the United States Army, Department of Defense, or the United States Government.

References

1. Jain, S., and S. Krishna, 2002, "Graph Theory and the Evolution of Autocatalytic Networks," John Wiley, New York. http://arXiv.org/abs/nlin.AO/0210070 (last visited 31 Aug. 2011).
2. Cares, J., 2005, *Distributed Networked Operations*, iUniverse, New York.
3. Chartrand, G., 1984, *Introductory Graph Theory*, Dover Publications, New York.
4. Jain, S., and S. Krishna, 1998, "Autocatalytic Sets and the Growth of Complexity in an Evolutionary Model," *Physical Review Letters*, Vol. 81: 5684–5687.
5. Wilenski, U., 1999, *NetLogo*, Center for Connected Learning and Computer-Based Modeling, Northwestern University, Evanston, IL, http://ccl.northwestern.edu/netlogo (last visited 31 Aug. 2011).
6. Deller, S., M. I. Bell, G. A. Rabadi, S. R. Bowling, and A. Tolk, 2009, "Applying the Information Age Combat Model: Quantitative Analysis of Network Centric Operations," *The International C2 Journal*, Vol. 3 No. 1; Online Journal of the Command and Control Research Program, http://www.dodccrp.org/files/IC2J_v3n1_06_Deller.pdf (last visited Mar. 2012).
7. Barabasi, A., 2002, *Linked*, Perseus Publishing, Cambridge.

14

An Exploration of Performance Distributions in Collectives

Jeffrey R. Cares
Captain, US Navy (Ret.), Alidade Inc., USA

14.1 Introduction

In the late 1990s, military innovators in the US began to espouse a concept of warfare to keep pace with a new, burgeoning, and poorly understood "Information Age" enabled by technology. In a flurry of books, articles, and briefings, this concept – Network Centric Warfare (NCW) – became in a very short time a dominant and transformative idea. Dominant, because NCW ideas were almost instantly adopted and embraced by nearly every major defense contractor and senior Pentagon leader (those who did not understand or outright rejected the ideas saw themselves instantly marginalized as luddites "who don't get it"). Transformative, because it spurred an incredible investment in enterprise-wide information technology (IT) systems that continues as a major share of worldwide defense budgets to this day. Now, after more than a decade and a half, the conceptual underpinnings of NCW have been long since cast aside while the hardware remains.

A new wave of innovation is washing through the international defense industry, ushering in what many are calling a "Robotics Age." This new age will be enabled by IT every bit as much as NCW was: interconnected swarms of self-organizing or remote-controlled unmanned vehicles are expected to take their place in a networked collective that includes legacy, manned systems. While industry and academia struggle to develop these new vehicles and control systems, their focus is narrowed just as much on engineering the vehicles as NCW enthusiasts were preoccupied with IT hardware. The hard conceptual work to figure out how forces will fight with these new platforms is taking a backseat to vehicular products – just as NCW theory and analysis did to IT investment.

Some view Robotics Age warfare as an outgrowth or maturation of NCW. Even if this is not strictly true, we would nonetheless be wise to revisit some of the initial NCW warfighting

Operations Research for Unmanned Systems, First Edition. Edited by
Jeffrey R. Cares and John Q. Dickmann, Jr.
© 2016 John Wiley & Sons, Ltd. Published 2016 by John Wiley & Sons, Ltd.

ideas to see if they are valid and can be applied to warfare characterized by increased automation, connectivity, and interdependence. Unfortunately, most of the claims of NCW were simply asserted, not tested. For example, the claim that networking necessarily improves performance throughout a force was accepted as a fundamental truism but never proven, as appropriate data from NCW forces under combat conditions did not and still do not exist. Since questions about the nature of performance of networked collectives in combat will no doubt resurface as unmanned technology matures, this chapter seeks to use human individual and networked performance data from another competitive pursuit – professional baseball – to make an initial inquiry into the subject.

14.2   Who Shoots How Many?

The US Air Force has long known about the "Ace Factor" – very few fighter pilots score high numbers of kills while a great many of them never notch a tally. Bolmarchic [1] put formal rigor to this street wisdom while investigating the distribution of outcomes for a great many, varied forms of modern combat. He noted, for example, that the distribution of kills in tank/anti-tank battles, of tonnage in submarine warfare, and of enemy kills in air-to-air combat all follow a similar statistical pattern. Not unlike other types of skewed distributions in human competition (such as power laws, Zipf's Law, or other Pareto-like distributions), he noticed a recurring distribution of combat outcomes such that a very small number of participants had a very high score, a moderate number had a moderate score, and a very large number had a very small (or even zero) score. Bolmarchic sought not just to describe the shapes of these distributions, but he was also looking for distributions that could be described by a single parameter, so that he could compare the imparity in one distribution to the imparity in another, perhaps even from a very different type of combat. To develop this imparity statistic he investigated the one-parameter form of the well-known multivariate homogeneous Pólya distribution, derived from an "urn scheme" assignment of outcomes.

Pólya's urn scheme works as follows. Assume there is a number of urns arrayed on a table, and each urn has an identical weight, β. The value of this β is some multiple of the weight of individual balls you will throw into these urns. As you throw balls into urns, you record the combined weight of each urn with its balls and bias the choice of your next urn by a probabilistic representation of the distribution of existing urn–ball weights, favoring the heavier urn–ball weights. This is a classic "rich-get-richer" scheme, and the shape of the distribution is driven by a single value, β. If β is small relative to the weight of each ball, then imparity is high, since early success by any urn will make future success more likely for that urn and less likely for others. If β is large relative to the weight of each ball, then imparity is low, since repeat success required to accrue strong weight bias toward any urn are nearly equiprobable events. Eventually one urn will receive bias, but not before weight is spread around the ensemble.

Mathematically, the probability that an urn gets the next ball is conditioned on where the balls are already,

$$\Pr\left[\text{Urn i gets the } k+1^{\text{st}} \text{ ball} | K_1 = k_1, K_2 = k_2, \ldots, K_n = k_n\right] = \frac{\beta + k_i}{n\beta + k} \qquad (14.1)$$

After a total of r balls have been tossed, the distribution of balls to urns follows the multivariate homogeneous Pólya distribution, with single parameter β,

$$\Pr[K_1 = k_1, K_2 = k_2, \ldots, K_n = k_n] = \frac{\binom{\beta + k_1 - 1}{k_1}\binom{\beta + k_2 - 1}{k_2} \cdots \binom{\beta + k_n - 1}{k_n}}{\binom{n\beta + \Sigma_i k_i - 1}{\Sigma_i k_i}} \quad (14.2)$$

In practice, of course, combat data aren't urns and balls but shooters and kills. We also don't care so much about naming our shooters Shooter 1, Shooter 2, and so on, than we care about how many shooters N_0 have no kills, how many have N_1, and so on. Equation (14.2) can be manipulated to show the distribution as a *profile probability*,

$$\Pr[N_0 = n_0, N_1 = n_1, \ldots, N_r = n_r] = \frac{n!}{n_0! n_1! \ldots n_r!} \frac{\left(\frac{\beta}{1}\right)^{n_1} \left(\frac{\beta+1}{2}\right)^{n_2} \cdots \left(\frac{\beta+r-1}{r}\right)^{n_r}}{\binom{nb+r-1}{r}} \quad (14.3)$$

Again, however, we can parameterize by β. In practice we can't observe β directly; we have data only on who shoots how many. But we can plug the data on who shoots how many into Eq. (14.3) and vary β until we get a maximum value. Using this maximum likelihood estimate, β^*, we can compare different data sets. Because of certain statistical regularities, Bolmarchic proposed to use the natural logarithm of the maximum likelihood estimate as the *imparity statistic*, $\kappa = \log(\beta^*)$. Both values are helpful in analysis: β for its analogy to weights and bias, and κ to allow for better comparison over a wide range of different β^* values.

14.3 Baseball Plays as Individual and Networked Performance

Bolmarchic computed and compared β^* values for a wide variety of combat data sets, including US Navy and US Air Force air-to-air kills of North Vietnamese aircraft (1965–1972), US anti-tank expenditures against Iraqi armor in the Battle of 73 Easting, Royal Navy expenditures in the Falklands, US submarine kills of Japanese shipping in World War II, and German U-Boat kills of Allied shipping in World War II. Some of these data sets were parsed into subsets, such as kills by wing/squadron/crew for air-to-air data, kills per Commanding Officer (CO)/boat type/region for U-Boat data, or expenditures by vehicles/crew for anti-tank data. He found that for all the standard statistical tests, the multivariate homogeneous Pólya distribution held up as a good fit for each these disparate data sets. These data are for classic, platform-centric warfare, and are unsuitable to answer questions about the impact of networking on performance distributions.

Data must meet two requirements to answer this question. First, there must be data from networked, collaborative competition. Second, the data must also have instances of individual performance, so that individual and collective performance can be compared. One very rigorously maintained database meeting these two requirements is the Major League Baseball (MLB) database. It contains scrupulously verified batting, fielding, and pitching statistics from 1871 to 2014. The data are hosted in many places, but they are open to public verification, maintenance, and free use online at the Lahman Baseball Database homepage [2]. Taking the second requirement first, there are many different tasks baseball players complete which can be categorized as individual performance, most notably, the "home run" (hr). An HR occurs when a player bats a pitched ball out of the ballpark in fair territory.[1]

Baseball also has what can be considered networked, collaborative tasks. The prime example is the "double play (DP)," which is by definition a continuous offensive play in which two "outs" are recorded. Teams take turns on offense, and each set of offensive turns is called an "inning." Each turn at offense last until three outs are recorded, which can occur by tagging a running player with a ball that is in play, catching a batted ball before it touches the ground, or by advancing the ball to a "base" (a filled fabric square attached to each corner of the dia-mond-shaped playing surface)[2] before a runner arrives there.

A DP is a very common baseball play, and can occur in a wide variety of ways involving anywhere from one to nine defensive players (nine is technically possible but more than four players involved in a double play is infeasible). There are usually six players who field inside the diamond, the first baseman (1B), second baseman (2B), third baseman (3B), shortstop (SS) (usually played on the line between second and third base), the catcher (who plays at home base), and the pitcher (who stands atop a mound in the middle of the diamond to throw balls to the batter, which the catcher catches if they are not batted). The classic DP occurs when one runner is at first base and the batter hits the ball to a fielder playing inside the dia-mond. The fielder can throw the ball to second base, causing the runner trying to advance from first to second to be out. The fielder catching this ball continues the play by throwing to first base, and if it arrives before the runner advancing from home base, that runner is out, too.

Double plays are also common when a batted ball is caught in the air. Players may not advance on balls caught in the air unless they wait at the base from which they want to advance until the ball is caught. The ball caught in the air is the first out, and a DP can occur if the player does not wait at the base (the player is out if the ball is thrown to the base before the running player returns) or if the player waits but does not beat the ball thrown to the base to which they are trying to advance. All nine players can participate in these types of DPs.

One more common DP occurs when a runner tries to advance to a new base in between pitches. This is called a "stolen base (SB)." Stolen base attempts result in DPs when a batter has two "strikes" (a strike is a hittable ball the batter does not hit; three strikes are an out). During this kind of DP, the batter does not hit strike three and the catcher throws the ball to the attempted SB and it arrives in time for the fielder at that base to catch and tag the runner before the runner touches the base. All infielders can participate in this "strike-em-out, throw-em-out" DP, but usually the catcher, pitcher, and the 2B or SS are involved.

Since all these plays require interdependent, interconnected collaborative effort from a col-lection of players under competitive pressure, they are sufficient proxies for the second requirement, "networked" performance data.

Baseball has many statistics, and more are introduced each season, fueled by the managerial imperative in the multibillion-dollar professional baseball industry to derive more return from their increasing player investments. For over 50 years, the Society for American Baseball

---

[1] There is the possibility of an "inside-the-park home run," during which a very fast player touches all bases before the outfield can return a batted ball that does not leave the park, but these are very rare.

[2] Home plate, the base at which players bat, is the fourth base, but is actually a five-sided hard rubber "plate."

Research (SABR) has kept watch over how to measure and analyze baseball, but in the last 10 years private sports research has become very big business. While data from a "pastime" might seem a frivolous source for proxies of combat data, baseball data analysis is in actuality very serious academic and intellectual work.

Some baseball statistics measure primitives; some are analytical composites of these primitives. An example of a primitive baseball measurement is the number of hits a player achieves during a season. Batting average is the number of hits per at bat the player achieves during a season. The multivariate homogeneous Pólya distribution – derived from an urn and ball scheme – obviously is applied to primitives (things you can count), not analytical composites. Furthermore, we are interested in direct measures of achievement, like hits, rather than indirect measures, such as the number of at bats. Certainly one will not continue to accrue the latter without production in the former, but that is the point: hits are a direct measure of how well a batter performs. Table 14.1 describes the primitive baseball statistics that were examined for this analysis.

14.4   Analytical Questions

We can state our analytical question more formally now that a valid single parameter statistical distribution and an appropriate data set have been identified. We know from Bolmarchic's study of combat data that individual performance is not only skewed but in many cases is well represented by the multivariate homogeneous Pólya distribution. We do not, however, have similar combat data for interconnected collectives (aggregations at the wing and squadron level for air-to-air data, for example, are just tallies for a group, not the result of highly integrated, connected performance). We have a data set, the MLB data set, which shows both individual and interdependent data. Early NCW concepts claimed that interconnection resulted in better performance for a group (although it was unclear if they were talking about individual performance within a group or better performance of the group as a whole). Since individual performance is skewed so that there are a few very high performers and a very large number of poor performers, what will happen when this mix of performers are connected? Will the talent of high performers contribute to make poor performers better ("a rising tide lifts all boats")? Or will connecting a small number of high performers to a very large number of poor performers drag the high performers down so that there are fewer performers in the "fat tail" of the distribution? Some optimistically call a skewed performance distribution a "distribution of aces," but it is also a "distribution of goats[3]" and there are many, many more goats than aces.

The rest of this chapter addresses collective performance in the following way. We will examine individual and interdependent primitive performance measures in baseball (things that can be counted) and compute the maximum likelihood estimate, β^*, and the imparity statistic, κ (if, in fact, the Pólya distribution is a good fit). We will next compare β^* and κ values to see if there is a difference in parity between individual and interconnected effort in MLB. Then, based on what has been hypothesized about the effects of networks and unmanned systems in future warfare, we will explore what the results from baseball suggest we should expect in networked combat performance.

---

[3] As early as the Old Testament (538–332 BC), goats have been identified as the source of failure in groups (Lev. 16–22).

Table 14.1 Measures of baseball performance.

Statistic	Description
Pitching statistics	
Wins (W)	A team's wins are attributed to the pitcher who pitched last when the winning team was leading
Losses (L)	A team's losses are attributed to the pitcher who pitched last when the losing team was behind
Complete game (CG)	A full nine-inning game pitched by a single pitcher
Shutout (SHO)	A full nine-inning game in which no runs are scored against a complete game pitcher
Hits allowed (H)	Number of hits allowed by a pitcher during a season
Earned runs allowed (ER)	An earned run is credited to the offense when no defensive errors occur while the batter is on base
Home runs allowed (HR)	Number of home runs allowed by a pitcher during a season
Base-on-balls (BB)	A pitcher is required to throw hitable balls ("strikes"). If a pitcher throws four non-hitable balls ("balls") to a batter, that batter is awarded first base, "on balls"
Strikeouts (SO)	A batter that does not hit three hitable balls ("strikes") has "struck out". Pitchers are awarded credit for striking out the batter
Intentional base-on-balls (IBB)	Some fearsome hitters are intentionally sent to first base, on balls, to prevent more damaging hits. Pitchers are awarded blame for IBB at bats
Wild pitches (WP)	A ball thrown so wildly that the catcher cannot catch it; awarded to the pitcher
Hit batsmen (HBP)	A batter struck by an ill-aimed pitch is advanced to first base; awarded to the pitcher
Double play ground balls (GIDP)	Some pitchers throw the ball to induce classic double plays from balls hit on the ground
Sacrific hits (SH)	Some balls are hit in such a way that the batter is out ("sacrificed"), but runners can advance
Sacrific flies (SF)	Some ball are hit high in the air to the outfield so that the batter is out but the runners can advance
Batting statistics	
Runs (R)	A batter who crosses home base from third base is credited with a run
Hits (H)	A hit is awarded when a player reaches base on a batted ball fielded without defensive errors
Doubles (2B)	A double is awarded when batters reach second base on their batted balls
Triples (3B)	A triple is awarded when batters reach third base on their batted balls
Home runs (HR)	A home run is awarded when batters reach home plate on their batted balls, usually because the ball was hit out of the park fairly
Runs batted in (RBI)	Number of runners a batter is credited with advancing to home plate
Stolen bases (SB)	Number of stolen bases credited to a runner
Caught stealing (CS)	Number of times a runner is caught stealing
Base-on-balls (BB)	Number of bases awarded on balls
Strikeouts (SO)	Number of times a batter strikes out in a season
Intentional base-on-balls (IBB)	Number of times a batter is intentionally put on first base
Hit-by-pitch (HBP)	Number of times a batter is hit by a pitch
Sacrific hits (SH)	Some balls are hit in such a way that the batter is out ("sacrificed"), but runners can advance

Table 14.1 (*Continued*)

Statistic	Description
Sacrific flies (SF)	Some ball are hit high in the air to the outfield so that the batter is out but the runners can advance
Grounding into double play (GDIP)	Number of times a batter is induced into a classic ground ball double play
Fielding statistics	
Put outs (PO)	Number of times a fielder causes a runner to be out
Assists (A)	Number of times a player participates in a play in which the runner is out
Errors (E)	Number of times a player misplays a thrown or batted ball
Passed balls (PB, catcher only)	Number of times a catcher misses a catchable ball thrown from the pitcher
Stolen bases (SB, catcher only)	Number of times a catcher fails to stop a runner from stealing a base
Caught stealing (CS, catcher only)	Number of times a catcher stops a runner from stealing a base
Wild pitches (WP, catcher only)	Number of times a catcher is in the game and the pitcher throws a wild pitch. This is not usually the catcher's fault, but very good catchers can prevent WPs, so baseball keeps track for both the pitchers and the catchers
Double plays (DP)	Number of times a fielder participates in a double play of any kind

14.5 Imparity Statistics in Major League Baseball Data

Table 14.2 shows the β^* and κ values for selected MLB statistics from the 2013 regular season (excluding statistics from the post-season World Series Championship tournament). Almost 90% (34 of 38) of the "countable" statistics listed in Table 14.2 are a good fit to the multivariate homogeneous Pólya distribution, using the standard statistical goodness-of-fit tests (chi-square, log-likelihood, Craven–von Mises, Kolmogorov–Smirnov, and likelihood ratio). The four statistics which do not have a good fit are discussed later.

Recall the urn–ball conceptual basis of β. We are not weighing urns and balls here; players are competing in a sport. As a precursor to understanding the results in Table 14.2, then, how should we interpret β^* (the estimate of β)? In the urn scheme, a very light urn means that early success is a strong driver of repeated future successes. Conversely, a very heavy urn means that early success is a very weak driver of repeated future successes. $\beta = 1$ is the breakpoint of "tending toward a strong driver" ($\beta < 1$ and more imparity is expected) and "tending toward a weak driver" ($\beta > 1$ and less imparity is expected). $\beta = 1$ is the point at which any future arrangement of balls in urns is just as likely as all others. One nice feature of using $\kappa = \log(\beta^*)$ as the imparity statistic is that for values of $\beta^* < 1$, κ is negative, and for values of $\beta^* > 1$, κ is positive. κ is more symmetric around 0 than β^* is around 1, so κ is very helpful in comparing very large and very small values of β^*. Under conditions of strong *im*parity, high performers and poor performers are both expected; under conditions of strong parity, high performers and poor performers are both *un*expected.

14.5.1   Individual Performance in Major League Baseball

We can use κ to answer two basic kinds of questions. The first is about the distribution itself (e.g., "Is there more parity in home runs this year than last?" or, "Is there more parity in home runs or second basemen?"). The second kind is about an individual player's performance within a distribution (e.g., "Is Babe Ruth a better home run hitter than Mark McGwire?").

Turning to the data in Table 14.2, let's see how we use the imparity statistic in both these ways. One might ask, "Is it harder for a player to be the league leader in SBs or in triples (3Bs)?" Baseball fans know that base-running speed plays a strong role in both SBs and 3Bs, but great hitting also plays a strong role in 3Bs, but not at all in SBs. From Table 14.2, we see $\kappa_{SB} = -1.64$, and $\kappa_{3B} = 0.03$; there is less parity in SBs and more parity in 3Bs. This means that there are a lot of players who perform poorly and a few who perform well in SBs, so one would expect a superbly fast base-runner to rise out of the base-running pack and accrue a lot of SBs. There are plenty of players who lack base-running speed, but who survive in the league through other talents, such as good defense for (notoriously slow) catchers or great hitting power. These talents can't help accrue SBs, so the distribution of SBs is really about parity of foot speed among baseball players. Triples require great speed plus great batting power, which is more rare than just great speed or just great batting power alone. Players with great speed and a decent bat and players with decent speed and a great bat will both vie for the lead in 3Bs, resulting in more parity in the distribution. As the κ values confirm, it is therefore harder to be a league leader in 3Bs because there is more parity, than to lead in SBs, because there is less.[4] In other words, while there might be keen competition for the lead in SBs between a handful of fast runners, most of the competition is weeded out early. The opposite is true for 3Bs: players can accrue 3Bs with more than just speed, so more contenders stay viable for the league lead longer.

Figures 14.1 and 14.2, which show the double cumulative distribution of triples and SBs, and the curve of β^* fit to the data, portray this graphically. The double cumulative distribution is useful in describing imparity data because it shows which percent of a population accounts for what percent of production. In Figure 14.1, for example, 68% of hitters hit only 20% of triples (which also means that 32% of hitters hit 80% of triples). In terms of actual numbers of triples, 80% of players hit two triples or fewer. Contrasting with Figure 14.2, 85% of runners account for 20% of SBs (and 15% steal 80%). In terms of actual SBs, 85% of runners steal six or fewer bases. Here we see the imparity graphically, and it comports with the previous conclusion. There is more parity in the distribution of triples so more players stay in the competition longer. Note also, that accruing triples is not as easy apparently as accruing SBs. The league leader for 2013 hit 11 triples but the league leader in SBs stole 52. In general, the more difficult a task, the more parity there is in its distribution among a population. So Figures 14.1 and 14.2 can be used to ask questions about individuals within a population. For example, should a team make an even trade of a player who hits six triples with one who steals 10 bases? The player with six triples is in the 98th percentile (the very top of the league) whereas the player with 10 SBs is in the 92nd percentile (still very good, but not as elite). The data tell us

---

[4] There is an important caution. Extraordinary performers are part of the distribution, and in certain cases their performance could be such an outlier that it affects the imparity statistic to make it look like leading the league was easier than they made it look. One should check the imparity of that statistic over previous years to confirm if this truly is "game-changing" performance.

Table 14.2 Imparity statistics for the 2013 MLB regular season.[a]

Statistic	β^*	κ
Pitching statistics		
Wins (W)	1.02	0.02
Losses (L)	1.10	0.09
Complete games (CG)	5.06	1.62
Shutouts (SHO)	8792.50	9.08
Hits allowed (H)	1.06	0.06
Earned runs allowed (ER)	1.03	0.03
Home runs allowed (HR)	0.89	−0.12
Base-on-balls (BB)	1.19	0.17
Strikeouts (SO)**	0.96	−0.04
Intentional base-on-balls (IBB)	2.25	0.81
Wild pitches (WP)	1.15	0.14
Hit batsmen (HBP)	0.70	−0.36
Double play ground balls (GIDP)	0.70	−0.36
Sacrific hits (SH)	0.97	−0.03
Sacrific flies (SF)	1.59	0.46
Batting statistics		
Runs (R)	0.51	−0.66
Hits (H)**	0.60	−0.51
Doubles (2B)	0.68	−0.38
Triples (3B)	1.03	0.03
Home runs (HR)	0.64	−0.44
Runs batted In (RBI)	0.56	−0.57
Stolen bases (SB)	0.19	−1.64
Caught stealing (CS)	0.57	−0.56
Base-on-balls (BB)	0.45	−0.79
Strikeouts** (SO)	0.50	−0.68
Intentional base-on-balls (IBB)	0.32	−1.14
Hit-by-pitch (HBP)	0.67	−0.41
Sacrific hits (SH)	0.65	−0.42
Sacrific flies (SF)	1.25	0.23
Grounding into double play (GDIP)	0.84	−0.17
Fielding statistics		
Passed balls (PB, catcher only)	2.21	0.79
Stolen bases (SB, catcher only)	0.98	−0.02
Caught stealing (CS, catcher only)	1.51	0.41
Wild pitches (WP, catcher only)	1.07	0.06

[a] Statistics identified with a double asterisk (**) were computed on a 10% random sample of that statistic. Computing over the full range of values for that data caused floating point over-runs in the statistical software employed for this study. There is a small probability that these β^* and κ might be missing important points in the fat tail, so are not definitive of the imparity of that statistic. They were computed to explore the extent to which a subset of this data could be represented by the multivariate homogeneous Pólya distribution and also to compare them with other statistics (not included in Table 1.2) for which the Pólya distribution did not fit.

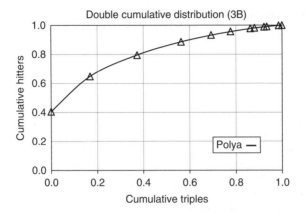

Figure 14.1   Percentage of batters hitting percentage of triples.

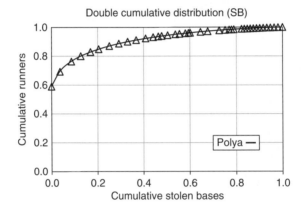

Figure 14.2   Percentage of runners stealing percentage of bases.

there are only 12 players who hit more triples than six but 75 more players who can steal more bases than 10 – so you'll likely find a similar base-stealer on the market but not likely that kind of triples-hitter.

The distribution of SBs has the most imparity of these 34 statistics. Next in imparity is intentional base-on-balls (IBB) awarded to batters ($\kappa_{\mathrm{IBB(B)}}=-1.14$), which stands in stark contrast to the strong parity of that statistic when attributed to pitchers ($\kappa_{\mathrm{IBB(P)}}=0.81$). This reflects baseball reality: there are only very few hitters that are so feared that they receive a high number of IBBs. Even the best pitchers, however, will be asked to put a feared hitter on-base for the opportunity to get a more likely out from the next hitter in line.

Very large values of β^* and κ tell a different kind of story. This chapter has discussed aces and goats in a somewhat simplistic light. Some goals in human competition are so difficult to achieve that almost no one can accrue great success, and most everyone performs at a very low level. Take, for example, the statistics for complete games (CGs, $\kappa_{\mathrm{CG}}=1.62$) and for complete game shutouts (SHOs, $\kappa_{\mathrm{SHO}}=9.08$). These reflect the fact that it takes incredible endurance and skill to complete a baseball game as the starting pitcher. The vast majority of starting pitchers will go an entire season without completing a single game before they are relieved by another pitcher. It is even more rare for a pitcher to complete the game without letting a single run across home plate. The two statistics are, in fact, correlated: most managers will let a pitcher pitch late into a game if they have been very effective (perhaps not even allowing a single hit, which is even rarer than an SHO), but as soon as their endurance appears to be breaking down, they will be quickly relieved if the score is close. So a pitcher in line for an SHO maybe be close to a CG, but lose both with only a few outs left in the game.

The imparity statistic can also be used to examine systemic conditions over time. Table 14.2 lists the imparity statistic for HRs, κ_{HR}, as −0.44. When calculated for previous years we find that $\kappa_{HR} = -0.70$ in 2012, $\kappa_{HR} = -0.65$ for 2011, and $\kappa_{HR} = -0.63$ for 2010, three values that are very consistent. What happened, then, in 2013? During this four-year period, MLB was in the final stages of a decade-long effort to ban performance-enhancing drugs (PEDs), culminating in very harsh penalties beginning with the 2013 season (even issuing a year-long ban to one of its major stars). Is this trend toward more parity an indication that a few PED abusers were the source of HR imparity, and that baseball is becoming free from widespread PED use? This, of course, is not directly related to unmanned vehicles, so is a question to take up in a different line of research.

Admittedly, this section contains quite a lot of baseball discussion, but while it may seem tangential to our analytical question, it is nonetheless very useful to establish the broad utility of β^* and κ to compare human performance data from a wide variety of countable measurements in a very competitive pursuit.

14.5.2   Interconnected Performance in Major League Baseball

There are four statistics for which the multivariate homogeneous Pólya distribution was a poor fit: putouts (POs), assists (A), errors (E), and double plays (DPs). Three of these statistics are closely related. A PO is awarded to a fielder who directly causes an out by catching a ball in the air, tagging a runner, or by stepping on a base to which a runner is trying to advance. An assist is awarded to a fielder who touches the ball prior to a PO. In a classic DP, suppose the ball is batted on the ground to the SS with a runner on first base. If the SS catches the ball and throws it without error to second base before the runner arrives, then the SS is awarded an assist and the 2B awarded a PO for the first out of the DP. If the 2B then throws the ball to first base without error, the 2B is awarded an assist and the 1B records a PO. So in that continuous play, all three players each receive a single DP credit, two players each receive a single assist credit, and two players each receive a single PO credit.

There are ways in which a PO can be credited individually, usually on a ball caught in the air. Assists, by contrast, come because of pairwise efforts (although sometimes more than one player can assist on a PO). Double plays are by definition two consecutive POs. If we are looking for interconnected performance in baseball, we find it here in the PO, A, and DP statistics.[5] While it is very inconvenient that the multivariate homogeneous Pólya distribution

---

[5] Errors are included in Table 14.1 but excluded from Table 14.2. They appear similar to individual statistics, but have some of the same computational difficulties that are discussed below with respect to double plays. One must be somewhat suspicious of errors anyway, since they are the most subjective of all baseball statistics – an official scorer decides what is a error and what is not, and it is widely conjectured that there is a strong "home team" bias among official scorers.

is not a good fit for these statistics, that fact in itself tells us that the nature of individual performance could indeed be different from interconnected performance. While it would be tempting (and disappointing) to end the analysis here with that single conclusion, the procedure for fitting the Pólya curve to the DP data unveiled some intriguing information that bears directly on the analytical question.

Once again recalling the urn–ball scheme we are reminded of something fundamental – there is an assumption of homogeneity in the weights of urns and balls. Without homogeneity in data, the Pólya distribution might be a poor fit. Bolmarchic found that there were some common kinds of inhomogeneity in combat data that could, however, be mitigated. A recurring type of inhomogeneity in combat data is a difference in opportunity among shooters over the time-span of the period analyzed. For example, a submarine captain who entered the war with his crew in the last month of the war did not have the same opportunity to perform as one who had been in theater for years. There are analogous sources of inhomogeneity in the baseball data: by rule, pitchers don't hit in half the major league ball parks, and there are plenty of near-ready minor league players who come up to the major leagues (usually early in the season or late in the season) and see very limited playing time before they return to the minor leagues.

Bolmarchic compensated for this lack of equal opportunity by conducting a two-parameter fit of the Pólya distribution, that is, optimizing over two parameters, β and N_0 (the number of shooters with no kills) instead of just the one original parameter, β. The assumption is that N_0 is suspect, but the other N_r are not, so we reduce N_0 until an optimal fit for β is achieved for some new N_0 and the original N_r. We can think of the new N_0 as a "natural" value for the number of shooters with no kills given the original N_r. This was the usual procedure for all 34 of the 38 statistics for which a good fit was achieved. Surprisingly, when this practice was applied to the DP data, over 30 000 *extra* players had to be added to N_0 before the likelihood values began to converge on an optimum. The process was close to converging at $\beta^* = 0$, $\kappa = -5.0$ when the statistical software failed due to floating point over-run errors. It is hard to conceive of a "natural" way to account for this N_0 in this case. One hypothesis might be some sort of externality from interconnectedness: that the highly inter-dependent performance of more than 2500 fielders looks like the distribution you might expect with an order of magnitude more performers and extreme imparity. This is an intriguing idea, but another natural way to think about the inhomogeneity of opportunity would be to look at "positional" inhomogeneity. For example, outfielders have fewer DP opportunities while SSs have many more.

To mitigate for positional inhomogeneity, the data were parsed in three ways, based on position: all infielders (1B, 2B, SS, and 3B), all outfielders (right field (RF), center field (CF), and left field (LF)), and individually (all positions including pitcher (P) and catcher (C)). While the previous attempt failed to provide a good fit, focusing just on the infielders provided better goodness-of-fit statistics. β^* was 0.17 (still a very light urn and stronger imparity than any individual performance statistic); κ was -1.76. N_0 had to be increased by 300 infielders, which was still unsatisfactory but left open the possibility of some kind of externality. Separating out the infielders by individual position showed that the fit improved (in fact, fit steadily improved as κ increased), but N_0 still required a larger population at 1B (77 extra 1Bs) and 2B (35 extra fielders). SS and 3B were good fits near the original N_0 (SS and 3B both needed three extra players). Table 14.3 shows the statistics for infield positions.

Table 14.3 Imparity statistics for MLB infielders, 2013 data.

Position	β^*	κ
1B	0.25	−1.39
2B	0.26	−1.33
SS	0.36	−1.03
3B	0.32	−1.13

Note that the κ for 1B (−1.39) is smaller (more imparity) than the κ for the entire infield (−1.76). This could be because 1B is primarily a recipient of other players' skill: he gets credit for a well-turned DP even if he is simply standing on the base for the second PO after exceptional skill by SS and 2B to execute a very difficult DP. 1B is almost always the least skilled of the fielding positions (catchers and 3Bs go there to extend their careers, and other backup fielders are routinely sent to first for a few games to cover for injuries rather than call a 1B up from the minors). This strongly suggests that there might indeed be some kind of "externality" received by 1B on almost all classic DPs. There are occasions when 1B has to initiate a DP, and sometimes he has to save a DP by snatching a poorly thrown ball out of the dirt, but there is a strong argument that the 1B gets more DP, A, and PO credit for doing less than anyone on the diamond. So on a team with great SSs and 2Bs, even an average 1B will look more exceptional than he actually is. Great SSs and 2Bs are hard to find, and it is possible that the relative imparity in infield performance is due strictly to the performance of these middle infielders (SS and 2B). Superb human performance, in this case, accrues not to the performers vital to the task, but to a player who is an adjunct to the action. Indeed, the top 10 league leaders in DPs in the 2013 season were all 1Bs.

It is a different case with 3B. He is almost always a DP initiator, and DPs are much harder from third than the classic DPs executed near the base at second (3B must execute longer throws, is often falling away from his throwing target, etc.). 3B is also not often at the end point of a DP. A similar argument about initiation can be made for SS. There are twice as many right-handed hitters in baseball than left-handed hitters, and right-handed hitters tend to hit to the SS side of second, so SS probably initiates the classic DP about twice as often as 2B. So one reason there is more imparity at the 1B and 2B positions than at SS or 3B could be that 1B and 2B both receive more credit for others' performance in DPs than they, themselves, give to others.

The distribution of DPs among outfielders, by contrast, was at the extreme of the distribution as well, but in the opposite direction: DPs are distributed with extreme parity among the outfield positions. The goodness-of-fit tests showed a very good fit for the Pólya distribution, and N_0 required adjustment in the usual way (reduction to account for inhomogeneity in playing opportunity). The outfield β^* was 18.23 and the κ_{OF} was 2.90, indicating much more DP parity among outfielders than infielders. Table 14.4 shows these statistics by each outfield position, displaying even more imparity at each individual position than for the outfield as a group.

There is no case in which an outfielder will receive credit for someone else's DP skill – they are always initiators. No outfielder was involved in more than four DPs in 2013. One might be tempted to look for a reason why the statistics decrease from LF around through CF to RF. A characteristic of cases in which there are both extreme parity and very small numbers of successes is that even a tiny bit of luck or extra success will dramatically change the imparity statistic. For example, no CF was involved in more than two DPs in 2013, but if one CF with

Table 14.4 Imparity statistics for MLB outfielders, 2013 data.

Position	β^*	κ
LF	27 271.57	10.21
CF	11 487.57	9.35
RF	700.76	6.55

two DPs executed just one more DP over the course of the 182-game season, then β^* would fall from 11 487.57 to 6.08, and κ_{CF} would drop from 9.35 to 1.81. Again note the relationship between initiation of DPs and higher parity.

Finally, we turn to pitchers and catchers. The highest DP credit given to pitchers in 2013 was six, which makes them about as rare as outfield DPs. Most of the outfielders who got DPs were credited with only either one or two, however. By contrast, there was a better DP distribution from one to six DPs among pitchers (11 got four or more DPs) than from one to four DPs among outfielders. N_0 was adjusted in the usual way to account for opportunity inhomogeneity and the fit was good. β^* was 1.12 and κ_p was 0.11 (which starts to look more like individual statistics from Table 14.2). Pitchers are strange defenders – sometimes they act like pitchers (initiating a strike-em-out/throw-em-out DP), sometimes like a 3B (initiating a P-2B-1B DP), and sometimes like a 1B (getting credit for covering first on a 1B-2B-P DP).

Catchers look more like pitchers than infielders. They can initiate (from a flyball/PO, or very short grounded ball C-2B-1B DP) or receive (2B-C-1B DP or strike-em-out/thow-em-out). N_0 had to be reduced in the regular way – interestingly, the population of catchers is the smallest of any position (there is only a starter and a backup on each team, like hockey goalies). The statistics showed a good fit, β^* was 2.00 (an urn twice as heavy as each ball) and κ_c was 0.69. These values, like those for the pitchers, are closer to individual performance statistics in Table 14.2 than the rest of the positional DP statistics.

We are still left with our challenge of comparing individual statistics to interconnected statistics. In our statistic tests we had to reject the hypothesis that the entire population of DP data were from the multivariate homogeneous Pólya distribution. We were able to see good fits (for all but 1B and 2B) once the data were separated out by position, but we still need to know something more about the entire population. Figure 14.3 compares the double cumulative distribution of DPs, A, and POs (interconnect plays) and SBs and wins (individual statistics). Stolen bases were selected for this chart because they have the most imparity of all statistics in Table 14.2. Wins are included since their distribution (along with bases stolen against catchers) is the closest to $\beta^* = 1$ (and $\kappa = 0$).

A first observation is the shapes of the data for POs, A, and DPs, which indicate some variant of a skew distribution and some kind of imparity. The fact that all points (except for a single point where zero POs are achieved by 16% of the population) are well below the double cumulative curve for wins shows that imparity in DPs is in a very different regime than the majority of individual statistics. Imparity for POs and A is comparable to SBs, but notice how there is a smaller population that performs poorly (the points of POs and A data to the left of about 10% cumulative success/77% of cumulative population are less than the points for SBs in that range). Notice also that there is a greater percentage of production from fewer players (the points of POs and A data to the right of about 10% cumulative success/77% of cumulative population are greater than the SB data in that range). This shows an extreme imparity where not only is there a "fatter tail," but a "skinnier head" as well. The entire DP data set (less the single point at zero DPs) lies above the SB data, which underscore how extreme the imparity is in DPs.

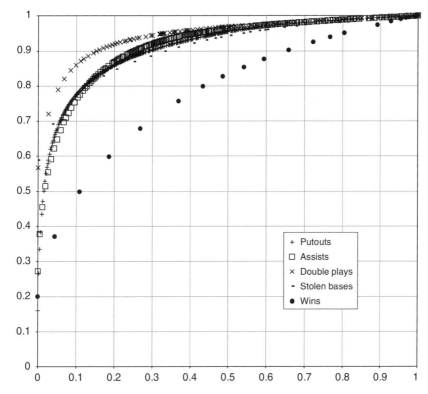

Figure 14.3 Comparison of the double cumulative distribution of various statistics. X-axis portrays percentage of cumulative success and the Y-axis is percentage of cumulative players.

14.6 Conclusions

This chapter has tried to compare interconnected performance with individual performance, as a way of exploring networked performance and non-networked performance in future warfare. Since this has been only an *exploration* using baseball as a proxy, we must be circumspect about what we should rightfully conclude; we are, after all, really talking about baseball, and extensions to future warfare must be made with care. But even with more than a modicum of caution we can confidently formulate some useful conjectures about interconnected combat performance that are still more advanced than anything that has yet been produced from the theoretical research of NCW or in research about the coming Robotics Age.

For example, we wanted to know if the talent of high performers would make poor performers better. If data from interconnected performance in baseball are an indication, then we should expect the kind of externality enjoyed by 1B and 2B to occur in networked warfare. When one usually talks about a rising tide lifting all boats, it is generally construed to mean that all boats rise by the same amount. What happens in baseball is that not all boats are lifted (there is still an N_0 population), the high performers are not necessarily lifted to the top (there were only three SSs in the 2013 top 20 ranking for DPs – at ranks 14, 18, and 20) and some modestly competent players get thrust to the very top (1B, of course). It is also interesting that those at the very top of the performance distribution do not return the favor – the tide keeps rising from the middle of the distribution, not from the top.

We also wanted to know if "goats" would hamper the heroic efforts of "aces." If data from interconnected performance in baseball are an indication, then we would have to say it depends on which goats and which aces. For example, an exceptionally incompetent 1B is not only preventing himself from being an ace, but he is bringing down the rest of the infield as well. So while the 2B or SS might not have had the league lead for all infield positions, a goat at first base will prevent them from being the ace among 2Bs or ace among SSs. What is interesting about the presence of the kind of externality we see in baseball data is that a goat at SS or 2B will also *fail* to create an ace at 1B where otherwise one would have emerged.

We can further say that although we had to reject the hypothesis that interconnected behavior in baseball was distributed according to the multivariate homogeneous Pólya family of distributions, we saw that the data was nonetheless similarly skewed, but with fatter tails and skinnier heads. This means that we can't use Bolmarchic's powerful imparity statistic like we can for 34 of 38 of the other statistics, but it does mean we can still interpret the double cumulative, for example, to compare performance. We lack, however, the conceptual analog of β and the direct comparison of κ, and we still have no explanation for why interconnected performance is different, why its distribution looks like it does, and if the apparent externalities are in fact real, operationally meaningful mechanisms.

Finally, this was an initial theoretical exploration using proxy data. A rigorous, definitive study of the fundamental concepts of NCW and warfare in the Robotics Age–to include performance distributions in collectives–is long overdue. It will require a substantial investment in a different kind of modeling and simulation than is currently used in the defense community to "prove" the worth of networks and robots, or analysts will have to wait for actual combat data. Given the poor understanding we still have about this topic, sending forces into combat still ignorant of these and other questions will deliver a twenty-first-century failure for the profession of operations research much more profound than its twentieth-century successes.

Acknowledgments

The author owes a deep debt of gratitude to Dr. Joe Bolmarchic, who provided invaluable help and assistance with the conceptual, mathematical, and statistical application of the multivariate homogeneous Pólya distribution to baseball data. He was exceedingly generous–even going so far as sharing his personal Pólya calculation routines. Special thanks to Sean Lahman, curator of the Lahman Baseball Database, for his assistance in interpreting the subtleties of the online MLB database.

References

1. J. Bolmarchic, "Who Shoots How Many?", unpublished briefing slides, Quantics Incorporated © 2000, 2003, 2010.
2. Lahman Baseball Database, http://www.seanlahman.com/baseball-archive/statistics/, accessed 20 Dec 2014.

15

Distributed Combat Power: The Application of Salvo Theory to Unmanned Systems

Jeffrey R. Cares
Captain, US Navy (Ret.), Alidade Inc., USA

15.1 Introduction

The US Navy's Littoral Combat Ship (LCS) was originally conceived as a small platform that would leverage networking and unmanned systems to fundamentally change how ships fight in some of the most demanding maritime scenarios, particularly in cluttered lethal environments just offshore. In early analyses, it was thought that severing the physical connection between this warship's combat power and its hull would be the main source of advantage for LCS in naval combat. As a fielded program, however, LCS has focused on different capabilities and leverage points, and while unmanned systems and modularity are still integral to platform, they have not been as important an element to LCS's operational prowess as originally intended. Part of the reason for this is that unmanned systems technology has not progressed enough for LCS to meet its full promise in this regard. As the technology improves, however, the opportunity still exists for LCS to become a revolutionary innovation in naval warfare. This chapter introduces a variant of Hughes' Salvo Equations to show how distributing and reconfiguring combat power among a group of LCS platforms can dramatically improve LCS's ability to fight, survive, and fight again in high-end anti-ship missile combat. These results extend beyond naval warfare to any operational concept for unmanned systems in which host platforms control large numbers of independent, distributable, unmanned vehicles.

Operations Research for Unmanned Systems, First Edition. Edited by
Jeffrey R. Cares and John Q. Dickmann, Jr.
© 2016 John Wiley & Sons, Ltd. Published 2016 by John Wiley & Sons, Ltd.

15.2 Salvo Theory

The classic form of naval warfare is the *gunnery duel,* a contest in which combatants apply *continuous fire* (increments of combat power applied over time) from large caliber guns against their foes. The primary physical process in a gunnery duel is an incremental depletion of the resistant force of the targeted hull, or *staying power.* Mathematical models of the gunnery duel were first developed more than 100 years ago and are well known to operations research analysts.

There are other types of naval weapons–rams, mines, torpedoes, bombs, and missiles–in which destructive power arrives instantaneously and in much larger increments. The form changes from continuous fire to *pulse fire.* Up until World War II, pulse fire weapons were considered secondary batteries; continuous fire from large caliber guns held center stage. World War II saw the end of big guns as a navy's main offensive capability. Air-delivered pulses of power–from both airplanes and missiles–became the central focus of fleets, but an accompanying mathematical model took decades to germinate. Although developed almost 30 years ago, this "new" model, Hughes' Salvo Equations, remains obscure to most naval operations research analysts, mostly because they favor probabilistic simulations of specific warfare scenarios over theoretical mathematical excursions. This chapter will show, however, that a theoretical exploration of naval combat using a variant of the Salvo Equations is very useful in identifying sources of advantage in a new age of naval warfare that employs large numbers of unmanned systems.

15.2.1 The Salvo Equations

The Salvo Equations, force-on-force equations for pulse combat, describe the damage inflicted by one side against another in a pulse weapon salvo (as a fraction of total pre-salvo force) as

$$\frac{\Delta B}{B} = \frac{\alpha A - b_3 B}{b_1 B}, \quad \frac{\Delta A}{A} = \frac{\beta B - a_3 A}{a_1 A} \tag{15.1}$$

where

A	= number of units in force A
B	= number of units in force B
α	= number of missiles fired by each A unit (*offensive combat power*)
β	= number of missiles fired by each B unit (*offensive combat power*)
a_1	= number of hits by B's missiles needed to put one A out of action (*staying power*)
b_1	= number of hits by A's missiles needed to put one B out of action (*staying power*)
a_3	= number of attacking missiles that can be destroyed by each A (*defensive combat power*)
b_3	= number of attacking missiles that can be destroyed by each B (*defensive combat power*)
ΔA	= number of units in force A out of action from B's salvo
ΔB	= number of units in force B out of action from A's salvo

Note that a two-way salvo exchange is expected (there is one equation for Side A and one for Side B) [1].

Consider two competing three-ship task groups, Side A and Side B. Assume these ships are identical in every way and have the following characteristics: $\alpha=\beta=4$, $a_1=b_1=2$, $a_3=b_3=3$. Then after a single salvo,

$$\frac{\Delta B}{B}=\frac{\Delta A}{A}=\frac{4(3)-3(3)}{2(3)}=\frac{12-9}{6}=0.50, \quad \text{or} \quad 50\%. \tag{15.2}$$

50% damage means each side will lose half its force, or the equivalent of 1.5 ships.

Alternatively, if there is only one ship on Side B while Side A still has three, then

$$\frac{\Delta B}{B}=\frac{4(3)-3(1)}{2(1)}=\frac{12-9}{2}=4.50, \quad \frac{\Delta A}{A}=\frac{4(1)-3(3)}{2(3)}=\frac{4-9}{6}=-0.83. \tag{15.3}$$

This means that B loses its single ship (actually, this one ship could be destroyed 4.5 times over) and A loses none (a negative amount of damage means less than 0% are damaged).

The equations can be iterated to explore sequences of salvoes during an ongoing battle. For example, assume $A=B=3$, $\alpha=3$, $\beta=4$, $a_1=3$, $b_1=2$, $a_3=3$, and $b_3=2$. Then after a single salvo,

$$\frac{\Delta B}{B}=\frac{3(3)-3(2)}{2(3)}=\frac{9-6}{6}=0.50, \quad \frac{\Delta A}{A}=\frac{4(3)-3(3)}{3(3)}=\frac{12-9}{9}=0.33. \tag{15.4}$$

B loses 1.5 ships and A loses only one. If the surviving ships attack each other with a second salvo, then

$$\frac{\Delta B}{B}=\frac{3(2)-2(1.5)}{2(1.5)}=\frac{6-3}{3}=1.00, \quad \frac{\Delta A}{A}=\frac{4(1.5)-3(1.5)}{3(2)}=\frac{6-4.5}{6}=0.25. \tag{15.5}$$

B loses its remaining force while A loses only half a ship (a quarter of a two-ship force).

15.2.2 Interpreting Damage

What are we to make of results like "destroyed 4.5 times over," "negative damage," or "losing half a ship"? Predicting damage at sea has always been difficult. In the era of big guns, analysts could calculate how many shells must impact an enemy's ship to exhaust its staying power, but there were frequently unexpected cases of, say, a single lucky round hitting a magazine, inability to control fires or flooding or disabled interior communications that contributed to a shorter combat life than the equations might prescribe. Given that catastrophic failures occurred with the relatively smaller increments of combat power from guns, it is no surprise that these effects are even more pronounced in salvo warfare. The exact outcomes from salvo exchanges are indeed exceptionally unpredictable.

Salvo Equations are best used, then, for *comparative analysis* rather than *predictive analysis*. Very useful aspects of war at sea can be explored with this perspective, such as force sufficiency, salvo size selection, the relative strength of offense against a certain defense, fractional exchange ratios, and so on. For example, inflicting 4.5 times more damage than required is one way to overcome the extreme variability in actual results: such a high level of overkill leaves much less to chance than a perfectly sized salvo. For reasons that will be discussed later, however, this also suggests that some missiles must be wasted in this effort to reduce uncertainty. Comparative analysis allows this tradeoff (overkill vs. wasted shots) to be quantitatively addressed.

The usual way that damage is assessed by the Salvo Equations is *pro rata*. For example, while in actual combat a three-ship force might incur 50% damage in a variety of ways (one and a half ships damaged, three ships each with 50% damage, two ships with 75% damage and one with none, etc.), for comparative analysis the analyst does not need to make that distinction. Explicitly, Eq. (15.1) is saying that when two forces exchange salvoes, then damage per ship ($\Delta B/B$ or $\Delta A/A$) stems from the interaction of three factors: offensive combat power (αA or βB), defensive combat power ($b_3 B$ or $a_3 A$), and staying power ($b_1 B$ or $a_1 A$). We decrement units of offensive combat power by units of defensive combat power and apply the remaining offensive combat power (if there is any) directly to staying power. Since combat power is resident in the hull, perhaps in tubes, launchers or magazines, any reduction in staying power also incurs a proportional reduction in combat power. So we assume that 50% damage means not only that staying power is reduced by half, but also that the ability to apply offensive and defensive combat power is reduced by half. The transition from Eqs. (15.4) to (15.5) shows how this approach to proportional damage assessment is applied in the calculations.

Since the analysis is not meant to be predictive, analysts have to be careful with proportionally attributing damage. In actual combat, a single missile hit on a ship with a staying power of two hits could very likely disable all the ship's combat power, yet the hull could steam on; alternatively the single hit could render the ship dead in the water, but the ship could retain full fighting strength. Exactly where a missile hits a ship's hull or superstructure determines what specific damage is incurred, but since substantial destructive power is delivered with each salvo weapon, a relatively small number of hits can put quite a large ship out of action. Once multiple hits have been absorbed by a ship, therefore, the results of comparative and predictive analysis converge.

15.3   Salvo Warfare with Unmanned Systems

Unmanned systems offer an unprecedented opportunity to once again change the character of naval warfare. In the very near future (and to a modest extent already today), ships will be able to package their combat power into autonomous offboard vehicles. We already see how sensors have literally taken flight in unmanned air vehicles (UAVs); soon anti-ship missiles and anti-missile missiles will be air-, surface-, and perhaps even subsurface-deployable by similar means. The US Navy has demonstrated how Cooperative Engagement Capability can allow one ship to control another ship's air defense missiles and routinely practices UAV handoffs between platforms. We should expect that advances in these capabilities will allow a wider range of ship–vehicle interactions, more handoffs between platforms, and a broader reconfiguration of combat power among a force. In short, vehicle autonomy will improve, and as it does vehicles will become more independent of their home platforms (perhaps to the point that the notion of a "home platform" becomes obsolete).

Let's look at another kind of sequential salvo exchange to explore how distributing combat power in this way can be a leverage point in modern combat. Assume two three-ship task groups as before, Side A and Side B. These ships have the following characteristics: $\alpha = \beta = 4$, $a_1 = b_1 = 2$, and $a_3 = b_3 = 3$. Side A, however, is able to distribute its offensive and defensive combat power among offboard autonomous vehicles, and the hull retains only sustainment and command and control functions. Side A is therefore not susceptible to proportional damage, yet Side B is configured in the usual way. After the first salvo,

$$\frac{\Delta B}{B} = \frac{4(3) - 3(3)}{2(3)} = \frac{12 - 9}{6} = 0.50, \quad \frac{\Delta A}{A} = \frac{4(3) - 3(3)}{2(3)} = \frac{12 - 9}{6} = 0.50. \quad (15.6)$$

Side B loses a ship and a half (of proportional damage). Side A's combat power is not reduced when its staying power is reduced by half, so Side A's one and a half ships now can fight like three (except of course, for staying power).[1] In the second salvo,

$$\frac{\Delta B}{B} = \frac{4(3) - 3(1.5)}{2(1.5)} = \frac{12 - 4.5}{3} = 2.50, \quad \frac{\Delta A}{A} = \frac{4(1.5) - 3(3)}{2(1.5)} = \frac{6 - 9}{3} = -1.00. \quad (15.7)$$

Side A receives no further damage while Side B is annihilated.

15.4   The Salvo Exchange Set and Combat Entropy

Section 15.2.2 introduced the idea of inefficiency in salvo exchanges. To better understand the source and effects of salvo inefficiencies, it is necessary to look at the set of outcomes for individual salvoes in greater detail. As will be seen, the Salvo Equations are an optimistic theoretical upper bound for the outcome of a salvo. Two concepts, the *Salvo Exchange Set* and *combat entropy,* show how things can be much worse for the combatants (and much, *much* worse for Side B during the salvo exchanges in Section 15.3).

The *Salvo Exchange Set* describes the different possible outcomes of salvo exchanges. It is defined as:

$$S \equiv \left\{ (H \cap D) \cup (H \cap D') \cup (H' \cap D) \cup (H' \cap D') \right\} \quad (15.8)$$

where **H** is the event that offense weapons are properly targeted and will hit their intended targets and **D** is the event that the defense is successful in destroying inbound weapons.[2]

Until weapons are used they have combat potential, which in a perfectly efficient system should equal the damage inflicted by their use. If salvo exchanges were perfectly efficient, the only member in the Salvo Exchange Set would be **(H ∩ D)**. Since examples of the other three subsets abound, the system is clearly not perfectly efficient. This loss of efficiency is called *combat entropy.* The four subsets resulting from the Salvo Exchange Set are defined below. A brief discussion of each subset's contribution to combat entropy is included.

---

[1] This assumes a modest magazine or a reload capability on the unmanned platforms, perhaps up to the original α and a_3 values.

[2] X′ denotes the complement of X, that is, "not X."

Subset 1.   **H ∩ D: The defense counters correctly targeted shots.**
              This is the most efficient case.[3]

Subset 2.   **H ∩ D′: The defense does not counter correctly targeted shots.**
              Combat power is wasted by the unsuccessful expense of counterfire. Still
              worse, unsuccessful "double-teaming" may occur.

Subset 3.   **H′ ∩ D: The defense counters incorrectly targeted shots.**
              Combat power is lost by ineffective offensive targeting and by expense of coun-
              terfire on a non-threat. Still worse, unnecessary double-teaming may occur.

Subset 4.   **H′ ∩ D′: The defense does not counter incorrectly targeted shots.**
              Here combat power is lost by ineffective offensive targeting and ineffective
              counterfire. Aside from simple misses, two effects, the "weapon-sump effect"
              (some targets are hit by more than their share of weapons while others do not
              receive enough hits) and "overkill" (the assignment of more weapons to all
              targets than are required) are often operative in this case.

The combat entropy defined above is a result of the physics of salvo exchanges, for example, radar inaccuracies or ineffective distribution of combat power. The sheer randomness of salvo exchanges also causes combat entropy. Even a "simple" scenario in which three units of side B each fire four missiles at one unit of side A will result in 4^{12} (almost 17 million) possible outcomes. Experiments have shown that as much as 30% of combat potential can be lost to entropy in this "simple" exchange [2].

Combat entropy increased as the numbers of shooters and targets increase. Re-examining the salvo exchanges in Section 15.3 from this perspective, we see that Side B's targeting challenge is much more complex than Side A's. While Side A need only consider the compli-cations of distributing offensive combat power among three targets, Side B could be shooting at as many as 21: 3 hulls (to be hit twice each), 12 offboard vehicles each carrying one offen-sive missile and 6 offboard vehicles each carrying one defensive missile. The combinatorics mean, of course, that Side B's targeting and coordination problem is not seven times worse, but many, many orders of magnitude more complicated, an additional leverage point that makes distributing combat power even more beneficial for Side A.

15.5 Tactical Considerations

Given that Side B now has 21 targets to consider (and perhaps another 15 or so after the first salvo if Side B attacks offboard vehicles in addition to hulls), Side B's β values would have to carry an usually large complement of anti-ship missiles. Alternatively, Side B's navy could invest in smaller ordnance, specially designed to combat smaller, offboard vehicles. Assuming Side B can overcome the daunting search and targeting challenges of 21 simultaneous engagements, it is instructive to see how distribution and reconfiguration allow Side A to employ a variety of tactical adaptations, yet another leverage point for forces that can achieve a high level of independence with their unmanned vehicles.

---

[3] Some combat power may be wasted if the defense has an unequal distribution of counterfire and "double-teams" inbound weapons.

The baseline tactical case is outlined in Section 15.3. Side B attacks Side A using traditional anti-ship missiles targeted at the hulls. As we have seen, Side A can rely on handoffs and reconfiguration to overcome Side B by the second salvo. What if Side B focuses its efforts on attacking the unmanned vehicles with smaller missiles? In this case, Side A is wise to disperse into three separate groups, so that Side B can find only one-third of the targets at once. Even if this third is lost, Side A will still have two-thirds of its force free to fire at Side B, which will now have its position established by the located one-third of Side A. All of Side A missiles can now attack faster than Side B can find the rest of the missing adversary force.

What if Side B, frustrated in its efforts to focus on first hulls and then offboard vehicles, develops a capability to electronically jam Side A's control frequencies (or links)? In this case Side A might mass its hulls to control its offboard platforms using shorter range directed links, similar to jam-resistant line-of-sight links used by shipboard helicopters today. Further frustrated in this non-kinetic attempt, Side B would be relegated to again attacking hulls or vehicles, and all the challenges of managing offensive fires against 21 targets.

One aspect of salvo warfare not addressed in Eq. (15.1), but covered in other versions of the Salvo Equations, is *scouting effectiveness,* the ability to find and fix the enemy for effective offensive shots. While even legacy warships have offboard vehicles for search (the most prominent being the ship-based helicopter), proportional damage will work much the same way with scouting effectiveness on these platforms as it does with offensive and defensive combat power. Some "cross-decking" of helicopters is possible, but shipboard flightdeck space is very limited. In addition, radars and sensors attached to the superstructure will be damaged when the ship is hit. Conversely, a force that can distribute or reconfigure search assets will have a decided advantage in targeting a second salvo, yet another leverage point for unmanned vehicles in modern naval combat.

15.6   Conclusion

This chapter has presented a theoretical basis for analyzing warfare between a legacy force and a force with a large number of independent, unmanned offboard vehicles. It has shown how the Salvo Equations can be used to identify tactical advantages for forces with unmanned systems, as well as the vulnerabilities inherent in the way forces are constituted today. The introduction suggested that this approach could deliver a revolutionary change in warfighting capability for the LCS programs. For this result to occur, of course, requires advances in autonomy, networking, and control that are not yet realized, as well as a fundamental shift in thinking – and investment – among senior naval leaders.

In some ways, however, this chapter has set up a false argument. If the US Navy has such a fundamental shift in thinking and investment, then others will follow suit, and the LCS's problem will not be a fight against a legacy force, but a fight against an adversary much more like itself. No one yet knows exactly how a distributed, networked force will fight a distributed, networked force, although some formulations have been attempted.[4] The US is not the only

---

[4] See, for example, the author's *Distributed Networked Operations* (2006, Alidade Press, Newport, RI).

nation pursuing advanced autonomy, networking, and control, and so it is not guaranteed to be the world leader in developing these capabilities. If it fails to do so, as it has already with LCS, then it will be quite ironic indeed when LCS cannot fight in either of the worlds of legacy or future naval combat.

References

1. Wayne P. Hughes, Jr., "A Salvo Model of Warships in Missile Combat Used to Evaluate Their Staying Power", Naval Research Logistics, Vol. 42, pp. 267–289 (1995).
2. Jeffrey R. Cares, "The Fundamentals of Salvo Warfare," pp. 31–41, Naval Postgraduate School Masters Thesis, Monterey, CA (1990).

Index

AAI Corp RQ-7 *Shadow 200* unmanned air vehicle, 244–246

AARS *see* after-action reviews

ace factor, 272, 275, 286

ACO *see* ant-colony optimization

ACSs *see* autocatalytic sets

activity allocation, 140

adaptive automation, 141

adjacency matrix, 257–258

AeroVironment *Puma AE* unmanned air vehicle, 243–244

AeroVironment *Raven* unmanned air vehicle, 243

AeroVironment *Wasp* unmanned air vehicle, 243

aerRobotix *CatOne* unmanned surface vehicle, 247

after-action reviews, 125

agent-base simulation, 259

AI *see* artificial intelligence networks

aided target recognition, 141

Airbus A380, 245

Air Force Institute of Technology, 7

Air France Flight 447, 247

AMAS *see* Autonomous Mobility Appliqué System demonstrations

Amazon.com, 243

ambulance drones, 243

analytical seminar wargaming, 137

analytical techniques

    agent-based simulation, 259

    analytical seminar war gaming, 137

    ant-colony optimization (ACO), 32, 44–46

    artificial intelligence (AI) networks, 240

artificial neural networks, 32

basic market algorithm, 35

closed-loop (CL) simulation, 119, 125

comparative analysis, 289

confirmatory triangulation, 137

consensus-based bundle algorithm (CBBA), 52

convex-hull with nearest-neighbor insertion (CHNNI), 36–37

deterministic model, 67

discrete-event simulation (DES), 236, 238, 251

dynamic programming, 32

finite distance estimation technique, 239

finite state machines, 199

frequency domain method, 239

fuzzy logic clustering, 32, 44–46

genetic algorithms, 32

GENIUS heuristic, 24

gradient estimation techniques, 239

gradient surface method, 239

graph partitioning, 39

heuristic methods, 24, 239, 248

heuristic search technique, 239

human-in-the-loop (HIL) simulation, 119, 125

integer linear program, 13

likelihood ratio, 239

Lin–Kernighan heuristic algorithm, 32, 37

live, virtual and constructive (LVC) simulation experiments, 194

market-based optimization techniques, 33–34

market-based solutions (MBSs), 34–40

Operations Research for Unmanned Systems, First Edition. Edited by
Jeffrey R. Cares and John Q. Dickmann, Jr.
© 2016 John Wiley & Sons, Ltd. Published 2016 by John Wiley & Sons, Ltd.

analytical techniques (*cont'd*)
   mathematical modeling, 161, 169, 171–172, 178–179
   Max-Pro optimization, 31–32
   meta-modeling, 239
   Min-Max optimization, 31
   Min-Sum optimization, 31
   mixed integer linear program (MILP), 27, 53–54, 98
   Monte Carlo simulation, 125, 161, 237
   multi-agent system (MAS), 31, 33
   multi-criteria analysis and ranking consensus unified system (MARCUS), 128, 132–133
   ordinal ranking, 268–269
   Parasuraman–Sheridan–Wickens model, 141
   Pareto distributions, 272
   particle swarm optimization, 115–116
   Perron–Frobenius Eigenvalue (λ_{PFE}), 255–256, 259, 263–270
   Perron–Frobenius theorem, 258
   perturbation analysis, 239
   power law distribution, 272
   predictive analysis, 289
   PRIMAL1 set covering heuristic, 24
   profile probability, 273
   random search formula, 68–70
   regression analysis, 161, 169, 171–172, 178–179, 195, 204, 212, 264, 266, 269
   response surface method, 239
   simulated annealing, 32, 239
   simulation optimization, 239
   skewed probability distributions, 272
   software agents, 28, 33, 259
   statistical thinking, 193
   steepest descent search method, 239
   subject matter experts (SMEs), 125
   Tabu search, 32
   three-index flow formation, 14
   toy problems, 45
   urn schemes, 272–273, 277, 282, 284
   visual interactive modeling systems (VIMSs), 238
ANDROS Wolverine disaster robot, 244
angular probability curve, 63
ant-colony optimization, 32, 44–46
anthropomorphism, 2
anticipatory spare ordering, 249
ARA eUGS unattended ground system, 244
ArcGIS™ mosaic tool, 87
ARDUPLANE autopilot software, 199
Arena™ logistics model, 238
armed robotic vehicle, 122
artificial intelligence networks, 240
artificial neural networks, 32
Association for Unmanned Vehicles and Systems International Driverless Car Summit, 247
Australian army's future operating concept, 120, 126
autocatalytic sets, 259

automatic target recognition, 140
automation continuum, 157–161, 180–186
Autonomous Mobility Appliqué System demonstrations, 156, 247
AUVSI *see* Association for Unmanned Vehicles and Systems International Driverless Car Summit

Babe Ruth, 278
BAE Taranis unmanned air vehicle, 245
basic market algorithm, 35
Battle of 73 Easting, 273
benefits of "unmanning", 2
bias, experiment, 190
Blackett, P. M. S., 1
Black Hornet nanocopter, 243
blocking factor, experiment, 190, 191
Bluefin-21 unmanned undersea vehicle, 247
Boeing 747, 245
Boeing 777, 245
Boeing A160 Hummingbird unmanned air vehicle, 244
Boeing Phantom Eye unmanned air vehicle, 247
boost arc, 116
BRANDO *see* breakpoint analysis with nonparametric data option software
breakpoint analysis with nonparametric data option software, 128
business logistics, 234
buyer agents, 34

C2 *see* command and control
capability maturity model integration, 219–222
CAS *see* commercially available software
CAST *see* convoy active safety technologies
CAT *see* combined arms team
CBBA *see* consensus-based bundle algorithm
census-taking methods
   classical, 81
   thermal infrared, 81
center runs, experiment, 203
centralized cognition, 3
central limit theorem, 165
CERS *see* cost-estimating relationships
Chernoff bound, 67
circular loitering path task, 106
C4ISR *see* command control surveillance and reconnaissance system
CL *see* closed-loop simulation, contractor logistics support
closed-loop simulation, 119, 125
CLS *see* contractor life support
CMMI *see* capability maturity model integration
coast arc, 116
COCMO II *see* Constructive Cost Model cost estimation software
combat cycle, 256

combat entropy, 291–292
combined arms team, 119, 123
command and control, 7
command control surveillance and reconnaissance
    system, 51
commercially available software, 237
commercial off-the-shelf, 250
comparative analysis, 289
comparison of performance CTPs, 22
components of wind acting on UAV, 111
computational efficiency of solutions, CTPs, 23
confirmatory triangulation, 137
consensus-based bundle algorithm, 52
Constructive Cost Model cost estimation software,
    218–220
Constructive Systems Engineering Cost Model's
    system life-cycle phases, 208–210
continuous fire process, 288
contractor life support, 224
contractor logistics support, 246
convex-hull with nearest-neighbor insertion, 36–37
convoy active safety technologies, 155
cooperative engagement capability, 290
cost-estimating relationships, 212, 213, 223, 227
    US Army, 213, 224, 227
costs of unmanning, true, 3
COSYSMO's see Constructive Systems Engineering
    Cost Model's system life-cycle phases
COTS see commercial off-the-shelf
counter-acquisition, 144
covering tour problems, 8–11
    comparison of performance, 22
    computational efficiency of solutions, 23
    infeasible solutions, 22
    measures of performance (MOPs), 21
    multiple vehicle variants, 24
    vigilant covering tour problem (VCTP), 8, 14
CTPs see covering tour problems
C-5 transport, 245
C-17 transport, 245
curse of dimensionality, 28
Curtiss-Sperry Aerial Torpedo, 242
customer service, 234

DA see driver assist vehicle operation
damage, 288
Danish Navy's Mine Countermeasure Mission mine
    hunters, 247
DARPA see Defense Advanced Research Programs
    Agency
Dassault nEUROn unmanned air vehicle, 245
Decider nodes, 256–270
Defense Advanced Research Programs Agency, 155
    TLAS infantry robot, 247
Defense Research and Development Canada, 60

defensive combat power, 288–291
definitive screening designs, 197
DEM see digital elevation model
demand forecasting, 234
DES see discrete-event simulation
design of experiments, 191
design point, experiment, 189
design power, experiment, 191
deterministic model, 67
DHL, Inc., 243, 249
digital elevation model, 85
discrete-event simulation, 236, 238, 251
distribution of goats, 275, 286
DOE see design of experiments
D-optimal experiment designs, 197–198
Dorado autonomous undersea vehicle, 60–61
DRDC see Defense Research and Development Canada
driver assist vehicle operation, 158, 161–163, 166–175,
    178, 182
driver warning vehicle operation, 158, 161–163,
    166–175, 178, 182
DSD see definitive screening designs
Dubins set, 97–113, 115–116
DW see driver warning vehicle operation
dynamic programming, 32

EADS Barracuda unmanned air vehicle, 245
easyJet Hexacopter, 243
edge covering matrix, 12–13
Elbit Hermes unmanned air vehicle, 244
ELIMCO, Inc., 84
engineering support, 235
ENVI™ mosaic tool, 88
Environmental Information Network of Andalusia,
    Spain, 82
environment and water agency of Andalusia, Spain, 84
ERDAS IMAGINE™ mosaic tool, 87
E-300 Viewer unmanned air vehicle, 84
evolutionary acquisition, 250
EXFOR see experimental force
expected flight time, 113
experiment
    bias, 190
    blocking factor, 191
    center runs, 203
    design point, 189
    design power, 191
    factor, 189
    held-constant factors, 195
    hypothesis construction, 191
    model misspecification, 191
    noise, 190
    nuisance factors, 195
    prediction variability, 191
    qualitative responses, 195

experiment (*cont'd*)
  randomization, 190
  replication, 190
  response, 190
  safety pilot, 199
  sample size, 197
  test engineer, 199
  test matrix, 197, 202
  type I error, 191
  type II error, 191
experimental design concerns, 191
experimental force, 120, 123, 128–132
experimentation, steps, 192
extended human sensing, 141
ExtendSim™ logistic model, 238

FA *see* full automation vehicle operation
factor, experiment, 189
false alarms, 144
FeDEx, 249
field service representatives, 245–246
FIFO *see* fly-in/fly-out servicing
finite distance estimation technique, 239
finite state machines, 199
2FIs *see* two-factor interactions
Fitts list of automation factors, 141
flat-earth approximation, 99
flight geometry flight checks, 84
Flyaway Unit Cost, 212
fly-in/fly-out servicing, 248
FOB *see* forward operating base
focal plane, 82–83
forward operating base, 245
frequency domain method, 239
FSRs *see* field service representatives
fuel burn rate model, 110
fuel consumption optimization, 116
full automation vehicle operation, 160–178, 186
fuzzy logic clustering, 32, 44–46

general dynamics *MUTT* unmanned ground system, 247
generic methodologies for verification and validation, 239
genetic algorithms, 32
GENIUS heuristic, 24
geodesic paths, 97
geo-referencing, 83–84
German U-boat kills of Japanese shipping (World War II), 273
glimpse rates, 144
Global Hawk unmanned air vehicle, 212–213, 247–249
global navigation satellite system, 83, 94
global test for curvature, 203
global war on terror, 242, 245, 250

GM *see* generic methodologies for verification and validation
GNSS *see* global navigation satellite system
GOCO *see* government-owned, contractor-operated
Google self-driving car, 247
government-owned, contractor-operated, 245
gradient estimation techniques, 239
gradient surface method, 239
graph partitioning, 39
ground mission planning system, 7
ground sample distance, 82–86, 93
ground surveillance radars, 245
GSD *see* ground sample distance
GSRs *see* ground surveillance radars
Gunnery Duel, 288
Gyrodyne QH-50 *DASH* unmanned vehicle, 242

hardware cost drivers, 215
HDT Protector unmanned ground system, 247
held-constant factors, experiment, 195
heuristic methods, 24, 239, 248
heuristic search technique, 239
hiding, 145
hierarchical market, 37–38
HIL *see* human-in-the-loop simulation
Hughes' Salvo equations, 287–291, 293
human input proportion, 156–157, 161–169, 171–180
human-in-the-loop simulation, 119, 125
hypothesis construction, experiment, 191
hypothesis of trust, 142

IACM *see* information age combat model
IAI *Heron* unmanned air vehicle, 244
IAI *Scout* unmanned vehicle, 242
IBM CPLEX®, 42–44, 48–49
IEDs *see* improvised explosive devices
image focal length, 82–83
image projection center, 84
image scale, 82–83
imparity statistic, κ, 273, 277–286
imperfect automation, 141
improvised explosive devices, 245
IMU *see* inertial measurement unit
inertial measurement unit, 85
infeasible solutions, CTPs, 22
influencer nodes, 256–270
information age, 271
information age combat model, 256–263
information entropy, 142
insitu RQ-21 *Blackjack* unmanned air vehicle, 244
insitu *Scan Eagle* unmanned air vehicle, 242, 244
integer linear program, 13
integrated logistics support, 235
intellectual property (IP) data, 242
intelligence, surveillance and reconnaissance, 123

interaction frequency, 157
inventory management, 234
I-optimal experiment designs, 197–198
iRobot *FirstLook* "throwbot", 244
iRobot *PackBot*
  unexploded ordinance (UXO) robot, 245
  unmanned ground vehicle, 122
Ishikawa diagram, 200–201
ISO/IEC 15288 systems engineering-system life-cycle
  processes, 208
ISR *see* intelligence, surveillance and reconnaissance

Java programming language, 107
Johnson function, 65, 69
joint offensive support team, 123
joint strike fighter, 250
JOST *see* joint offensive support team
JSF *see* joint strike fighter
just-in-time logistics, 250

Kettering *Bug* unmanned vehicle, 242
key installations, security protection of, 7
2^k factorial design, 196–197
kill chain, 7
KINs *see* key installations (KINs), security protection of
Kongsberg *Remus 100, 600* and *6000* unmanned
  undersea vehicles, 247

Lahman Baseball database, 274
land capability analysis branch, Australian DSTO, 119
LaserMotive, Inc., 243
lawn-mowing search patterns, 62–64, 71
LCA *see* land capability analysis branch, Australian
  DSTO
L-COM *see* logistics composite model
LCS *see* Littoral Combat Ship
leader-follower vehicle operation, 159, 161–178,
  183–184
levels of experimental factors, 195
LF *see* leader-follower vehicle operation
likelihood ratio, 239
Linfox, Inc., 249
Lin–Kernighan heuristic algorithm, 32, 37
Littoral Combat Ship, 287, 293–294
live, virtual and constructive simulation experiments, 194
Lockheed Martin *MULE* infantry support vehicle, 247
Lockheed Martin *Stalker* unmanned air vehicle, *243*
logistics, business, 234
logistics communication, 234
logistics composite model, 237
logistics doctrine, 235
logistics functions, 235
loitering task, 105
low rate initial production, 242
LRIP *see* low rate initial production

LSA autonomy *RAP* unmanned ground system, 247
LVC *see* live, virtual and constructive simulation
  experiments

main effects, 197
maintenance planning, 235
maintenance robots, 249
Major League Baseball database, 273, 275
Malaysia Airlines Flight, 370, 247
manned-unmanned teaming, 241
manned vehicle operations, 121
manpower cost avoidance, 2
MARCUS *see* multi-criteria analysis and ranking
  consensus unified system
Mark McGwire, 278
market-based optimization techniques, 33–34
market-based solutions, 34–40
MAS *see* multi-agent system
mathematical modeling, 161, 169, 171–172, 178–179
MATLAB®, 42–43, 48–49, 100, 112
maximization of observability in navigation for
  autonomous robotic control, 7, 9
maximum thrust arcs, 116
Max-Pro optimization, 31–32
MBSs *see* market-based solutions
MCM *see* mine countermeasures reconnaissance
  operation
MDVRP *see* multi-depot vehicle routing problem
measures of baseball performance, 276–277
measures of effectiveness, 62, 126
measures of performance, CTPs, 21
MEs *see* main effects
meta-modeling, 239
middle east representative town, 120
military off-the-shelf, 250
MILP *see* mixed integer linear program
mine countermeasures reconnaissance operation, 59
mine density distributions, 72–74
minimum thrust arcs, 116
MINITAB software, 109
Min-Max optimization, 31
Min-Sum optimization, 31
Miricle Camera LVDS 307K, 84
mixed integer linear program, 27, 53–54, 98
MLB see Major League Baseball database
model misspecification, experiment, 191
modular unmanned scouting craft, littoral, 212–213
MOEs *see* measures of effectiveness
MONARC *see* maximization of observability in
  navigation for autonomous robotic control
Monte Carlo simulation, 125, 161, 237
MOPs *see* measures of performance, CTPs
Morse and Kimball, 1
MOTS *see* military off-the-shelf
MQ-4C Triton unmanned air vehicle, 246, 248–249

multi-agent system, 31, 33
multicollinearity, experiment, 191
multi-criteria analysis and ranking consensus unified
    system, 128, 132–133
multi-depot vehicle routing problem, 30–33
multiple vehicle variants, CTPs, 24
multi-UAV2 simulation, 99
multivariate homogenous Pólya distribution, 272–273,
    275, 282
MUM-T *see* manned-unmanned teaming

National Research Council report on test and
    evaluation, 193
NCW *see* network centric warfare
negative damage, 289
network centric warfare, 2, 271, 275, 286
network-enabled systems of systems, 3
network logistics models, 238
noise, experiment, 190
non-confounding experiment designs, 197
normal probability distribution, 238
Northrop Grumman
    CaMEL unmanned ground system, 247
    ExTASS surveillance system, 245
    MQ-8B *Fire Scout* unmanned air vehicle, 244
    MQ-9 *Reaper* unmanned air vehicle, 244
    X-47B *UCLASS* unmanned air vehicle, 244–245
NRC *see* National Research Council report on test and
    evaluation
NREC *Gladiator* unmanned ground system, 247
nuisance factors, experiment, 195
null-agent, 41

OEMS *see* original equipment manufacturers
offensive combat power, 288–291
operational environment, 214
operator control units, 240
OPFOR *see* opposition force
opposition force, 120, 128–132
order processing, 234
ordinal ranking, 268–269
organic logistics support, 246
organic spares manufacture, 250
orienteering problem, 32
original equipment manufacturers, 233, 242, 245, 249
orthogonal array experiment designs, 197
orthographic restitution, 86
ortho-rectification, 84

packaging, handling, storage and transport, 234, 236
Paketkopter logistics drone, 243
parallel search patterns, 62
Parasuraman–Sheridan–Wickens model, 141
Pareto distributions, 272
Parrot *AR.drone* unmanned air vehicle, 243

particle swarm optimization, 115–116
parts and service support, 234
pattern search, 239
Perron–Frobenius eigenvalue (λ_{PFE}), 255–256, 259,
    263–270
Perron–Frobenius theorem, 258
personnel and manpower, 236
perturbation snalysis, 239
photographic mosaics, 86–92
photomosaic post-classification process, 90
photomosaic texture filter, 90
pioneer unmanned air vehicle, 108–109, 112
pixel, 82–83, 86
PlackettBurman experiment designs, 197
planar logistics model, 238
plant and wholesale site selection, 234
PNOA *see* Spanish National plan of aerial
    ortho-photography
power law distribution, 272
prediction variability, experiment, 191
predictive analysis, 289
PRIMAL1 set covering heuristic, 24
prize-collecting TSP, 32
probability of acquiring a target during a time
    interval, 145
probability of detection as a function of
    range, 62–63
probability of not acquiring a target during a time
    interval, 146
probability of target recognition, 144
procurement, 234
profile probability, 273
ProModel™ logistics model, 238
pseudo-random number generation, 237
pulse fire process, 288

QinetiQ *Dragon Runner* "throwbot", 244
QinetiQ *TALON* UXO robot, 245
Qinetiq *Zephyr* unmanned air vehicle, 247–248
qualitative responses, experiment, 195
quality control of National and Regional
    photogrammatic flights program, 86

radius of avoidance task, 108
radius of sight task, 103
RAM *see* reliability, availability and maintainability
random search, 239
random search formula, 68–70
randomization, experiment, 190
rational allocation of activities, 141
RC *see* remote control vehicle operation
recognized ground situation picture, 7, 9
ReconRobotics Throwbot, 244
REDIAM *see* Environmental Information Network of
    Andalusia, Spain

regression analysis, 161, 169, 171–172, 178–179, 195,
    204, 212, 264, 266, 269
reliability, availability and maintainability, 242
remote control vehicle operation, 158, 161–163,
    166–168, 170–175, 178, 181
replication, experiment, 190
response, experiment, 190
response surface method, 239
reverse logistics, 234
RGSP *see* recognized ground situation picture
Riverwatch *Nacra* unmanned catamaran, 241
robotic decision-making ability, 156–157, 169–180
robotics age, 2, 271, 285–286
robustness, 255, 267–268
Rockwell Collins patrol persistent surveillance
    system, 245
ROE *see* rules of engagement
Royal Navy ordnance expenditures (Falklands), 273
RQ-1 *Predator* unmanned air vehicle, 242, 246
rules of engagement, 7
Ryan BQM-34 *Firebee* unmanned vehicle, 242

Saab *Double Eagle* unmanned air vehicle, 245
Saab *Leopard* unmanned air vehicle, 245
Saab *Seaeye Falcon* unmanned air vehicle, 245
SABR *see* Society for American Baseball Research
safety pilot, experiment, 199
Salvo Exchange Set, 291–292
sample size, experiment, 197
Schiebel *Camcopter,* 244
scientific method, 193
scouting effectiveness, 293
screening experiment, 197
searching, 145
search patterns
    aspect angle degradation, 64–65
    maximum angular probability of detection, 65
    measures of effectiveness, 62
    uneven "lawn-mowing", 72
    zigzag, 72
SEER-H cost estimation software, 215–216
seller agents, 34
sensor nodes, 256–270
side-scan sonar, 59, 61
    measures of performance, 62
simplest complete combat network, 258
Simul8™ logistics model, 238
simulated annealing, 32, 239
simulation model development process, 239
simulation optimization, 239
Simulink simulation, 101
skewed probability distributions, 272
Slocum *Glider* unmanned undersea vehicle, 247
SMDP *see* simulation model development process
SMEs *see* subject matter experts

SMSS™ *see* Squad Mission Support System
Society for American Baseball Research, 274
software
    ArcGIS™ mosaic tool, 87
    ARDUPLANE autopilot software, 199
    Arena™ logistics model, 238
    breakpoint analysis with nonparametric data option
        (BRANDO) software, 128
    constructive cost model (COCMO II) cost
        estimation software, 218–220
    COSYSMO's system life-cycle phases, 208–210
    ENVI™ mosaic tool, 88
    ERDAS IMAGINE™ mosaic tool, 87
    ExtendSim™ logistic model, 238
    IBM CPLEX®, 42–44, 48–49
    Java programming language, 107
    logistics composite model (L-COM), 237
    MATLAB®, 42–43, 48–49, 100, 112
    MINITAB software, 109
    multi-UAV2 simulation, 99
    ProModel™ logistics model, 238
    SEER-H cost estimation software, 215–216
    Simul8™ logistics model, 238
    Simulink simulation, 101
    visual basic, 67
software agents, 28, 33, 259
software cost drivers, 215
sojourns, cumulative, 146
Spanish National plan of aerial ortho-photography, 85,
    87, 89
spiral development, 250
SQ *see* status quo vehicle operation
Squad Mission Support System, 214, 226, 247
statistical thinking, 193
status quo vehicle operation, 158, 161–163, 166–168,
    170–175, 178, 180
staying power, 288–291
steepest descent search method, 239
subject matter experts, 125
substitution myth of applying automation, 141
supply (warehousing) support, 236
support and test equipment, 236
systems engineering and program management cost
    drivers, 215

Tabu search, 32
tactical wheeled vehicles, 157, 180–186
tagged image file world, 87
target glimpses, 144
target nodes, 256–270
target recognition, 140
TBF *see* time between failure
T&E *see* test and evaluation costs
technical support network, 250
technology refresh, 250

test and evaluation costs, 230
test engineer, experiment, 199
test execution strategy, 198–199
test matrix, experiment, 197, 202
test schedule, 198
Textron *CUSV* unmanned surface vehicle, 247
Textron Systems *MicroObserver* unattended ground system, 244
TFW *see* tagged image file world
Thales IAI *Watchkeeper 450* unmanned air vehicle, 244
Thermotechnix Systems, Ltd., 84
three-index flow formation, 14
time between failure, 238
time to repair, 238
Titan *Solara* unmanned air vehicle, 247
TO *see* vehicle tele-operation
TOC *see* total cost of ownership
Toll, Inc., 249
total cost of ownership, 2
toy problems, 45
traffic and transportation, 234
training and training support, 236
traveling salesman problem, 8, 31–32, 115
    Concorde TSP solver, 32, 37
    prize-collecting TSP, 32
triangular distribution, 163
TSP *see* traveling salesman problem
TSP solver, Concorde, 32, 37
TTR *see* time to repair
two-factor interactions, 197–198
type I error, 191
type II error, 191

UK Ministry of Defense architecture framework, 140
UK Ministry of Defense Phoenix unmanned air vehicle program, 245
ULE/R *see* ultra-long endurance/range systems
ultra-long endurance/range systems, 247–248
unexploded ordinance robots, 245
uniform probability distribution, 170, 238
unit costs, 212
unmanned air vehicles
    delivery systems, 243
    flight time minimization, 30
    history, 188–189
    minimizing fuel consumption, 30
    mission "profit" maximization, 30
    requirements for assignment to targets, 28
    test planning process, steps, 193–198
unmanned systems training costs, 225
unmanned vehicle control modes, 226
unreliable automation, 141

urn schemes, 272–273, 277, 282, 284
US Air Force air-to-air kills (Vietnam), 273
US Air Force Project RAND, 237
US Air Force Research Laboratory, 99
US Army Robotics Rodeo, 247
US Department of Defense 5000 system life-cycle standard, 208–209
US Department of Defense architecture framework, 140
US Department of Defense master plan for unmanned air system development, 189
US Department of Defense RTD&E, 189
US Department of Defense RQ-2 pioneer unmanned air vehicle program, 245
US Department of Defense unmanned systems integrated roadmap FY2013-2038, 241
US Navy air-to-air kills (Vietnam), 273
US Submarine kills of Japanese shipping (World War II), 273
UXO *see* unexploded ordinance robots

VCTP *see* vigilant covering tour problem
vehicle routing problem, 11, 31–33
vehicle routing problem with time windows, 96
vehicle survivability, 143
vehicle tele-operation, 158, 161–163, 166–168, 170–175, 178, 181
verification and validation, model, 238–239
vertex covering matrix, 13
vigilant covering tour problem, 8, 14
VIMSs *see* visual interactive modeling systems
Virtualrobotix *R Brain 4* Hexacopter, 241
visual basic, 67
visual interactive modeling systems, 238
VRP *see* vehicle routing problem
VRPTW *see* vehicle routing problem with time windows

WA *see* waypoint vehicle operation
warehousing and storage, 234
warehousing models, 238
waypoint vehicle operation, 159, 161–178, 185
WBSs *see* work breakdown structures
weapons sump effect, 292
Weibull probability distribution, 238
WGS84 *see* World Geodesic System 1984 reference database
WHOI *see* Woods Hole Oceanographic Institute Nereus unmanned undersea vehicle
Woods Hole Oceanographic Institute Nereus unmanned undersea vehicle, 247
work breakdown structures, 212
World Geodesic System 1984 reference database, 85

Zipf's law, 272